TOWARDS WORLD HI

Heritage, Culture and Identity

Series Editor: Brian Graham,
School of Environmental Sciences, University of Ulster, UK

Towards World Heritage
International Origins of the Preservation Movement,
1870-1930

Edited by
MELANIE HALL
Boston University, USA

Routledge
Taylor & Francis Group

LONDON AND NEW YORK

First published 2011 by Ashgate Publishing

2 Park Square, Milton Park, Abingdon, Oxon OX14 4RN
711 Third Avenue, New York, NY 10017, USA

Routledge is an imprint of the Taylor & Francis Group, an informa business

First issued in paperback 2016

Cover image: 'St. Sophia, from the Bosphorus'. Watercolour. After a drawing by David Roberts for *The Life and Works of Lord Byron, 1833–4*, Vol. 1. © Victoria & Albert Museum, London.

British Library Cataloguing in Publication Data
Towards world heritage : international origins of the
  preservation movement 1870-1930. -- (Heritage, culture and
  identity)
  1. Historic preservation--History--19th century--
  Congresses. 2. Historic preservation--History--20th
  century--Congresses. 3. Historic preservation--
  Cross-cultural studies--Congresses.
  I. Series II. Hall, Melanie.
  363.6'9-dc22

Library of Congress Cataloging-in-Publication Data
Hall, Melanie.
  Towards world heritage : international origins of the preservation movement 1870-1930
/ by Melanie Hall.
    p. cm. -- (Heritage, culture and identity)
  Includes bibliographical references and index.
  ISBN 978-1-4094-0772-0 (hardback) -- ISBN 978-1-4094-0773-7 (ebook)
  1. World heritage areas--History. 2. Cultural property--Protection--History. 3. Historic
preservation--History. I. Title.
  G140.5.H35 2011
  363.6'909034--dc22

                                                                              2011012732

ISBN: 978-1-4094-0772-0 (hbk)
ISBN: 978-1-138-27499-0 (pbk)

Contents

List of Figures

List of Abbreviations

CAB	Cabinet Office Papers
CO	Colonial Office Papers
FO	Foreign Office Papers
GStA PK	Geheimes Staatsarchiv Preussischer Kulturbesitz
MARBL	Gregory Family Papers, Manuscript, Archives and Rare Books Library, Emory University
HLG	Housing and Local Government Papers
ICBS	Incorporated Society for the Promotion of the Enlargement and Construction of Churches and Chapels
ISA	Israel State Archives
MAP	Médiathèque de l'Architecture et du Patrimoine
NAS	National Archives of Sweden
SPAB	Society for the Protection of Ancient Buildings
T	Treasury Papers
TNA	The National Archives, Kew
WORK	Office of Works Papers
UNESCO	United Nations Educational, Scientific and Cultural Organization

Notes on Contributors

Anne M. Blackburn is Professor of South Asia Studies and Buddhist Studies, and Director of the Sinhala Program in the Department of Asian Studies at Cornell University. She studies Buddhism in South and Southeast Asia, with a special interest in monastic culture and southern Asian Buddhist networks. Her publications include *Buddhist Learning and Textual Practice in Eighteenth-Century Lankan Monastic Culture* (2001); *Approaching the Dhamma: Buddhist Texts and Practices in South and Southeast Asia*, co-edited with Jeffrey Samuels (2003); and *Locations of Buddhism: Colonialism and Modernity in Sri Lanka* (2010).

Raymond Cohen graduated from Oxford University, and received his PhD from the Hebrew University of Jerusalem. He recently retired from the Chaim Weizmann Chair of International Relations at the Hebrew University, where he also served as head of department and director of the Leonard Davis Institute for International Relations. He has held research fellowships at the United States Institute of Peace, and the Watson Institute for International Studies, Brown University, and was Corcoran Visiting Professor at the Center for Christian-Jewish Learning, Boston College (2008-10). He has published widely on international relations with an emphasis on conflict resolution, diplomatic negotiation, and intercultural communication. His latest books are *Isaiah's Vision of Peace in Biblical and Modern International Relations* (2008) and *Saving the Holy Sepulchre: How Rival Christians Came Together to Rescue their Holiest Shrine* (2008). He is currently working on a study of reconciliation between the Jewish people and the Catholic Church with special reference to the normalization of relations between the Holy See and Israel.

Erik Goldstein is Professor of International Relations and of History at Boston University. He was previously Professor of International History at the University of Birmingham and a member of the Centre for Byzantine, Ottoman, and Modern Greek Studies. He has published widely on the history of British foreign policy and was the founder-editor of *Diplomacy & Statecraft*. His research interests include the evolution of Anglo-American relations, and British policy in the eastern Mediterranean. He is also interested in the role of both religion and culture in diplomacy. He is a Fellow of the Royal Historical Society.

Melanie Hall is Associate Professor and Director of Museum Studies, Boston University. She previously worked for English Heritage (the Historic Buildings and Monuments Commission for England) for which she wrote seventeen *Lists of Buildings of Special Architectural or Historic Interest*. She writes on Anglo-American preservation at the turn of the twentieth century, and the origins of the National Trust. Her publications include: 'Affirming Community Life: Preservation, National Identity and the State, 1900', in Chris Miele (ed.), *From William Morris: Building Conservation and the Arts and Crafts Cult of Authenticity 1877-1939* (2005), and 'Plunder or Preservation? Contesting the Anglo-American Heritage in the Later Nineteenth Century,' in Peter Mandler and Astrid Swenson (eds), *From Plunder to Preservation. Britain and the Heritage of Empire, 1800-1950* ('Proceedings of the British Academy', forthcoming). She is a Fellow of the Society of Antiquaries, London.

Michael Holleran runs the Graduate Program in Historic Preservation in the School of Architecture at the University of Texas at Austin. He worked previously at the University of Colorado as Associate Dean of Architecture where he founded the Colorado Center for Preservation Research. He earned his PhD in urban studies at the Massachusetts Institute of Technology. He is a historian of the preservation of both buildings and landscapes and his publications include *Boston's 'Changeful Times': Origins of Preservation and Planning in America* (1998) which won the Antoinette Downing Award from the Society of Architectural Historians.

Peter Mandler is Professor of Modern Cultural History at the University of Cambridge and Bailey College Lecturer in History at Gonville and Caius College. His most recent books are *History and National Life* (2002) and *The English National Character: The History of an Idea from Edmund Burke to Tony Blair* (2006). He is currently completing a book tentatively titled 'Return from the Natives: How Margaret Mead Won the Second World War and Lost the Cold War' and editing with Astrid Swenson *From Plunder to Preservation. Britain and the Heritage of Empire, 1800-1950* ('Proceedings of the British Academy', forthcoming).

Chris Miele is an architectural historian and chartered town planner. He has written widely on the history of the British preservation movement, including editing a collection of essays, *From William Morris: Building Conservation and the Arts and Crafts Cult of Authenticity 1877-1939* (2005). His recent publications include *The Supreme Court of the United Kingdom: History, Art, Architecture* (2008); and *Danson House: The Anatomy of a Georgian Villa* (2008). He is a Partner in the Planning and Development Department, at Montagu Evans LLP where he advises on major development and regeneration proposals. He recently advised the Trustees of the British Museum in support of their World Conservation and Exhibitions Centre, and he is currently advising the Dean and Chapter of Westminster Abbey on their 2020 Vision proposals. He is a Fellow of the Royal Historical Society and of the Society of Antiquaries, London.

Thomas Otte is a Senior Lecturer in Diplomatic History at the University of East Anglia. He has published widely on diplomatic and international history of the late nineteenth and early twentieth centuries. His most recent publications include *The Foreign Office Mind: The Making of British Foreign Policy* (2011) and *Coalitions in British Politics: From Glorious Revolution to Cameron-Clegg* (2011). He has also written on the impact of railways in international politics, as well as on religion's role in international relations.

Astrid Swenson is a Lecturer in the Department of Politics and History at Brunel University. She studied History, History of Art, Romance Languages and European Studies in Mainz, Dijon and Cambridge. She was a Research Fellow of the Cambridge Victorian Studies Group's interdisciplinary project 'Past versus Progress in Victorian Britain,' funded by the Leverhulme Trust. Her publications include *The Rise of Heritage. Preserving the Past in France, Germany and England, 1789-1914* (forthcoming) and with Peter Mandler (eds), *From Plunder to Preservation. Britain and the Heritage of Empire, 1800-1950* ('Proceedings of the British Academy', forthcoming).

Ola Wetterberg is Professor of Conservation and Chair of the Department of Conservation at Göteborg University. His research covers the conservation movement in Sweden in the beginning of the twentieth century and its relation to contemporary development of town-planning practice, as well as the transformations and morphology of the urban fringe, and connections between nature and culture in environmental planning. He has published extensively in Swedish on preservation, including most recently: *Malmberget: Structural Change and Cultural Heritage Process, A Case Study* (2009), *The Significance of History in Heritage Management* (2009), and *History's Role in kulturmiljörvaltning* (2008).

Foreword
Facing the Past

Charles Dellheim

I was both surprised and elated when my colleague Melanie Hall invited me to participate in the conference she was planning on the 'International Origins of the Preservation Movement, 1880-1920.' Historic preservation had figured prominently in my first serious historical work, but I had moved on long ago. The conference took place in Boston on April 1, 2006, thirty years to the date after my Yale PhD oral exam in History, an event that I perversely enjoyed (much to the amusement of my friends who had teased me about undergoing this ritual on April Fool's Day of all days). At the time, all I knew about historic preservation came from some impassioned essays by John Ruskin and William Morris. I had no idea that I would ever do any research of my own on the subject. But my move to Britain soon changed that.

When I arrived in London in October 1976 to do my dissertation research, historic preservation seemed more relevant to the state of the nation than the condition of its buildings. The International Monetary Fund's bailout of nearly bankrupt Britain (which had helped found the organization less than three decades before), dramatized the political, economic, and cultural ills known as 'the British disease.' All the talk focused on decline, but it was easier to attribute blame than to find solutions. Some kept their sense of humor anyway. The newspaper seller in Russell Square greeted me in the morning as 'Guv' when I bought the establishment paper, *The Times*, and as 'Comrade' when I opted for the left-wing *Guardian*).

I was fascinated by the political spectacle that surrounded me, but I had come to interpret British history not to change it. The historical problem that concerned me was the fascination with the medieval past in the increasingly industrial, progressive society of Victorian England. Historic preservation didn't figure at all in the string of prospectuses that I handed to my generous and wise adviser, Franklin Le Van Baumer. He tore each iteration to shreds, with a huge desk blotter balanced on his lap and a gently encouraging smile on his face. My plan was to study medievalism from the point of view of 'high culture,' examining how poets, historians, artists, and architects used and abused the forms and symbols of the Middle Ages.

I spent much of my first months in London exploring the capital and trying to get a feel for a culture that I had studied but never visited. I took my lead from Marc Bloch and Henri Pirenne, two great historians who made a habit of walking around the towns they visited before hitting the local archives. So I made the rounds at the usual tourist sites from the Tower of London to the Palace of Westminster and walked around neighborhoods from Hampstead to the East End. All the while, I was aware, perhaps too much so, that I was following in the path of American expatriates such as Henry James. The falling pound was a godsend to me, albeit a disaster for the country-at-large. It enabled me to spend many evenings on the South Bank at the National Theatre or Royal Festival Hall, attending the latest Tom Stoppard play or listening to Alfred Brendel work his way through Beethoven. London was as inexpensive then as it is expensive now. My small room at the Crescent Hotel, a modest bed-and-breakfast in Bloomsbury run by a charming Italian widow, cost me about £4.00 per night (plus the coins I fed into the electric fire) – the current price of a double latte (sans mocha). I bought loads of books at Dillon's from Trotskyite book clerks who deigned to accept my money while readying themselves for the coming revolution. While doing my best to learn about the ins and outs of British society by day, I amused myself at night by reading about the varieties of alienation anatomized in Saul Bellow's fiction and Georg Simmel's social theory.

Britain may have been falling apart, but I could hardly have been happier. My work wasn't going well, however. I struck out in the manuscript collections of the (old) British Library and (old) Public Record Office. I decided to try my luck at the Society for the Protection of Ancient Buildings, which William Morris had founded nearly a century before. When I arrived at its headquarters on Great Ormond Street late one afternoon, I was greeted by its Secretary, Mrs Dance, a genteel county lady (or so she seemed) who was a far cry from its firebrand socialist founder. She could hardly have been more gracious or helpful however. She showed me into a small room, which housed the SPAB archives. The good news was that Mrs Dance very kindly granted me the extraordinary privilege of looking at whatever I liked; the bad news was that SPAB had no finding aid and I had no idea what I was looking for.

And so I simply began going through the tattered, old case files one by one, trying to find anything that would help me understand Victorian attitudes to the past. Trained as an 'analytical' historian, I looked down on narrative as "mere description." In the SPAB archives, however, I discovered just how difficult it sometimes was to reconstruct what actually happened. Going through file after file of faded pages in hard-to-read handwriting, I remembered what Peter Gay had told me during my first year of graduate school: 'History isn't always a champagne dinner,' he had warned (even though it had seemed just that for him). This injunction, which had annoyed me at the time, kept me going when everything seemed to be going wrong with my research. Fortunately, I gradually found a number of rich stories, including those of Kirkstall Abbey, Leeds, and the York city churches.

The approach that I took in what eventually became *The Face of the Past* was shaped by the experience of living in a society soaked in history's physical remains. The presence of the past was almost always on my mind as I traveled through the Continent and then Britain. It was my first trip to Europe and I couldn't get enough of the architecture that I came across day by day. None of this might have amounted to much, though, without the amazing guidance and support that I received from Asa Briggs, preeminent scholar of Victorian Britain, Provost of Worcester College, Oxford, and probably one the few historians in Britain who would have had any interest in my project. He encouraged me to do 'visual cultural history' and sent me off into the provinces to look at archives and architecture that I otherwise would have missed. Always innovative, Asa believed in theory well before theory was fashionable, but he was enough of a traditionalist to think that it should inform rather than dominate historical writing. At his suggestion, I delved into a number of new fields, including city planning, urban sociology, social geography, and mental mapping. I learned a lot more about how buildings shape the sense of place from reading Kevin Lynch's *The Image of the City* than from the heartfelt pleas or nostalgic cries of preservationists decrying (usually for good reason) threats to historic survivals.

Historic preservation was only part of the story of how Victorians regarded, treated, and rehabilitated medieval artifacts. I wanted to understand what attitudes to the past revealed about the larger culture and society. What interested me was the place of old buildings and remnants in the Victorian imagination; the various forces that shaped their fate; and how preservation served provincialism, nationalism, and professionalism.

The Face of the Past died quietly. My approach got me into a certain amount of trouble with formalist architectural historians who were mainly interested in aesthetic style and technique. They didn't like intruders (at least this intruder) from another discipline. The book, which was published by Cambridge University Press in 1983, during the heyday of the 'new social history,' had little, if any, impact on British historians who were more concerned with class than culture. By then, I was teaching at Arizona State University, painfully far from my native city, New York, or my adopted city, London. It took no time at all to figure out there was about as much interest in historic preservation in Phoenix (where the 1920s may as well have been the Middle Ages) as in snow removal.

My interests changed, too, as I began to focus on the interaction between culture and business. From time to time, I delved into some of the literature on 'heritage,' above all David Lowenthal's landmark work, *The Past is a Foreign Country*, which I regularly used when I taught historiography.

But I can't say that I thought much about historic preservation until I chaired a session at the Boston University conference on the International Origins of the Preservation Movement. For me, it was a sort of homecoming, an occasion to think about ideas and issues that I had left behind, not without a certain regret. Like the narrator in T.S. Eliot's 'Burnt Norton' I felt as if I had arrived where I had started and knew the place for the first time.

I was grateful for the chance to have an early look at the present volume, *Towards World Heritage*, which originated in the conference. It brings together the work of specialists from different fields and shows the fruits of cross-disciplinary research, which combines history, politics, international relations, and religion with art and architecture. Although the authors do not doubt that historic preservation is an aesthetic and social good, this is a work of scholarship rather than advocacy. It is fortunately free of the notes of self-congratulation that informed many of the early works in the field. In fact, Peter Mandler's chapter exposes some of the Whiggish assumptions, which shaped the work of early preservationists. Raymond Cohen's examination of the Church of Nativity in Bethlehem demonstrates that we can learn a great deal from failed attempts to preserve revered buildings. *Towards World Heritage* is notable too because it expands the study of preservation to include landscape as well as architecture. Melanie Hall's piece shows that some of the same considerations that drove campaigns to save historic buildings informed efforts to protect, and indeed sacralize, Niagara Falls too.

Towards World Heritage is most exciting because it truly delivers on its title, focusing on the interplay between politics and culture internationally. Erik Goldstein's study of Hagia Sophia, for example, analyzes both what this great church meant to different individuals and groups and how conflicting political interests shaped its modern history. Britain and America get their share of attention in this volume, but not at the expense of other societies. Astrid Swenson looks at preservation legislation in France, Germany and England; Ola Wetterberg takes us to Sweden; and Anne Blackburn to Sri Lanka. By the journey's end, we have a much richer sense of how preservation movements developed in various societies, what ideals and interests they promoted, and to what effect.

Towards World Heritage goes a long way towards answering questions about the international origins of preservation. Even so, much remains to be done in the field, especially on the issues of ethnicity, gender, class, and race. There has been relatively little sociological and cultural analysis of which groups have – and haven't – been involved in preservation. Guarding historic buildings and remains naturally seems more important to those who identify with the world they represent. The fact that certain Christians have argued for the preservation of their religious monuments is unsurprising. Their efforts may (or may not) be admirable, but it's clear that they serve the interests, and stoke the self-regard, of elites or those who aspire to join their ranks. But little, if anything, has been said about other groups, notably Jews, who became important players in the arts as both producers and brokers in the late-nineteenth and twentieth century. How did various Western societies deal with the monuments (such as they were) of those they regularly persecuted or repressed? To paraphrase Ghandi: Historic preservation may be a very good idea, but whose culture was worth memorializing? Whatever directions new scholarship takes we can be fairly certain of one thing: We will continue to see our own image, for better and worse, in the face of the past.

Acknowledgements

This volume has grown out of a conference organized to explore the International Origins of the Preservation Movement, 1880-1920 held at Boston University in April, 2006. The conference, one of a series that have explored issues relating to historic preservation and house museums internationally, was generously funded by a grant from the Humanities Foundation at Boston University, and it is a pleasure to thank its then Director, Professor Katherine O'Connor, whose enthusiasm and personal support for this series have been invaluable. The conference was a collaboration between the Departments of the History of Art and Architecture, and International Relations, the Preservation Studies Program, and the International History Institute.

Several colleagues generously assisted in the project. Special thanks go to Keith Morgan, Richard Candee, Alice Tseng, Claire Dempsey, Kim Sichel, and Fred Kleiner as well as to the several contributors to the volume and the conference. I am also grateful for the assistance provided at various stages by the following graduate students, some of whom have gone on to careers in museums and preservation: Melissa Renn, Danielle Cavanna, Bridget O'Riley, Ginger Smith, and Alexandra Polemis.

Thanks also go to the many colleagues, curators, custodians, archivists and other interested parties who attended the conference and provided stimulating discussion. Additionally, to all of those who have cared for the sites and archives included in these chapters, without whom the material we are considering would not have been preserved. Finally, I am pleased to thank Val Rose and her editorial team at Ashgate, who have helped to steer the project to publication.

Melanie Hall
Boston University

Introduction
Towards World Heritage

Melanie Hall*

The idea of protecting sites as examples of 'world heritage' developed during the twentieth century but had its antecedents in the nineteenth century. We are now familiar with the idea of 'world heritage'; and the inter-governmental World Heritage Committee which was formed after the United Nations Educational, Scientific and Cultural Organization (UNESCO) adopted the Convention Concerning the Protection of the World Cultural and Natural Heritage (World Heritage Convention) (1972, enacted 1975) designates sites of World Heritage.[1] This formula has proved successful; there are currently almost one thousand World Heritage Sites in one-hundred-and-fifty-one countries, inscribed for their 'cultural and natural' interest. While it has been accepted that the idea of international cooperation in preservation became established in the nineteenth and early-twentieth century, there is little understanding of how this actually happened; the phenomenon of 'world heritage' has been neither adequately examined nor explained.[2] This volume explores and explains the contexts for the emergence of 'international heritage' during the important formative period from 1870 to 1930 when the protection of specific places, or sites, moved from a cause of national and imperial concern to one of international concern.

* Particular thanks for their comments on this section are due to Raymond Cohen, Erik Goldstein, Howard Malchow, Thomas Otte and John Walton.

1 Francesco Francioni, *The 1972 World Heritage Convention, A Commentary* (Oxford: Oxford University Press, 2008).

2 See, for example, Michael Fry, Erik Goldstein, Richard Langhorne, *Guide to International Relations and Diplomacy* (London: Continuum, 2002), 40-54; Armin von Bogdandy, Rüdiger Wolfrum, Jochen von Bernstorff, Philipp Dann, Matthias Goldmann, eds, *The Exercise of Public Authority by International Institutions Advancing International Institutional Law* (Heidelberg: Springer, 2010), 303-04; UNESCO World Heritage Centre, *World Heritage: Challenges for the Millennium* (2007), 26-8, online at http://whc.unesco.org/documents/publi_millennium_en.pdf. Recent considerations of the institutionalization of World Heritage that refer to its deeper intellectual origins include, Craig Forrest, *International Law and the Protection of Cultural Heritage* (London and New York: Routledge, 2010), 1-30. See also, Peter Mandler and Astrid Swenson, eds, *Plunder to Preservation: Britain and the Heritage of Empire, 1800-1950.* 'Proceedings of the British Academy' (Oxford: Oxford University Press, forthcoming).

Why 1870-1930?

The ascendancy of international cooperation on matters of 'heritage protection' occurred in the context of empires and during a period of enormous change. The date range covered here begins in the 1870s, when societal and economic conditions and industrial development were beginning to prompt more focused, more organized, and broad-based efforts to reserve 'cultural' and 'natural' resources for representational leisure in developed countries. Such efforts, institutionalized within societies and in law are noticeable in Europe and in its imperial territories as well as in the United States of America which, following its Civil War, was taking off as an industrial power and becoming more integrated into the 'international community.'

Preservation of 'cultural and natural' resources was taking place beyond the local level and at government level in several countries by the 1870s; legislation was usually framed around categories of 'nature' or of 'culture'; international preservation initiatives were also beginning to gain momentum, important precedents for this are discussed below. By the 1890s, 'heritage' conceived in terms of 'nature *and* culture' was being instituted in several preservation societies notably in the Massachusetts Trustees of [Public] Reservations (founded 1890); Britain's National Trust for Places of Historic Interest and Natural Beauty (instituted 1894-95 but in planning earlier), and the [New York, later] American Scenic and Historic Preservation Society (1896 and 1901). All three societies collected multiple properties; all were to some extent interlinked. The National Trust aspired to act internationally, affiliating with organizations in the United States, Canada and elsewhere, and forming a short-lived branch in Washington DC (1899-1904).[3] Their initial aims were comparatively broad and their functions still developing; however, none were 'modern' in John Breuilly's definition of 'institutions with specialized functions and extensive societal reach.'[4] They relied on group philanthropy and self-selecting support. Such institutions provide early precedence for the collection of *World* Heritage sites. Legal debates began to be framed around the nature-culture paradigm, although 'nature' was understood in 'cultural' terms as scenic, beautiful and sometimes, of literary and historic interest.

In the 1930s the world-wide depression, the trend toward national autarchy, and the drift into another world war may have worked against the agendas of international preservationism. 1945, however, would see its significant return in the creation of UNESCO, following the Second World War to 'contribute to peace and security by promoting collaboration among nations through education, science and culture in order to further universal respect for justice, for the rule of

---

3   Melanie Hall, 'The Politics of Collecting: The Early Aspirations of the National Trust, 1883-1913', *Transactions of the Royal Historical Society*, 6th series, vi (2003), 354-56.

4   John Breuilly, 'Modernisation as Social Evolution: the German Case, c1800-1880,' *Transactions of the Royal Historical Society*, 6th series, xv (2005), 121.

law and for the human rights and fundamental freedoms which are affirmed for the peoples of the world, without distinction of race, sex, language or religion, by the Charter of the United Nations.'[5] Considering how and in what ways such values underlie early internationalizing preservation initiatives raises contextual questions that intersect as underpinning themes throughout this book.

As the international heritage phenomenon is complex, involving people from different countries, professions and disciplines, including politicians and diplomats, practitioners, authors, artists, and philanthropic advocates, so this volume is multi-disciplinary and international in scope. Some chapters can be read as case studies, while themes work synchronically and diachronically to provide the reader with a rich and textured understanding of an important aspect of the global culture with which we live. One of the volume's strengths is in making connections that, hitherto, have not been apparent. To understand better how heritage resources came to be perceived as of international importance, we consider examples of successful campaigns as well as those that were unsuccessful or partially successful. Usually, studies of preservation involve success stories; 'failures' are included in so far as they rouse action. This is not the case here. Some considered sites are now sanctioned by World Heritage status; some involved international efforts but did not gain world heritage recognition; others have faded into obscurity, their use now limited. All help to contextualize the criteria and circumstances in which international heritage developed. Some chapters use case-studies, others take a more exploratory angle in opening out areas for further research. Thus, the volume provides much for existing scholars while suggesting scope for new scholars.

Themes and chapters in this book can and should be read and understood synchronically and diachronically. The book is divided into two sections; the first section has case studies, each from a different disciplinary perspective. However, politics, art and architecture, and religion intertwine. The book opens with what was arguably the first truly inter-national and trans-national preservation effort at Niagara Falls. It then moves to Constantinople [Istanbul] where preservation of the Hagia Sophia mosaics by the Sultan led to British attempts to *redeem* the building for Christendom; this building has subsequently become part of a World Heritage Site. Preservation in Ceylon [Sri Lanka] took place within the context of the British Empire, though religion and aesthetics were important factors for Governor Gregory; here, too sites are now inscribed on the World Heritage List. At the Church of the Nativity, Bethlehem neither the 'religious', the Ottoman Empire, or the British Mandate were willing to intervene to 'preserve' the building, despite its religious status. At Sulgrave Manor, an 'ancestral site' in Britain was preserved due to British and American group interests with common political objectives. Part two considers the growing *international* awareness of the need for heritage protection and the interactive ways in which this is achieved. The next chapter considers the legal processes that resulted in preservation's rise as a facet of state*craft* in Europe; what look like national activities in fact occur in a continental

---

5   Article I, UNESCO Constitution, quoted in Forrest, *International Law*, 31.

context. Aspects of preservation practice in England, the United States and Sweden reveal the various influences of religion, of complex political, educational, and professional group interests, and the need for planning. This too takes place within the context of a growing awareness of the efforts of other countries. Finally, an 'anti-history' of preservation considers the ways in which preservation has been considered and opens up areas for further investigation.

Notions of Heritage and Preservation

The idea of 'heritage' has both cultural and legal connotations that pertain broadly and specifically. Heritage is generally understood as something handed down from the past which has value in the present; at its best, it provides an imaginary realm where the possibility of a common good can be imagined and, perhaps, discussed without intentional or overt suspicion of bias.[6] As more recently the criteria of 'intangible heritage' has been added to UNESCO's adoption of the 'culture and nature' taxonomy, it is clear that the 'heritage' realm is expanding; this suggests a useful time to examine earlier contexts. While 'cultural' heritage or, rather, architectural heritage gains prominence in this volume, natural heritage is also considered.[7] The nature-culture dialectic was largely self-reinforcing and can be understood in the context of similar oppositional dichotomies with which the Victorians explained their world: town and country, metropolis and wilderness, hub and periphery.[8] The dialectical approach is familiar to diplomacy particularly in diplomatic gift-giving and reciprocity which, as Harold Nicolson observes, has to take 'the correct dialectical form'; however, dialectics shift according to need and circumstance.[9] World heritage sites are, in one sense, each nation's 'gifts' to the world.

The term 'preservation' used in this volume's title likewise has cultural and legal connotations. It is more common to speak of heritage in Britain and preservation in the United States, for example. Such distinctions may owe much to the law and patterns of settlement. Preservation's terminology was far from fixed during the period in question. Sites were variously reserved, preserved, protected, and even redeemed by different groups. Terminology carried differing connotations

---

6   See, for example, Robert Hewison, *The Heritage Industry: Britain in a Climate of Decline* (London: Methuen,1987), 9; David Lowenthal, *Possessed by the Past: The Heritage Crusade and the Spoils of History* (New York: The Free Press, 1996), 94-95; Raphael Samuel, *Theatres of Memory: Past and Present in Contemporary Culture*, vol. 1 (London, Verso. 1994), 292.

7   The term 'cultural heritage' is similarly problematical and has been discussed by Forrest, *International Law*, 1-30.

8   See Raymond Williams, *The Country and the City* (New York and Oxford, 1973).

9   Harold Nicolson, *Diplomacy* (London, New York: Oxford University Press, 1950), 23-28, 106.

that formed part of then-current political debates. It still can. Terminology is used differently by different disciplinary interests, in different countries, and can depend for its nuanced meanings on counter-prevailing oppositional terms. Hence, the title of this book references both preservation and heritage; there is a similar elision in the wordage of the Convention.

Countries and Disciplines Covered in this Book

While there is a focus on the British Empire, the United States and Sweden are also included, as are mainland France and Germany. The Ottoman Empire is present in two of the chapters. Put another way, preservation in the following countries is considered: Canada, England, France, Germany, Sri Lanka, Sweden, Turkey, and the United States. Not all of these countries existed within their current boundaries or under their current names; some had complex relationships. Bethlehem is also included but what state that is currently part of is a matter of debate. Until 1905, Sweden meant the Kingdom of Sweden and Norway, and included parts of what is now Finland, as well as Lapland. Sri Lanka was known as Ceylon, a British colony. The United States and Britain shared a history and, so, heritage resources but the British Empire shared a border with the United States via Canada. However, within these countries preserved sites could be recognized as expressing difference, while suggesting a common exhibitionary culture that helped to suggest an internationally-unified civilization.

The Question of Community

Questions of community are central to several chapters and were topical in the nineteenth century. Membership of a community was, for preservationists, equated with the rights and responsibilities of citizenship. However, communities could be imagined in multiple ways. Benedict Anderson, writing on nationalism, has considered how a sense of community might be communicated and defined, with what, and by whom. As David Cannadine, echoing Anderson observes, 'Nations … are in part imagined communities, depending for their credibility and identity both on the legitimacy of government and the apparatus of the state, and on invented traditions, manufactured myths, and shared perceptions of the social order that are never more than crude categories and oversimplified stereotypes.'[10] So, too, are international communities. Heritage sites were visited by travellers and tourists as real and imagined groups. Preservationists themselves also formed loosely-affiliated groups, influenced by one another's activities and achievements.

---

10   Benedict Anderson, *Imagined Communities: Reflections on the Spread and Origin of Nationalism* (London and New York: Verso, 2006); David Cannadine, *Ornamentalism: How the British Saw their Empire* (New York: Oxford University Press, 2001), 3.

International preservation initiatives emerged in complex contexts of colonialism, imperialism, and nationalism. Relationships and communities were considered and represented geographically and historically; that is, in terms of 'nature and culture' and of 'past and present.' Creating a range of international sites helped to demarcate acceptable norms in distinct social and industrial arenas. Simultaneously, such sites provided structures of legitimate meaning that brought a system of shared values to the social experience and allowed complex and shifting group identities to be re-imagined, at least by some.

Most preservation studies divide the built and landscape environments, or cultural artifacts and nature. That this is not the case here accords with late nineteenth century preservation practice, with well-established categories of collecting, and with the taxonomies of World Heritage itself. While establishing heritage of (and as) *place* was very much a focus of preservation activity during this period, a seemingly 'intangible heritage' was often covertly represented. As becomes apparent in this volume the role of religion, the arts, education, commerce, and science were inter-related agencies. Several chapters discuss the impact of then current thinking among the preservationist community upon its choices.

Religion and Preservation

'Modernization', industrialization, and urbanization are well-understood to have prompted preservation efforts. These factors not only altered the face of nations, they also brought secularization and an increasingly scientific understanding of the world.[11] But how far is the relationship between these factors and preservation understood? As this volume reveals, religion was a factor in national and international preservation; several of the most important targets of international preservation were religious buildings; some of its seminal texts, including John Ruskin's *The Stones of Venice* (1851-53), were inspired by religious buildings.[12] Religion's role in determining cultural responses to preservation targets cannot easily be ignored. Religions often use aesthetics and affect to win hearts and minds as well as to impart normative values and so too, in many cases, does preservation. Nevertheless, aesthetics and religion were subject to hierarchical systems of value; several chapters in this book question and examine the components of these values.

---

11   Some aspects of preservation move in more 'scientific' directions, notably in the social science of archaeology or, in the nascent attempts to protect rare species and habitats; these have come to be viewed in terms of conservation and are not concerns here.

12   For the complexities of Ruskin's approach to buildings see, Robert Hewison, *Ruskin on Venice: 'The Paradise of Cities'* (New Haven and London: Yale University Press, 2009), esp. 117-21; his writing is often a poetic plea for preservation.

The function of religion in nationalism and national identity formation has been minimized by writers who themselves subscribe to modernist views.[13] More recently, Adrian Hastings has reminded us that religion should be seen as a fundamental component of nationalism. Hastings asserts religion's profound importance during the medieval period and to the identity-formation of ethnic groups; it is also important to the cultural traditions of nations. Hastings argues that, 'while the role of religion has been far from single-faceted in its relationship to ethnicity and the construction of nations, it has been integral to this wider history, perhaps even determinative. The history of religion can never be best understood within a box of its own and that is evidently particularly true in a field such as this where religion, politics and culture so obviously interact.'[14] While *national* communities could be imagined around the church in England and Sweden, as Chris Miele and Ola Wetterberg demonstrate, through the chapters of this book it becomes inescapably apparent that *international* romanticized-Christian communities were re-imagined around significant, monumental Byzantine and medieval churches. More modernizing trends are discernable both in the approach of preservationists to a slightly-expanding range of religions, and in the different religious groups that engaged in the act of preservation. As Peter Mandler argues, we need to better understand preservation's complex religious dimensions.

In preservation, the economic and social interests of tourism and publishing, education, histories of various kinds and especially art and architectural histories, and more modern cultural productions intersected with those of politics and religion. The interests of religion interrelate with those of preservation in multiple and sometimes unexpected ways. The earliest international example discussed here is a campaign to *redeem* the Hagia Sophia in Istanbul; as Erik Goldstein demonstrates, politics overlapped with preservation. Preservation issues at that site are complex; the building was made a mosque in the 1453 conquest of Constantinople thereby protecting it from destruction. That the Hagia Sophia is now part of a World Heritage Site is testament to the possibility of cooperation. Religious interests do not always coincide with those of preservation as becomes clear in the case of the Church of the Nativity in Bethlehem discussed by Raymond Cohen. Despite several factors that could have made it attractive to preservationists, the Church did not become an international preservation target and languished.

Anne Blackburn and Chris Miele consider the complex relations between religion and preservation with respect to Buddhist sites in Sri Lanka, and the Anglican (Episcopal) Church in England. As Ola Wetterberg indicates, protecting religious buildings as indicative of parish life was important in Sweden.

---

13 See, for example, Eric Hobsbawm, *Nations and Nationalism Since 1780 Programme, Myth, Reality* (Cambridge: Cambridge University Press, 1992); John Breuilly, *Nationalism and the State* (New York: St Martin's Press, 1982); Ernest Gellner, *Nations and Nationalism* (Ithaca: Cornell University Press, 1983); Anderson, *Imagined Communities.*

14 Adrian Hastings, *The Construction of Nationhood, Ethnicity, Religion and Nationalism* (Cambridge: Cambridge University Press, 1997), 2.

Anuradhapura and Kandy are now both World Heritage Sites; in England, three significant religious sites have been inscribed on the World Heritage list; Gammelstad Church Town is a World Heritage Site, one of only sixteen remaining Swedish church towns. Nevertheless, acts of preservation as commemoration are integral to certain religions; others take an approach of continuous renewal.

The Sacralization of Nature and Culture

Some sites carried culturally sacralized, rather than overtly religious meanings. As religious impulses were romanticized in nature, so certain sites became transcendentalized.[15] The Niagara Falls campaign demonstrates aspects of this development, as Melanie Hall explains. It falls into the category of sites of the 'sublime' which, as a philosophical and aesthetic construct, induces awe. Such sites provided encompassing sensory experiences where sights, sound, and smell could be joined by touch, and motion. The protection of sites for environmental or proto-ecological purposes was rare.

A similar sacralizing process is found at other, secular, cultural sites including birthplaces, which have religious overtones, and ancestral homes that exhibit similarities to forms of ancestor-worship. Thomas Otte's discussion of Sulgrave Manor, George Washington's ancestral home in England demonstrates that this was a significant shrine for specific but influential American groups speaking in the name of the nation. Such impulses underpin new nationalistic cultural norms that were represented around heroic figures and inspiring places with which transformative events in a grand narrative of the life of either a nation or international community might be associated.

Both natural and cultural sites accrued international meaning through various means. As people travelled and migrated with industrialization, urbanization, imperial expansion and colonization individuals, families, and community groups acquired more complex and multi-national identities. So, too, texts and images became increasingly familiar to wider groups with developments in printing and educational subjects such as art history and literature. The role of art and, particularly its reproduction, publication, and dissemination in print (with or without text) as an agency in internationalizing preservation is a theme that runs through several chapters here, notably Erik Goldstein's and Melanie Hall's. Some sites acquired iconic, episodic or *monumental* status which, Chris Miele explains, was observed by William Morris. Monumentality was emphasized by legislation.[16] The competitive legislative processes that helped internationalize heritage protection forms the theme of Astrid Swenson's chapter.

---

15   Tony Tanner, *The Reign of Wonder, Naivety and Reality in American Literature* (Cambridge: Cambridge University Press, 1965), 19-25.

16   The idea of the monument as a political, legal, variously religious or quasi-religious, and experiential entity was also topical during the nineteenth century when the preservation

International Relations, the Church, and Preservation

International preservation sites have an important context within and representational values for international relations. Invariably, international historians concentrate on the general political framework within which diplomatic activities took place; scholars of preservation rarely resort to international relations.[17] However, politics cannot be divorced from representation and political activity is negotiated and communicated through public representation, symbolic actions, and public interests. Religious institutions had long developed sophisticated techniques for representing temporal and 'spiritual' power that helped maintain stability and loyalty among the faithful by spreading 'normative' and 'unitary' values across a 'society,' whether in one or more countries. The preponderance of religious sites covered in this volume speaks to the association between preservation, nationalism, and religion or, at least, certain religious impulses and – then – culturally-'European' values. A significant point of origin for the intertwining of preservation, international representation, and the power of religious and commemorative sentiment is to be found at the Congress of Vienna and, as it is little considered, is worth outlining.

The great powers increasingly sought agreement and alliance during the nineteenth century in systems that provided both state representation and bureaucratic practices of governance. It was desirable to represent the state and its practices as one and the same thing, both autonomous and indivisible, at specific sites. The metaphorically-named Concert of Europe was a system and a structure wherein great powers came together to achieve international cooperation and agreement that was then represented in matters of state. Cooperation was often achieved incrementally through the diplomatic congresses that helped to foster the developing sense of international interest following the Congress of Vienna.

---

process involved defining the monument legally; its seminal text, written to promote preservation, was Alois Riegl, *Der modern Denkmalkultus, sein Wesen, seine Entstehung* (Vienna: W. Braumuller, 1903) ('The modern cult of monuments: its character and its origin', in Nicholas Price, M. Kirby Talley Jr., and Alessandra Vaccaro, eds, *Historical and Philosophical Issues in the Conservation of Cultural Heritage* (Los Angeles: the Getty Conservation Institute,1996), 69-83. For more recent discussions of built monuments see, Robert S. Nelson and Margaret Olin, eds, *Monuments and Memory Made and Unmade* (Chicago: University of Chicago Press, 2003), esp. 1-10; Jukka Jokilehto, *A History of Architectural Conservation* (Oxford: Butterworth-Heinemann, 1999), 215-18 et passim; Miles Glendinning, 'The Conservation Movement: A Cult of the Modern Age, *Transactions of the Royal Historical Society*, 6th series, vi (2003), 361-62. For the monument idea in nature see, Alfred Runte, *National Parks, The American Experience* (Lincoln: University of Nebraska Press, 1997), 14-17, 30-33 et passim.

17   For an inquiry into the sub-structures of diplomacy see, Melanie Hall and Erik Goldstein, 'Writers, the Clergy, and the 'Diplomatization' of Culture: Sub-structures of Anglo-American Diplomacy, 1820-1914,' John Fisher and Antony Best, eds, *On the Fringes of Diplomacy: Influences on British Foreign Policy* (Aldershot: Ashgate, 2011), chapter 3.

The rise of international cooperation and international preservation is synonymous with so-called congress and conference diplomacy which mushroomed during the nineteenth century.[18]

The idea of the preservation of cultural artifacts as a heritage of place of origin first emerged as a public and political issue at the Congress of Vienna. This Congress sought to settle a raft of international issues that established certain international principles pertaining to 'natural' and 'cultural' resources. The first significant example of *heritage* artifacts being used to suggest international cooperation occurred when the Second Treaty of Paris (1815) required the return of some of the antiquities and works of art that comprised the so-called Bouquet de Napoleon, looted from the Papal States and elsewhere during the Napoleonic Wars.[19] Although few works were in fact returned, this representational peace project by European Allies set an important precedent for the claims of national rights over significant examples of cultural property in times of war. The return of these cultural artifacts had representational value in affirming the Concert's respect for national sovereignty and the heritage of *place*, which gained importance as a means of affirming specifically national identities through geographical and cultural references both to antiquity and to current notions of the beautiful (suggestive of stability) and the sublime (suggestive of awe).[20]

Effectively, the return resulted in the affirmation and designation of Rome as a *site-museum*. It is not insignificant that the Papal States were the first to receive heritage recognition. Rome, also the former epicentre of the Roman Empire, was already regarded as religiously and culturally distinctive by grand tourists; here empire and history could be re-imagined as destiny. Advocacy for the return of the artifacts had been eloquently voiced by French pamphleteer, Quatremère de Quincy who saw the entire area of Rome as an interlinked geographical, geological and cultural studio. He described 'the true *museum* of Rome' in panoptic terms as made up not only of artifacts but of all the associated geographical and cultural features such as quarries, stones, and transport systems, and diverse cultural monuments and sites that had helped give rise to the artifacts.[21] Thus, he both expanded and united the taxonomic categories of artificial and natural artifacts used in cabinets of curiosities and encyclopædic museums, while also amplifying

---

18   Fry, Goldstein, Langhorne, *Guide*, 111-19; Erik Goldstein, 'The Origins of Summit Diplomacy', in D.H. Dunn ed., *Diplomacy at the Highest Level: The Evolution of International Summitry* (London and New York, 1996), 26-29.

19   Margaret M. Miles, *Art as Plunder: The Ancient Origins ofDebate about Cultural Property* (Cambridge: Cambridge University Press, 2008), 319-27.

20   It is worth remembering that the artefacts were returned to museums and collections, not to archaeological digs.

21   Quoted in Francis Haskell and Nicholas Penny, *Taste and the Antique, The Lure of Classical Sculpture, 1500-1900* (New Haven and London: Yale University Press, 1981), 110, et passim [my italics]; see also, Jokilehto, *History of Architectural Conservation*, 72-75.

inherent ideas of the museum as *theatrum mundi* by tying them to a sense of *genius loci*.[22] De Quincy considered that such contexts provided a better understanding of the origins and modes of production of cultural artifacts. Perhaps unsurprisingly, Rome is now a World Heritage Site; so, too, is the Vatican City.

Rome was important symbolically because it brought together Christian and Classical traditions, celebrated in the art, architecture and artifacts of the Vatican. The Church, however, represented a continuity of diplomatic practices as the origins of international diplomatic congresses are found in the Great Councils of the Church, convened to resolve issues of spiritual well-being that also threatened political stability.[23] Since at least the Renaissance important diplomatic conferences were invariably followed by festive celebrations and commemorated with works of art and by more permanent cultural productions that existed to bolster new agreements.[24] For example, redrawn maps and group portraits represented and legitimized new state boundaries and affiliations. These all manifested power, wealth, status, stability, continuity and cooperation in varying degrees. Acts of preservation, intended to encourage political and economic stability and (for advocates) the right to the pursuit of social well-being, were not dissimilar from those of the Church Councils. The act of returning artifacts to a church-state by Europe's Concert effectively suggested harmony between the great powers and the Church while at the same time asserting and demarcating the limits of secular authority. Economic restrictions were placed on the return of the artifacts by making the Italian states responsible for the cost of transportation.

Demonstrating agreement over moveable cultural artifacts was less controversial than over 'moving' natural resources and so doubtless helped pave the way for international agreement on more contentious and difficult issues. The Second Treaty of Paris also considered a shared natural resource that crossed national boundaries, the River Rhine; this was made an internationalized waterway administered under a Central Commission for Navigation on the Rhine.[25]

---

22   For the development and meaning of museums a useful place to start is Paula Findlen, 'The Museum: Its Classical Etymology and Renaissance Geneology,' *Journal of the History of Collections* vol. 1, no. 1 (1989), 59-78; Eilean Hooper-Greenhill, *Museums and the Shaping of Knowledge* (London: Routledge, 1992); Oliver Impey and Arthur MacGregor, *The Origin of Museums: The Cabinet of Curiosities in Sixteenth and Seventeenth Century Europe* (Oxford: Clarendon Press, 1985).

23   For modern diplomacy's development from Church and secular Italian practices see, M.S. Anderson, *The Rise of Modern Diplomacy, 1450-1919* (London: Longman, 1993) esp. 2-11; R.T.B. Langhorne, 'The Development of International Conferences, 1648-1830', *Studies in History and Politics* iii (1981-2), 65-75.

24   See, for example, J.R. Mulryne and E. Goldring, *Court Festivals of the European Renaissance: Art, Politics and Performance* (Aldershot: Ashgate, 2002) and J.R. Mulryne, Helen Watanabe O'Kelly and Margaret Shewring, eds, *Europa Triumphans: Court and Civic Festivals in Early Modern Europe*, 2 vols, (Aldershot: Ashgate, 2004).

25   Concerns about moving natural resources on and in the Rhine continued throughout the nineteenth century; free navigation on the Rhine (as well as other European rivers)

Although at issue were different states' riparian rights, the river's 'sublime' Falls of Schaffhausen were becoming an international cultural site popularly ratified by artists, poets, and tourists.

Preservation and Site-museums

The development of the idea of a *site-museum* as an important international theatre of significant cultural productions that were deemed worthy of preservation (rather than development or destruction) *in situ* as elements in a larger landscape came with the protest against looting of antiquities by Napoleon's troops. While they can be sites of political activity, as a representational theatre a site-museum is also a container of wonders and values among which the spectator is a voluntary actor.[26] Site-museums are associated with monumentalized icons of culture and nature; however, they are highly complex. In a site-museum a monument can be an individual item, a series of items, the entire area, or some combination of these. Site-museums thus seemingly combine the attributes of a fixed place and a collection. Each site-museum acts effectively as an episode in an exhibition or collection of similar sites found, for example, in guide books or tour itineraries. But it is not simply visual, it is an occurrence and experience as well as an incident or installment. Thus, the values of one and of the group are complex and compounding.

While De Quincy's pamphlet has been regarded by later advocates as an appeal for site-based preservation, the artifacts' return had representational value as a ceremonial act of cultural diplomacy. Parades of newly discovered or acquired antiquities heading towards museums or demarcating routes through landscaped gardens formed part of the cultural imagination of the period. Indeed, the practice of parading looted treasures through Imperial Rome had been emulated by Napoleon when the looted art works were paraded through Paris, an event that was subsequently commemorated on painted Sèvres vases made for the royal palaces.[27] However, the performative and ceremonial aspects of preservation have

---

was included in the Vienna Final Act of 1815 (Annexe XVI), and concerns about natural resources *in* the Rhine resulted in the 1886 treaty concerning salmon stocks.

26 Tanner, *Reign of Wonder*, 9-11, 258-61, although Tanner writes about American literature, his comments on wonder and art forms have wider application; Steven Greenblatt, 'Resonance and Wonder', in Peter Collier and Helga Geyer-Ryan, eds, *Literary Theory Today* (Cambridge: Polity, 1990), 74-90.

27 Miles, *Art as Plunder*, 321-24; D.I. Kertzer, *Ritual, Politics, and Power* (New Haven: Yale University Press, 1988), esp. 1-14, 159-61, participating in preservation can be considered as a form of national and, increasingly, international ritual; Haskell and Penny, *Taste and the Antique*, 76-77, 110-11, figs. 40 and 62; Simon Schama describes the river-roads of Italy, including the statue-infused cascades at Caserta, Simon Schama, *Landscape and Memory* (New York: Vintage Books, 1996), 342-45; Castle Howard's garden design begins with a statue walk which leads, ultimately, to the mausoleum, David Watkin, *The*

been largely overlooked. Nevertheless, preservation can be seen both in terms of a static site, and an active theatre that comprises staging, scripted performances, and spectatorship.[28] The return of artifacts by diplomatic agreement suggests recognition that there were objects of international heritage that should be protected or exempted from looting and perhaps, as many were owned by the Papal States, of the sacralized values of commemoration that could be tied to place and notions of stability through beauty.[29]

Where the state is involved, political interests coincide overtly with those of preservation, as Erik Goldstein analyzes in his discussion of the campaign to *redeem* the Church of St. Sophia, a focal point of the lost Byzantine Empire and an important landmark of the then-ailing Ottoman Empire, as a symbol of support for oppressed Christianity in that region of the world. A monumental religious symbol thus can be associated with broader ideologies and concerns. It is a tribute to the site that respect for Islam, for Christianity, for history and the arts has resulted in its site-museum status and incorporation within a World Heritage Site. In contrast the Church of the Nativity in Bethlehem, discussed by Raymond Cohen, failed to become a site-museum because the respective Ottoman and British governing powers were unwilling to intervene to protect this religious building. Neither the secularizing forces of aesthetics nor tourism were sufficient to counter prevailing religious tensions over ownership of the site. Instead, religious and secular 'pilgrimage' needs could be better met by the more monumental Church of the Holy Sepulchre. Now a World Heritage Site, the ancient city of Jerusalem has multiple attractions and, so, compound interest. These attractions include landscapes that were, perhaps, more accessible to Protestants and others.[30] As Peter Mandler argues, there is much greater need for an understanding of the nuances of religious sensibilities and, in turn, their impact on ideas of production and stewardship, ideas that differ not only between religious groups but also within broad religious communities.

English Vision, The Picturesque in Architecture, Landscape and Garden Design (London: John Murray, 1982), 8.

28   Elements of the complex theatrical production of site-seeing are described by Henry James in *Niagara*, an account that was itself a work of description and an advocacy for preservation. 'Niagara, 1871,' in Henry James, *Portraits of Places* (Boston: Houghton, 1884), 64-76. In preservation, as with theatre, verb and noun are interactive.

29   It is worth noting that the ancient Greek 'Horses of St Mark,' returned to the Basilica in Venice, had been taken by Venetian forces in the thirteenth century from Constantinople, which may, in turn, have acquired them from Chios. Napoleon intended them for the Arc de Triomphe.

30   For contemporary accounts see, Eitan Bar-Yosef, *The Holy Land in English Culture 1799-1917* (Oxford: Clarendon Press, 2005), 78-79. Sensibilities can include visual as well as other sensory factors such as smell, sound and general ambiance. It is perhaps not insignificant that the Anglicans chose to build an alternative church. A further site, The Garden Tomb, discovered and investigated during the 1860s and 1870s, is run by a multi-denominational, evangelical association.

Taste and taste-making are also important factors in preserved site-museums. Visuals and ideals of beauty are important to preservation as to some religious expression. Nineteenth-century artistic and poetic interest in sensory experience also permeate preservationism; that is, a supposition (almost an intuition) that ideas and understanding originate in the senses, incorporating not only aesthetic but, also, a range of sensory experiences that are often met in preservation sites. To some extent this was underpinned by science and philosophy. Taste found further expression in the emerging discipline of archaeology which reveals, displays, and interprets sites in a seemingly 'true' or 'scientific' form.[31] The sensory element of site-visiting is often affirmed by nineteenth-century guides to such sites that incorporate poetry and pictures, and make experiences of them both synaesthetic and individualized, even if that experience is re-imagined in sacralized terms. That perception is often given pre-eminence is affirmed by the adoption of the term 'site-*seeing*'.[32] The taste for classical aesthetics combining with Christian cultural expression helped shape investigations of sites further afield than Rome; Anne Blackburn demonstrates its influence on Governor Gregory's preservation choices in Sri Lanka. Efforts to redeem Santa Sophia came with knowledge of its Byzantine mosaics and, perhaps, with the frisson of an alternative religious experience, occasionally tinged with cross-dressing, as Erik Goldstein shows. Interest in Niagara Falls was for many a highly charged, sensory experience enhanced by a 'sublime' painting, as Melanie Hall demonstrates.

Preservation Advocacy, Professionals, and the Concert System

Internationalizing preservation gained momentum by the 1870s. Thus, the transformation of heritage initiatives to include international sites followed the transformation of European politics and the development of Concert diplomacy that sought to achieve harmony between the powers through international conferences and symbolic, ceremonial acts.[33] International preservation initiatives were further assisted by the close association between political life and publishing, by the

---

31   Any attempt to 'create' a site-museum requires intervention. Archaeology is its own kind of interpretation but, one that gives priority to the physical (natural) context of the artifacts.

32   The difference between site-seeing, site-experiencing, and site-analysis is a difference in levels of wonder, as defined by Tony Tanner and Steven Greenblatt (op cit at note 22). It is possible to wonder at something, or to wonder about something. Nevertheless, at site-museums the element of wonder retains its resonance, no matter how far revealed by archaeological 'science' or organized by creative curating.

33   P.W. Schroeder, 'The Nineteenth Century System: Balance of Power or Political Equilibrium?' *World Politics* xxxix (1986-7), 135-53; P.W. Schroeder, *The Transformation of European Politics, 1763-1848* (Oxford, New York: Oxford University Press, 1994).

extension of the public sphere of politics, and the rise of professional society.[34] In turn, print techniques allowed the use of more sophisticated images and so sought illustrative subjects. The extension of the public sphere was acknowledged by the established political elites of Europe who also utilized it extensively so as to ensure their political and social survival and win public opinion. A similar confluence of exhibitions and conferences was found at the international fairs which burgeoned from the mid-century, where trades, industries, and the arts were displayed and discussed.

Preservation was among those interests that attracted overlapping and interconnected professional and voluntarist interests. Interests included politics, education, religion, publishing, writing and the arts and, sometimes, banking. Politicians often acted in a personal (hobbyist) capacity while understanding the publicity platform that preservation activism and sites provided. James (later Viscount) Bryce, Regius Professor of Law at Oxford University, Liberal Member of Parliament and, from 1907-13, Britain's Ambassador to the United States, a prolific author, active mountaineer, member of several amenities societies, was a keen observer of public opinion and an active supporter of preservation as several chapters in this volume attest.[35] Artur Hazelius, the Swedish ethnographer and professor of Swedish dialects, gained an international reputation by exhibiting his tableau of Swedish life at the 1878 Paris and the World Exhibition, and subsequent worlds' fairs before he established the open-air museum of Skansen, outlined by Ola Wetterberg.[36]

As preservation professionalized, so too professional organizations and literature developed. Michael Holleran and Ola Wetterberg in particular deal with this aspect of preservation's development. However, certain seminal figures make appearances in several chapters as either advocates or practitioners. Ruskin's contribution is well known but that of his Harvard counterpart, Charles Eliot Norton becomes apparent in the Niagara campaign. The influence of practitioner and advocate William Morris is also known but not his involvement in the Eastern Question, as Erik Goldstein reveals. Morris's organization, the Society for the Protection of Ancient Buildings, significantly extended his influence abroad; his

---

34 Jurgen Habermas, *Strukturwandel der Öffentlichkeit: Untersuchung zu einer Kategorie der bürgerlichen Gesellschaft* (Frankfurt, 1962) (*The Structural Transformation of the Public Sphere: An Inquiry into a Category of Bourgeois Society*, trans. Thomas Burger Cambridge, Mass: MIT Press, 1988); Craig Calhoun, ed., *Habermas and the Public Sphere* (Cambridge, Mass: MIT Press, 1992); Harold Perkin, *The Rise of Professional Society: England since 1880* (London: Routledge, 1989), esp. 9-26; 141-74. Preservation intersects with public, bourgeois, and professional interests, and particularly with concerns about land ownership, leisure, and social welfare.

35 Bryce wrote several times on public opinion; see, for example, James Bryce, 'The Nature of Public Opinion,' in *The American Commonwealth*, vol. 2, 1900, (New York: Macmillan, 1960), 247-54.

36 Ralph Edenheim and Eivor Martinus, eds, *Skansen* (London and Stockholm: Scala, 1995), 7. trans. Neil Smith.

follower, Charles Robert Ashbee, figures in both Peter Mandler's and Raymond Cohen's chapters. It could be added that Ashbee also made tours of the United States on behalf of England's National Trust.[37] The influence of French architect, Eugène-Emmanuel Viollet-le-Duc has importance for both Sweden and the United States.

That small but influential politically-orientated individuals and groups invariably act in preservation with varying agendas and varying degrees of reach is demonstrated in many of the chapters, once again emphasizing preservation's modernizing but not entirely modern tendencies. It is also clear that preservation happens at multiple levels of interest. Nevertheless, as Anne Blackburn's and Thomas Otte's work demonstrates, central to the making of a monumental site of lasting significance is its compound interests and continued support. The sacred cities of Anuradhapura (1982) and Kandy (1988) have been inscribed as World Heritage Sites; Sulgrave Manor has limited resonance beyond a core group of Americans.

The formal structures of concert diplomacy were underpinned by a raft of legal, voluntary, and cultural activities. Issues of cultural heritage were among the subjects covered by international unions and conferences that considered such matters as trade, communication, navigation, and copyright. Often, specific lobbying groups formed with international agendas. An international artistic and literary society formed in Paris in 1878 to promote protection of international heritage; its aims were institutionalized in the Berne Convention of 1886 for the Protection of Literary and Artistic Works and, subsequently, in the League of Nations' Committee on Intellectual Cooperation.[38] Similarly, the Anglo-American Society for the Preservation of Wild Flora and Fauna of the Empire was established in Africa in 1903 and is noted by UNESCO as among the turn-of-the century societies and international treaties that assisted the development of an international heritage framework.[39]

Astrid Swenson elucidates the growing professional conferences that led to the implementation of legislation by more than one nation state, tracing particularly the inter-national influences on this emerging facet of statecraft. The institution of preservation legislation by the European powers, France, Britain and Germany was concurrent and intersecting.[40] Both Michael Holleran and Ola Wetterberg explain influences on the development of preservation practice respectively in the United States and Sweden; these are predominantly French, British, and German and are both ideological and practical. Thus manifestations of status were also matters of statecraft and diplomacy, in turn intersecting with considerations of

---

37   Hall, 'Politics of Preservation', 354-55.

38   Paul S. Reinsch, 'International Unions and their Administration', *American Journal of International Law*, i:3 (1907), 597.

39   UNESCO World Heritage Centre, *World Heritage*, 26.

40   R.T.B. Langhorne, *The Coming of Globalization, Its Evolution and Contemporary Consequences* (Basingstoke: Palgrave, 2001), 118-22.

taste and trade, notably the publishing and tourism trades.[41] However, respect for professional integrity and scholarship are also important factors, as Ola Wetterberg makes clear.

Tourism, Publishing, and Spectacle

The needs of publishing for new sites as illustrations and of tourism for new destinations contributed to the rise in importance of visual arts. An imagined consumer- and educational-community of English literature readers has been noted as bolstering the sub-structures of diplomacy.[42] Printed sources could highlight the desirability of site-seeing, preservation's causes were also overtly supported by press barons, as Erik Goldstein, Melanie Hall, and Thomas Otte make clear. Similarly, tourism also helped in the suggestion of both individual and community identities. This view was endorsed by James Bryce who envisaged a 'community of travellers', brought together by their common experiences of international sites whether as elite travellers, tourists, or armchair-tourists. Newer depictions of world communities were also being displayed; these included worlds' fairs, as well as the increasingly familiar circus shows and spectacles; all intersected, supported, and clashed with preservation's interests, as Mandler, Hall, Otte, and Wetterberg indicate.

The industrialization with which worlds' fairs is associated has a more complex relationship to preservation than has been recognized. These new preservation sites were, after all, accessed by roads and railways that invariably skirted their peripheries. The process of associating beauty and 'the sublime' with international places as well as with divinity was doubtless further enhanced by the awe-inspiring, steam-powered modes of transport that made them accessible, and by illustrated guide books where text was amplified by poetics.[43] In site-museums, statis and a sense of deep time provided explicatory oppositional dichotomies to motion and progress in which the active tourist could position themselves as if in a self-directed production. This sense of engagement in a panorama is most compellingly conveyed in Henry James's description of a visit to Niagara and underpins a voluntaristic sense of national or international identity.[44]

Grand narratives and tourist itineraries were important to the development of preservation. International touristic, artistic, architectural, archaeological, and literary interest in Rome, the destination point of The Grand Tour, and its importance

---

41   This was not new, see Hall and Goldstein, 'The Diplomatization of Culture'.

42   Bradford Perkins, *The Great Rapprochement, England and the United States, 1895-1914* (London: Gollancz Ltd, 1969), 119-56.

43   See, Michael J. Freeman, *Railways and the Victorian Imagination* (New Haven: Yale University Press, 1999).

44   James, *Niagara*. I am grateful to Bernard Richards for discussions of Henry James.

to western artistic taste is well-documented.[45] This was followed by tours further afield to a broadening range of sites, including Istanbul. By the second half of the nineteenth century, Eastern Tours were a feature of packaged itineraries purveyed by Henry Gaze, Thomas Cook and others, that revolutionized travel to the Holy Land.[46] As the Bible Tour supplemented the Grand Tour, sites of Christendom intersected with sites of Classicism in broader explorations of the Ancient World. Similarly, touring in Asia and Africa was increasingly possible for those with trade links and dependent territories. All areas were punctuated with significant landscape sites of history, culture, geology, geography, and ethnography that provided points of reference and comparison. Some of these became monumentalized illustrations in grand-narrative discourses. With the United States as a tourist destination came a New World tour, emphasizing distinctions with an *Old World* as opposed to an Ancient World.[47]

The management of a new society through politics, manufacture and trade in particular were added as American interests. Thus, incrementally, the preservation (or reservation as it was often known in the States) and management of site-resources gained importance. The institutionalization of a site goes beyond preservation to management as becomes apparent at Niagara Falls. Debates about preservation and conservation are also about representation and management; however, these have developed somewhat differently to serve different national political agendas and differing interests in landscapes, natural resources, and buildings or other artifacts.[48] This has obscured an understanding of the underlying principles and values. More sophisticated ideas of conservation were developing within highly politicized preservation debates to ensure the viability and continued and multiple use of industrially-important sites such as Niagara Falls and other tourist sites. Thus, a seemingly more 'scientific' management of a site was better able to incorporate the values of industrial use, but its ability to display 'protective,' pastoral or 'whiggish' values, (to reference Peter Mandler's usage) was diminished. Although there are many waterfalls and water features within World Heritage Sites these are not used industrially and, predominantly, are contained within national parks as is the case with Yellowstone and Yosemite.

---

45   Haskell and Penny, *Taste and the Antique*.

46   Bar-Yosef, *Holy Land in English Culture*, 64: John K. Walton, 'Thomas Cook', in Richard Butler and Roslyn Russell, eds, *Giants of Tourism* (Wallingford: CABI, 2010), 87-88.

47   Indicatively, the gateway to the New World was New York. For introductory observations on this sense of difference see, Christopher Mulvey, *Anglo-American Landscapes: A Study of Anglo-American Travel Literature* (Cambridge: Cambridge University Press, 1983), esp. 249-67.

48   The conservation of game is another area in which the preservation/conservation oppositional was worked out, as well as in archaeology, which represented professional, preservational, educational, and conservational interests as a social science.

Niagara Falls is not a World Heritage Site.[49] Ola Wetterberg demonstrates the rise of a more scientific approach to 'cultural' preservation in Sweden influenced by the natural sciences.

The entry of the United States into the international preservation arena was, however, rather more complex as it took place representationally at a border site it shared with British Canada. Here, the desire to maintain voluntary peaceful cooperation against imperialistic ambitions is perhaps a sub-text as this United States-driven preservation initiative also helped to demarcate that country's desire to harmonize with the concept of the Concert of Europe, albeit as the Concert was beginning to deteriorate.

Conclusion

As Chris Miele explains, the concept of community underpinned and continues to underpin much heritage discourse. So too, as Peter Mandler argues, do concerns about the impact of the ethics and aesthetics of industry and commerce. The expanding range of sitescapes and site-museums has similarities with branding, as Miele indicates. These reflected a more permanent, varied and sophisticated practice than the displays of ethnographic and village life that characterized the Worlds' Fairs but, like them suggested hierarchy, meaning, and values. By placing collective events in a cohesive unity that includes past, present and future it suggests mastery over time, a construct radically challenged by steam and rail travel, as well as of easier movement between time zones.[50] However, site-museums and targets of preservation were more than social experiences and cultural reference points. They represent the ability of different cultural powers to collaborate peacefully over access to resources that have meaning and appeal for more than one community. Hierarchical and structured places as they may be, they are also spaces set aside for recreation, pleasure, and the imagination and indicative of a deeper desire for peaceful harmony and a sense of stability in an increasingly changeable and chaotic world. That such sites survive at all remains witness to the agency of random chance. As internationalizing forces put ever greater pressures on their chances of survival, efforts began to find a systematized approach to both defining and preserving an international heritage which, ultimately, became the World Heritage scheme.

---

49   By the end of the nineteenth century the idea of a monument had developed to include a specific area that could include more than one monument and, indeed, a planned landscape as with the canals of Amsterdam. Thus, it can be more easily considered as a site-museum.

50   Robert Rydell, *All the World's a Fair: Visions of Empire at American International Expositions 1876-1916* (Chicago: University of Chicago Press, 1987), 2.

PART I
Case Studies

Chapter 1

Niagara Falls: Preservation and the Spectacle of Anglo-American Accord

Melanie Hall*

> The suggestion ... that an object of this class, so unparalleled as the Falls of Niagara should be placed under the joint guardianship of the two governments ... is a proper concern of the civilized world.[1]

Niagara Falls is a significant, naturally-occurring spectacle on the long border between the United States and Canada, and the northern hemisphere's largest waterfall. As the scenic icon of the North American continent and regarded as the eighth wonder of the world, Niagara Falls lent itself to a demonstration of cultural diplomacy. Cultural expressions of alliance, diplomatic representation, and the relationships between well-governed states and their citizens have deep trajectories in western civilization through recognizable types of temporary and permanent spectacles. Water festivals as spectacles of state date from the Renaissance and earlier, when 'stupendous scenic effects [were cast] with a powerful political and social role ... [as] manifestations of the virtue of magnificence.'[2] Natural and man-made water-features were brought together in grand gardens where technology, design, and landscape harmonized with seemingly natural ease to represent ownership, stewardship, and governance. Suffused with symbolism of ordered, well-governed worlds, examples remain on show from Tivoli to Versailles. These destination points on the European tour were made familiar through paintings, panoramas, and engravings.[3] By referencing such scenic traditions a nation's place in the progress of western civilization could be signaled to those familiar with the type. Representations of governance and diplomacy encompassed spectacle and pageantry in which hierarchical roles were established on lines of class, gender,

* I am grateful to Raymond Cohen, Erik Goldstein, Howard Malchow, Peter Mandler, Stephen Mosely, Thomas Otte, Bernard Richards, and John Walton who have all made helpful suggestions and comments.

1   Memorial Addressed to the Governor of New York, and the Governor-General of Canada, March 2, 1880, quoted in Charles Mason Dow, *The State Reservation at Niagara, a History* (Albany: J.B. Lyon, 1914), 18-20.

2   Roy Strong, *Art and Power, Renaissance Festivals 1450-1650* (Berkeley: University of California Press, 1984), 11, 39-41.

3   Simon Schama, *Landscape and Memory* (New York: A.A. Knopf, 1996), 337-52. John Vanderlyn's famous Panoramic View of Versailles (1818-19) toured the United States.

and race and made seemingly fixed by divine authority.[4] Several cultural traditions came together at Niagara Falls, for which Britain still spoke in matters of foreign policy, to make this a suitable site for a spectacle of Anglo-American governance and diplomatic accord.

The significance of Niagara Falls' preservation as an international effort has been largely overlooked.[5] Although the campaign to reserve its 'natural beauty' began in the United States, preserving the scenic splendours of the spectacular waterfall and its surrounds required the cooperation of Britain and Canada.[6] Failure or success in bringing order to the site showcased Anglo-American governance in the new world. The border and waterways were especially at issue during and just after the Civil War, which saw the so-called Fenian raids, that is, free-booting Irish-American incursions northwards. The site's maintenance, at once reserved, restored, conserved, and monumentalized, was a symbolic representation of a cooperative, imperial governing order. The cultural and diplomatic campaign can be seen as a celebratory display of unity, and the preservation of the Falls' aesthetic power was in large measure due to successful statecraft. It figuratively and literally set the scene for further cooperative efforts.

From the late 1860s the United States and Great Britain entered a new period of rapprochement in their relationship.[7] As they did so celebrating peace, friendship, and a common heritage in the name of Anglo-American civilization became an area of activity.[8] Festivities included temporary, parallel pageants that expressed

---

4   Eugene Jones, *Native Americans as Shown on the Stage, 1753-1916* (Metuchen: Scarcrow Press, 1988), 1-8.

5   The preservation of Niagara Falls spans decades; numerous accounts cover different aspects. Many are written from an American perspective and deal with particular aspects. The most comprehensive remains Gail Edith Hallett Evans, 'Storm over Niagara: A Study of the Interplay of Cultural Values, Resource Politics, and Environmental Policy, 1670s-1950' (PhD diss., University of California Santa Barbara, 1991). George A. Seibel, *Ontario's Niagara Parks, One Hundred Years, A History* (Ontario: Niagara Parks Commission, 1985) brings a Canadian perspective. Two accounts consider aspects of the international relations at the site, N.F. Dreisziger, 'The Campaign to Save Niagara Falls and the Settlement of United States-Canadian Differences, 1906-1911,' *New York History* 55 (1974), 437-58, and Peter Neary, 'Grey, Bryce, and the Settlement of Canadian-American Differences, 1905-1911,' *Canadian Historical Review*, vol. 49, no. 4 (1968), 357-80.

6   Alfred Runte, 'The Role of Niagara in America's Scenic Preservation,' in Jeremy Elwell Adamson, *Niagara: Two Centuries of Changing Attitudes, 1697-1901* (Washington, DC: Corcoran Gallery of Art, 1984), 117-18. For scenic values in American preservation see Alfred Runte, *National Parks, The American Experience* (Lincoln: University of Nebraska Press, 1997), 11-18, 20-22, 31-32, 38-39, 43-47, 71-73.

7   A significant starting point was the settlement by arbitration of the so-called Alabama claims. See also, Bradford Perkins, *The Great Rapprochement, England and the United States, 1895-1914* (London: Victor Gollancz, 1969).

8   Roland Quinault, 'The Cult of the Centenary, c.1784-1914,'*Historical Research*, vol. 76, no. 176 (1998), 303-23.

the similar ideas and ideals of two nations united and divided by a common language and traditions. While such celebrations *asserted* friendship, preserving the scenic splendour of Niagara Falls was a powerful *operational* representation of collaboration. This demonstrated not only the wish for cooperation but also the will to bring that about, and it should be regarded as the first successful example of an international preservation effort.

Political, legal, industrial, and representational factors at Niagara Falls were complex and interlinked. The boundary between the Dominion of Canada, still a part of the British Empire, and the United States ran along the Niagara River; however, the riparian rights of the Niagara's waters had not yet been clarified when efforts began to preserve the Falls' banks and islands. Attempts to suggest an *international* viewing park within and around the Falls from the 1870s reflected and presaged a need to bring order and control to the spectacular shared resource. Efforts culminated during the following decade in several border and island parks that lent the *aura* of internationalism to the site. As a *visual* order was instituted, the Falls' representational power was harnessed in the interests of international governance. Nevertheless the viewing platforms were national territories and the waterfalls themselves still lacked international protection.

The aesthetic character of this iconic waterfall and water rights were interlinked. Although the waterfall's image of pristine wildness had been compromised, Niagara Falls had gained a reputation as a beautiful water-feature with both a definite and indefinite structure and appearance that encompassed agitated motion and monumentality.[9] Its dynamic character was still considered by many to be effectively contained by its situation although natural forces, a spirit of Nature, or 'acts of God' brought endless variation and fugitive effects. Furthermore it was those very indeterminate, effervescent and, for some, transcendental qualities that were considered most able to evoke a sensory, even a spiritual, response.[10] As science and technology developed God and Nature were not the only forces able to command the Falls' characteristic appearance or, indeed, its physical properties and energies. Although industrial use altered the appearance of the banks, Niagara's Falls were becoming subject to international industrial and economic forces that sought to utilize their waters to produce hydro-power. Curbing industrial activity in the surrounding area paved the way for restraining developments at the Falls that took aesthetics into account. Canal construction further upstream raised the prospect that the very character of the waterfall might be altered by man's intervention. This prospect intensified with developments in the transmission of electricity from hydro-power. Demonstrating cooperation in the surrounding environment of the Falls undoubtedly paved the way for agreement in potentially

---

9   Elizabeth McKinsey, *Niagara Falls, Icon of the American Sublime* (Cambridge: Cambridge University Press, 1985).

10   Barbara M. Stafford, *Voyage into Substance, Art, Science, Nature, and the Illustrated Travel Account, 1760-1840* (Cambridge, Mass: MIT Press, 1984), 185-87 *et passim*, especially 240, 242-49.

more contentious and complex negotiations about the waters themselves. Literally setting the scene at Niagara in which monumental beauty and rarity were given status imposed a *sensibility of preservation* at the site, to use Raymond Cohen's evocative phrase. This, in turn, required an ethos of stewardship in economic development which could be introduced as a factor in the subsequent negotiations of a boundary waters treaty in 1907. The reward for cooperation was the retention of nature's beauty alongside the use of its bounty.

i

The campaign to preserve Niagara Falls began in 1869, two years after Canadian Confederation. Cultural and political relationships at the Falls intersected in several ways; between the United States and Canada, Canada and Britain, and Britain and the United States. There were regional, bi-lateral, and national interests; however, there was also a will to cooperate and respect for the unusual magnificence of the resource. To varying extents all three countries engaged in representational culture in response to internal, nation-building imperatives that included external considerations such as international tourism, migration, and status in the ranks of 'civilization'. As external events and shared interests in North America brought Britain and the United States closer together, the political, diplomatic, and cultural interests of preservation coincided.

That Britain and the United States had already engaged in preserving similar sites provided a framework for joint efforts. Several factors helped foster a spirit of cooperation including a shared language and literature, history and traditions. Following Thomas Carlyle's influential essay, 'On Heroes, Hero-Worship, and the Heroic in History' (1841) homes of heroes were preserved as museums in both countries. Carlyle's ideas found sympathetic reception among the educated American elite and helped in the formation of a transatlantic culture.[11] Shakespeare's Birthplace in Stratford-upon-Avon became a site-museum in 1847, although it was already a destination on American as well as English tourist itineraries.[12] George Washington's Virginia home, Mount Vernon followed in 1851; Washington would gradually emerge as an Anglo-American, as well as an American, hero. Such representational symbols of 'civilization' vied with competing models of the use and display of the resource; American circus-owner and show-business entrepreneur, P.T. Barnum made a well-publicized attempt to

---

11   Carlyle's friendship with Ralph Waldo Emerson is well documented; see Samuel Slater, ed., *The Correspondence of Emerson and Carlyle* (New York: Columbia University Press, 1964).

12   Christopher Mulvey, *Anglo-American Landscapes, A Study of Nineteenth-century Anglo-American Travel Literature* (Cambridge: Cambridge University Press, 1983), 74-92; Michael Dobson, *The Making of the National Poet: Shakespeare, Adaptation and Authorship, 1660-1769* (Oxford: Oxford University Press, 1992), 228-31.

acquire Shakespeare's Birthplace, a symbol of America's theatrical heritage, when it came up for auction. Barnum wanted the house for his private museum and not for the civic-national domain.

For Gladstonian Liberals, the idea of a supra-national community of 'English-speaking peoples' gained attraction from the 1860s.[13] James Bryce MP was the foremost advocate of Anglo-American unity and Ambassador to the United States (1907-13). Bryce explained in 1886 to a sceptical Edward A. Freeman, Regius Professor of Modern History at Oxford University, that his object was 'to maintain our English citizenship and nationality over the whole world ... I wish we could bring in the USA too: we ought never to have let them go out; nor do I despair of some sort of permanent arrangement with them in the future'.[14] Bryce's *History of the American Commonwealth* (1888) established him as 'the foremost authority on the new civilization across the Atlantic', and had significantly improved understanding between these two powers.[15] His interest in preservation is less well-known but was a tool of his diplomacy that would serve Niagara's cause.[16] An English-speaking identity found support (not necessarily for the same reasons) among 'high culture' groups in the United States. Walter Hines Page, American Ambassador to Britain (1913-18), wrote to Bryce in 1898, 'English-speaking men are friends and friends for the largest purposes of civilization'.[17] Preservation at Niagara Falls was in part a manifestation of their ideals of civilization against Barnum-style entertainment and Barnum himself, as we shall see.[18]

In the long-settled British Isles preservation occurred in the voluntarist sector, sometimes with government interest behind the scenes.[19] Strategies for acquiring and managing natural resources for public, scenic-leisure purposes developed through voluntarist societies from the 1860s; Bryce was among the activists.

---

13   Hugh Tulloch, *James Bryce's American Commonwealth: The Anglo-American Background* (Woodbridge: Boydell Press, 1988), 7, 10. Francis H. Herrick, 'Gladstone and the Concept of the "English-Speaking Peoples,"' *Journal of British Studies*, vol. 12, no. 1 (1972), 150-51.

14   Quoted in Christopher Harvie, 'Ideology and Home Rule: James Bryce, A.V. Dicey and Ireland, 1880-1887,' *English Historical Review* 91 (1976), 313.

15   Keith Robbins, 'History and Politics: the Career of James Bryce,' *Journal of Contemporary History*, vol. 7, no. 3-4 (1972), 43.

16   For Bryce and preservation see Robert A. Lambert, *Contested Mountains: Nature, Development and Environment in the Cairngorms Region of Scotland, 1880-1980* (Cambridge: White Horse Press, 2001), 60-72; Melanie Hall, 'The Politics of Collecting,' *Transactions of the Royal Historical Society*, 6th series, xiii (2003), 345-57.

17   Page to Bryce, April 23, 1898, in Burton J. Hendrick, *The Training of an American: The Earlier Life and Letters of Walter H. Page* (Boston: Houghton, 1928), 263-64.

18   Lawrence Levine, *Highbrow, Lowbrow, the Emergence of Cultural Hierarchy in America* (Cambridge, Mass: Harvard University Press, 1988), 13-81, 85-168 discusses both Shakespeare in America and cultural divisions during the nineteenth century.

19   Calls for legislative protection for built ancient monuments came from Sir John Lubbock, MP (Liberal), FSA.

The Lake District (known as Wordsworthshire in America) was a particular area of interest. Calls for its designation as a 'national property' had begun with Wordsworth's *Guide to the Lakes* (1810), though there was then no model for translating rhetoric into action. By the 1870s industrial and urban challenges to the District's scenic-leisure facility included railways and non-local water-use as Thirlmere, one of the smaller lakes, was converted to a reservoir for the distant industrial hub of Manchester during 1879-94.[20] Water resource issues were topical as social amenity, industry, and scenic representation competed for the same resources.

Appreciation of the scenic characteristics of Niagara Falls and its surrounding banks was voiced by international visitors and East Coast arts advocates from the early nineteenth century. Niagara was both a national and an international destination. As early as 1834 English Congregational Ministers, Andrew Reed and James Matheson deemed 'Such spots … [to be] the property of civilized mankind'; no means for protecting the area had yet been devised and Alexis de Tocqueville, the French political philosopher, advised seeing it before industry encroached further.[21] Concern for 'natural' scenery sparked in East Coast society circles by the utilitarian spoliation of Niagara helped to motivate the reservation of large-scale tracts of land in western territories at Yosemite (1864) and Yellowstone (1872).[22] The aftermath of civil war and westward expansion towards areas of economic and governmental underdevelopment brought imperatives for establishing federal government identities and for strengthening northern state identities. Urban and 'wilderness' park schemes linked by railways and art helped to suggest physical and cultural unity between the states and the federation, and between *metropoli* and 'wilderness'.[23]

American 'wilderness' parks in areas still untouched by modern society have *ancient and monumental* scenery with a strong aesthetic component. Their 'natural' wonders have geological, geographical, botanical, and arboreal values. These resources supported educational interests in the natural sciences and provided subjects for visual and literary representations of the nation. The model simultaneously exploited a *new* type of *world* identity around 'natural' resources in juxtaposition to 'cultural' resources of the *old world*, that is architecture and other built monuments. This was an adroit and diplomatic oppositional coupling that showcased America's distinctive natural resources and growth potential. [24]

---

20   Harriet Ritvo, *The Dawn of Green: Manchester, Thirlmere, and Modern Environmentalism* (Chicago: University of Chicago Press, 2009).

21   Quoted in Charles M. Dow, ed., *Anthology and Bibliography of Niagara Falls* (2 vols, Albany, NY: J.B. Lyon, 1921), vol. 2, 1070-71.

22   Runte, 'The Role of Niagara,' 117-18.

23   Runte, *National Parks*, 14, 24-25.

24   For diplomacy's use of oppositional couplings see Harold Nicolson, *Diplomacy* (London, New York: Oxford University Press, 1950), 23-28, 106. For nature as a cultural construct see Schama, 7-19.

As these 'national parks' were apparently less subject to European-style, 'old world' hierarchies of ownership and access, they were also construed as a more democratic representational form.

While these areas were set aside for leisure and sport, rights to natural materials were reserved.[25] Their representational value was cultural, political, and fiscal and marked government's arrival and claim; governance, stewardship, and the social hierarchy were represented by possession, access, and control. Parks were destination points on the transcontinental railroads from which they were accessed at their peripheries thus, nature and technology were represented in harmony. All could be visually re-presented as a national collection in guide books and museum galleries.[26]

ii

Although the Falls had a monumental status and were regarded by some as an 'object', their 'identities' were liminal. At once changing and not changing, the waterfalls were an apt and dramatic representational analogy for awe, power and change; their symbolism was difficult to harness in any specific cause. Identities at Niagara Falls were complex as the Falls had accrued symbolic meanings. The site held deep memories; the name Niagara was an Anglo-French variation of Onguiararonon referring to the Iroquois inhabiting the area. The truncated version was applied to the river.[27] The British took the territory from the French who retained an interest in Niagara's representational value. After Independence, Niagara Falls accrued a sub-diplomatic identity as a sublime, powerful and awe-inspiring symbol of a natural and sacralized order between two Anglo-Saxon powers that were readjusting their relationship. Identities realigned as political, economic and scientific forces challenged both the order of power and the power of order at the site.[28]

On one level, and especially visually, the massive waterfall symbolizes Deluge, disorder, wildness and untameability. Maintaining and representing

---

25   W. Turrentine Jackson, 'The Creation of Yellowstone National Park,' *Mississippi Valley Historical Review*, vol. 29, no. 2 (1942), 199. For the monumental aspect of national parks see Runte, *National Parks*, 14-47. For nationalism, nature and national parks see Eric Kauffmann, '"Naturalizing the Nation:" The Rise of Naturalistic Nationalism in the United States and Canada,' *Comparative Studies in Society and History*, vol. 40, no. 4 (1998), 666-95. Protection of built ancient monuments in the United States lies beyond the scope of this chapter.

26   Richard Grusin, *Culture, Technology, and the Creation of America's National Parks* (Cambridge: Cambridge University Press, 2004), 1-15.

27   John Rydjold, *Indian Place Names, Their Origin, Evolution and Meanings* (Norman: University of Oklahoma Press, 1968), 267-68.

28   For an explanation of the relationship between scientific understandings of such phenomena and their visual representation see Stafford, 242-45, 367-75, 467-71.

order, meaning, and political power at the Falls in the face of change was strongly symbolic, and would require diplomatic cooperation. Diplomacy attempts to maintain relations 'between identities that are continuously under construction in conditions that, if not anarchic, are characterized by very thin social contexts'.[29] The waterfalls' initial symbolic associations of peace and divine order between English-speaking peoples were endorsed by such writers as Charles Dickens and Harriet Martineau. United States nationalism and its changing power relationship with the British Empire was expressed when Senator John Quincy Adams (former President and minister to Britain) suggested in a public speech in Buffalo that 'Heaven had considered this vast natural phenomenon too great for one nation'. Adams adroitly suggested concord and joint ownership while asserting national identity at the site.[30]

Economic and industrial change brought cultural change. Construction of the Erie Canal enhanced the area's popularity as a tourist attraction. Tourism increased with near-by Buffalo's expansion as an industrial metropolis and railway hub. Industry at the Falls stood for celebratory alliance when harnessed, or inadequate control if unregulated. The 1848 railway bridge across the Niagara River suggested to London's *Times* 'an iron link of civilization between the two ruling Powers of the world, [which] will never be severed'.[31] Alongside new technology came well-publicized modern mayhem. Brawls, narrow escapes, fatalities, tourist exploitation, and even murders at the privately-owned viewing platforms reported in Buffalo and Toronto newspapers were disseminated by *The Times*.[32] As sensationalism and exoticism catered to tourists and day-trippers, *The Times* dubbed the area an urbanized 'pleasure-ground'.[33] Dare-devil entrepreneurship interrupted genteel sight-seeing views; in 1859 François Gravelet, known as Blondin, made the first tight-rope walk across Niagara Gorge. Tourist 'fungus' included 'Chinese pagodas, menagerie, camera obscura, museum watch tower, wooden monument and old curiosity shops'.[34] Nature's pastoral associations receded further as timber deforestation turned the banks to industrial wastelands. Although the Falls topped Anthony Trollope's 1862 list of the world's 'wonders of art ... and ... beauties of nature', ahead of Giotto's tower, the Apollo Belvedere, the peaks of the Alps, and valleys of Jamaica's Blue Mountains, he found their environs as potentially turbulent as 'life in Paris', and 'the full tide of trade round the Bank of England'.[35]

---

29   Paul Sharp, 'For Diplomacy: Representation and the Study of International Relations,' *International Studies Review*, vol. 1, no. 1 (1999), 36.

30   Dow, *Anthology*, vol. 1, 233.

31   *The Times* (Jan. 18, 1848), 5.

32   *The Times* (Sept. 19, 1860), 7; (Nov. 1, 1861), 9; (Jan. 5, 1864), 7.

33   *The Times* (Aug. 23, 1859), 10.

34   G.D. Warburton, *Hochelaga; or England in the New World* (London, 1847), quoted in Dow, *Anthology*, vol. 2, 1073.

35   Seibel, 135-36; Anthony Trollope, *North America* (2 vols, London, 1862), vol. 1, 136-52, quoted in Dow, *Anthology*, vol. 1, 296-305. For industry, tourism, and design at the

Trollope's urban, fashionable, and commercial analogy was followed by Thomas Carlyle's vitriolic essay *'Shooting Niagara: And After?'* (1867). Written on the eve of Disraeli's reform bill, Carlyle saw extending the franchise as a suicidal political, economic, and cultural leap. For the obstinate and cranky social critic (whose brother emigrated to the Niagara peninsular), the culture of the Falls epitomized popular 'American style' mass democracy: it was materialistic, un-Christian, unbounded, individualistic and racially inclusive. Niagara, as he would have it, was a place of the 'swarmerai'. Nevertheless, his essay presumed a relationship between the representational power of culture and the representation of political power; so, too, did Niagara 'reservationists' as they became known.

iii

Preservation requires monuments with a strong visual element. The representational power of the Falls gained a new dimension with Frederic Church's magnificent *Niagara* (1857). The painting provided a catalyst for preservation. Skillful art dealers displayed *Niagara* in New York and Boston before touring it around Britain; reproduction prints were sold as souvenirs. Waterfalls held a fascination and challenge for art; as science investigated the composition of matter, waterfalls attracted artists who were concerned with more diffuse substances such as airs and currents. For John Ruskin, whom Church admired, waterfalls presented

Figure 1.1   Frederic Edwin Church, 'Niagara' (1857). Oil on canvas, 40 x 90.5 inches. Courtesy of The Corcoran Gallery of Art, Washington, DC. Museum Purchase, Gallery Fund. Accession no. 76.15

site also see, William Irwin, *The New Niagara, Tourism, Technology, and the Landscape of Niagara Falls 1776-1917* (University Park: University of Pennsylvania Press, 1996).

a challenge both to depict realistically the nebulous phenomena of nature such as fume or sun-bows, and to retain the specific and individual, divinely-created character of particular falls.[36] When Ruskin was reported by American critic Henry T. Tuckerman to have praised the treatment of the atmospheric effect of light upon the waterfall, Church acquired a reputation as J.M.W. Turner's heir.[37] Although the Falls were already regarded as a *natural* monument, capturing their natural 'wonder' had challenged writers and artists until *Niagara*. This was realized in Church's work not only by the realistic treatment; the painting with its mass of water also carried powerful associations of the Biblical Deluge and Creation, while rainbows symbolized God's promise to Noah not to destroy the world by Flood. However, in Church's painting, the triumph of realism seems to be suggested by the rainbow being broken. This triumphant and widely-applauded representation of the Falls established their visual image and for the United States created a national, *cultural* monument.

As the work's reach extended so Church's reputation grew and with it a new, international status for American art. The painting acted as a cultural ambassador for the United States as it toured London, Liverpool, Manchester, and Glasgow to great acclaim. *The Times* recognized that 'the most defiant of all ... the many scenes which the New World offers' had been mastered concluding, 'American originality and grasp are too great to be long confined to the fields of industrial or mechanical activity'. It anticipated 'a new and national development of painting also'.[38] In response, Americans agreed that Church's Canadian viewpoint was the best for observing the Falls. *Niagara* triumphed again at the Paris 1867 *Exposition Universelle* and found a permanent site in W.W. Corcoran's Washington DC gallery, bought (1876) for $12,500, the most to date paid for an American painting.[39]

As the propagandistic role of *Niagara* the painting became inter-linked with viewing the waterfall, Church advocated at the influential New York Century Club for polite viewing platforms at the Falls. Influential advocates combined high-minded idealism and professional aesthetic motives in promoting scenic interests. Church, a commissioner of New York's Central Park, was joined by his distant cousin and famous Park designer, Frederick Law Olmsted. A Niagara Falls Association was formed by Olmsted and Charles Eliot Norton, the editor of *The North American Review* and *The Nation*, Harvard University art history professor,

---

36   Stafford, 44, 320, 426, 467.

37   Henry T. Tuckerman, *Book of the Artists* (New York, 1867), 371, in Jeremy Elwell Adamson, 'Frederic Edwin Church's "Niagara:" The Sublime as Transcendence' (PhD diss., University of Michigan, 1981), 33. Turner was known to have been fascinated by images of the Falls, John Gage, *Color in Turner: Poetry and Truth* (New York: Praeger, 1969), 121.

38   *The Times* (Aug. 7, 1857), 12.

39   Clara Marburg Kirk, *W.D. Howells and Art in His Time* (New Brunswick: Rutgers University Press, 1965), 252-67; McKinsey, 249; Adamson, 'Frederic Edwin Church's "Niagara," 32-45, 561.

and well-known as a friend of Ruskin and Carlyle. They were joined by Unitarian Reverend Jonathan Harrison who made extensive lecture tours of New England campaigning for parks at the site.[40] The Association attracted powerful support from politicians including two future New York governors and United States presidents, Grover Cleveland (Democrat) and Theodore Roosevelt (Republican).[41] Cleveland's friend William Dorsheimer, a district attorney for northern New York then prominent on the Republican Party's liberal wing, saw the Falls' representational potential for New York State and his own ambitions. As head of the Capitol Commission in Albany, Dorsheimer built his reputation by promoting city beautification schemes. He commissioned Olmsted to design Buffalo's park system and complete plans for Albany with architect H.H. Richardson. In Dorsheimer's words, these three shared a vision of 'higher civilization' and of park-planning as a means to achieve it.[42]

Public park projects are aspirational, concerned to provide scenery for a society brought together in an idealized, voluntaristic civic-national pageant. For Olmsted the Falls, surroundings and viewing practices were inter-related and the Falls' affect had been compromised by activities along the banks. He observed, 'the idea that Niagara is a spectacular and sensational exhibition, of which rope-walking, diving, brass bands, fire-works and various side shows are appropriate accompaniments, is so presented to the visitor that he is forced to yield to it, and to see and feel little else than that prescribed to him'.[43] Olmsted sought a relationship between public display and performance whereby spectacle would work as a voluntaristic form of civic pageantry and aspired to effect this through design. To accomplish the desired effect an 'illusionistic stage' is required so that, 'the eyes of the spectator could be almost forced to look at things in a certain way'.[44] The technique had been well-employed in Renaissance court masques, a clearer form of representational culture in which roles, costumes and a script were also ascribed. In a public park, the eyes and movements of a tourist-spectator can be choreographed, albeit lightly. Olmsted envisaged parks laid out with polite walks and rides, with well-placed stations to encourage orderly, awe-inspired

40   'Niagara Falls, The Scenery and the story of the Falls, with an account of their redemption, together with the geological history of Niagara River and an account of the Water power of the Falls, etc. etc. compiled by J.B. Harrison, Secretary of the Late Niagara Falls Association,' (n.d.) Charles Eliot Norton Papers, Library of Congress, Washington, DC, MMC-1114.

41   Evans, 154, at note 60. Kathleen Dalton, *Theodore Roosevelt, A Strenuous Life* (New York: A.A. Knopf, 2002), 247.

42   William Dorsheimer, *The Life and Public Service of the Honorable Grover Cleveland* (Philadelphia, 1884), 36, quoted in Francis Kowsky, 'Municipal Parks and City Planning: Frederick Law Olmsted's Buffalo Park and Parkway System,' *Journal of the Society of Architectural Historians* 46 (1987), 49-64.

43   Harrison (n.d.), ch. 1, 4, n.d., Charles Eliot Norton Papers, Library of Congress, Washington, DC, MMC-1114.

44   Strong, 32-38.

viewing. Henry James concurred, identifying Goat Island between the American and Horseshoe Falls as the most appropriate viewing venue. For James, visiting the Falls was analogous to participating in a series of panoramas where 'the tourist effectively playing at showman' produces his individual (though similar) self-directed, sight-seeing experiences. Reminding *The Nation's* readers (1871), 'We place a great picture, a great statue, in a museum: we erect a great monument in the centre of our largest square, and if we can suppose ourselves nowadays to build a cathedral, we should certainly isolate it as much as possible and expose it to no ignoble contact', he proposed that '... the State buy up the precious acres ... [for which] ... no price would be too great to pay'.[45] An effective American festival of the Falls required British and Canadian cooperation and so would necessitate a celebration of this alliance.

Cooperation between Anglo-American preservationists was facilitated by compounding, multi-layered factors including family ties, friendship, language and literature, religion and race; professional and amateur associations, the churches, universities, clubs and societies; by government and legal affiliations. Preservation's protagonists were socially-motivated, highly-educated activists often in significant public positions. Their political affiliations spanned political parties, though progressive Republicans and Liberals are notable among them. Their sense of order and place was both ideological and material. They shared a sense of property ownership, and ideas of culture and cultivation. They had an ethos of cooperative group action in the public interest, a respect for diplomacy in both internal and international relations, and a sense that beauty and rarity were of value and worth protecting. In Canada preservation was necessarily promoted at the highest levels as national arts and scientific societies, which formed the campaigning constituencies elsewhere, were as yet still forming. In Canada's first governor-general, the Earl of Dufferin and Ava, American 'reservationists' found a willing advocate. American East Coast artists and photographers were already active at the site. Church's contemporary, Alfred Bierstadt acted as intermediary; it was a sub-diplomatic role he had played before.[46] Bierstadt's paintings of sublime scenery in the United States and Canada (often commissioned by transcontinental railway companies), including Niagara, were also instrumental in national-identity building and preservation.[47] William H. Hurlbut, *The New York World* editor and a

---

45   'Niagara, 1871,' in Henry James, *Portraits of Places* (Boston: Houghton, 1884), 365, 372-76. James's essay was published originally in *The Nation*.

46   Allan Pringle, 'Albert Bierstadt in Canada,' *American Art Journal* vol. 17, no. 1 (1985), 2-27; Melanie Hall and Erik Goldstein, 'Writers, the Clergy, and the "Diplomatization" of Culture: Sub-Structures of Anglo-American Diplomacy, 1820-1914,' in John Fisher and Antony Best, eds, *On the Fringes of Diplomacy: Influences on British Foreign Policy, 1800-1945* (Aldershot: Ashgate, 2011).

47   Runte, *National Parks*, 14. Bierstadt's brother, Charles, had a thriving photography business at Niagara Falls.

mutual friend of Church, Olmsted and Dufferin, assisted in a sub-diplomatic role he would play again.[48]

Politically-motivated preservationists were aware of economic and scientific developments and of democratic, industrial, and international change. During his final weeks of office, in October 1878, Dufferin proposed at the Ontario Society of Artists' Annual Meeting (of which he was President) the formation of 'a small public *international* park ... carefully preserved in the picturesque and unvulgarized condition in which it was originally laid out by the hand of nature', urging the governments of New York State and Canada 'to acquire whatever rights may have been established against the public'.[49] For many, the relationship between culture, politics, economics, nature and development indicated a successful balance between civilizations, societies, and economies.[50] Turbine engines with improved efficiency had already appeared at the Falls in 1877, when British engineer Sir William Siemens also raised the prospect of long-distance transmission of electricity from Niagara.[51] The suggestion that the Falls be placed under joint guardianship of the Canadian and United States governments demonstrated prescient awareness of industrial and economic change. Ontario's Premier Oliver Mowat attended the meeting, and it is inconceivable that Dufferin had not consulted widely beforehand. Protocol obliged him to have discussed his speech with the Canadian Prime Minister (and Crown loyalist), John A. MacDonald.[52]

High-level support in New York State and Canada coincided. Dufferin, with extensive political contacts in Washington and East Coast society circles, was well-placed to further the project in conjunction with New York's Governor Lucius Robinson.[53] Following Dufferin's address, Robinson expressed his 'trust that the good sense of both governments may finally abolish the system of nuisances which has to [sic] long defiled the greatest natural ornament of the continent ...'.[54] Robinson recognized the Falls' international status and that for New York State it was both a tourist asset and an advertisement. His 'personal' view was that, 'If the annoyances which surround the Falls of Niagara were once removed its

---

48 Olmsted to Thomas V. Welch, superintendent of the New York State Reservation, May 28, 1888, quoted in Dow, *State Reservation*, 12.

49 'Address given to the Ontario Society of Artists,' Sept. 28, 1878, in William Leggo, *The History of the Administration vof the Right Honorable Frederick Temple, Earl of Dufferin, K.P., G.C.M.G., K.C.B., F.R.S., Late Governor General of Canada* (Montreal: Lovell, 1878), 822-24 (my italics).

50 M. Daunton, 'Britain and Globalisation since 1850: I. Creating a Global Order, 1850-1914,' *Transactions of the Royal Historical Society*, 6th series, xvi (2006), 1-15.

51 Evans, 210-11, 235-36.

52 Seibel, 23-24.

53 Leggo, 274-75, 633-34, 745-46.

54 Dufferin to Robinson, Oct. 1, 1878 reprinted in Leggo, 824 verso. Robinson to Dufferin, Oct. 7, 1878, Papers of the Marquis of Dufferin and Ava, Public Record Office of Northern Ireland, Belfast (hereafter PRONI), D1071/H/B/R/270/2.

great natural advantages would soon number it with the most popular and pleasant resorts of the world'.[55] Niagara Falls already hosted an international society season. *The Pall Mall Gazette,* referencing *The New York Herald* exploited the theme of family alliance, envisaging the resort as a bridge towards 'Anglo-American union' by facilitating 'matrimonial unions which again will lead to international union'.[56] This referenced traditions of maintaining state alliances through marriage at a time when Anglo-American high-society marriages were topical.[57]

Aligning political support proved challenging. Robinson pushed a proposal through the New York State legislature in January 1879 for a Commission to study the feasibility of an '*International Park*' urging in a public address published in *The Nation* that 'both governments ... protect such travelers [from all quarters of the globe] from improper annoyance on either side'.[58] Visiting the Falls that month the Marquess of Lorne, Canada's new Governor General and Queen Victoria's son-in-law, told *The New York Times*, 'I heartily concur in the idea, and it should be carried out', though left vague how this might happen.[59] New York sought solutions; James T. Gardiner, the New York State Survey director, and Olmsted were charged to consider whether part of a mile-wide strip along the riverbank known as the Niagara portage, formerly state property, might be compulsorily repurchased and redesigned. They proposed to remove mills along the banks and 'restore and preserve' a more naturalistic and picturesque landscape.[60] The Survey Commissioners opened discussions with the Ontario cabinet. Lack of clarity about national and provincial jurisdiction over the riverbank, known as the Crown or Chain reserve, caused bureaucratic obstacles over control and funding between the Province of Ontario and the Dominion government. Canadian economic interests focused more on industrial development at the Falls. A New York regime change stalled the project; Governor Alonzo Cornell was more sympathetic to emerging industrial wood pulping interests. [61]

The Falls' campaign continued, bringing together powerful political, cultural, and economic interests. Signatories to a petition in March 1880, addressed jointly to Governor Cornell of New York and Lord Lorne included: W.A. Wheeler, Vice-President of the United States, and Alex Ramsey, Secretary of War; the chief justices of the United States and of Canada; Whitelaw Reid (future ambassador to Britain,1905-12); and seven United States Supreme Court associate justices.

---

55   Robinson to Dufferin, Oct. 7, 1878, Papers of the Marquis of Dufferin and Ava, PRONI, D1071/H/B/R/270/1.

56   *Pall Mall Gazette* (Oct. 15, 1878), 8.

57   Richard W. Davis, '"We are all Americans Now!" Anglo-American Marriages in the Later Nineteenth Century,' *Proceedings of the American Philosophical Society*, vol. 135, no. 2 (1991), 140-99.

58   *The Nation* (Feb. 6, 1879), 28, 101-102.

59   *New York Times* (Jan. 23, 1879), 1.

60   Dow, *State Reservation*, 17-18.

61   Evans, 142-59; Seibel, 1-2, 23-26.

American senators and congressmen; British members of parliament; bishops from all three countries; university presidents, academics and the directors of the British and American geological surveys signed. Banking magnate J.P. Morgan accompanied cultural figures Church, Bierstadt, Ralph Waldo Emerson, Oliver Wendell Holmes, Henry Wadsworth Longfellow, John Greenleaf Whittier, Carlyle, and Ruskin among others. The petition expressed the opinion that, 'Objects of great natural beauty and grandeur are among the most valuable gifts which Providence has bestowed upon our race. The contemplation of them elevates and informs the human understanding. They are instruments of education. They conduce to the order of society. They draw together all races, and thus contribute to the union and peace of nations'. Suggesting Anglo-American imperial hegemony and cultural harmony on the continent, it requested that 'the Falls of Niagara ... be placed under the joint guardianship of the two governments'.[62]

The proposed sale of Goat Island stimulated action. Just weeks after the appeal Gardiner and the press raised the prospects of 'a row of factories' such as those already lining the river banks, 'a military camping ground [and] ... 'a rifle range', or the purchase of the entire island by P.T. Barnum and its complete disappearance under a big top.[63] This suggested the possibility that a 'monster hippodrome' where 'the usual "great combination" circus companies may give performances through the visiting season' would remove the island from view.[64] Circuses brought a different vision of the social order and doubtless emphasized the wild and untameable associations of the site; Wild Bill Hickok and Buffalo Bill already showed Amer-Indian war-dance spectaculars at Niagara Falls in which customary rituals performed for the crowds presented alternative cultures of exotic wildness than those traditionally depicted in English and American poetry and drama.[65] The Falls and the agreements about them were of most interest to an American elite with its affinity for English culture, its sensibility grounded in a literary pastoral or romantic wildness, and as well its interest in an ordered landscape not intruded upon by either trade and industry or the *hoi poloi* (unless well-confined on viewing platforms). The reservationists' desired representation of the natural order at Niagara was of Anglo-American stewardship of resources peacefully not militaristically achieved, orderly and respectful viewers, craft-trading 'Indian squaws with bead trinkets', 'ornaments and curiosities', and other elements kept out of sight.[66]

---

62   Memorial Addressed to the Governor of New York, and the Governor-General of Canada, Mar. 2, 1880, quoted in Dow, *State Reservation*, 18-20.

63   *Huddersfield Daily Chronicle* (May 5, 1880), 4, quoting Gardiner.

64   Ibid.; *New York Times* (April 21, 1880), 4.

65   Alan Trachtenberg, *Shades of Hiawatha. Staging Indians, Making Americans* (New York: Hill and Wang, 2004), xi-xxv.

66   William Morris, *Letters sent Home* (London: Frederick Warne, 1875), 225-26. Morris should not be confused with William Morris, Arts and Crafts designer and preservationist.

The International Park project, as it was known, had diplomatic and cultural aspects that were not in tandem. The project stalled on the Canadian side but persevered in New York during Grover Cleveland's governorship with what would be significant landmark legislation for preservation and more elite class interests. In New York in 1883 'An Act to Authorize the Selection, Location and Appropriation of Certain Lands in the Village of Niagara Falls for a State Reservation; and to Preserve the Scenery of the Falls at Niagara' was approved. The acknowledgement of scenic interests in law provided precedence for future debates. Elite, scenic interest was more difficult to harness in Canada. Ontario's Premier Oliver Mowat introduced park legislation in 1880, though it failed to attract broad public support. With title to the Chain Reserve lands unclear, neither the Provincial nor Dominion government would accept financial responsibility for a park.[67]

Financed by a limited issue bond, New York State instituted parks on Bath and Goat Islands and the riverbank designed by Olmsted with polite serpentine walks and carriage rides. At the opening of the Niagara State Reservation on 15 July 1885, Governor Hill trumpeted New York's munificence to the assembled crowds, highlighting its political unity, cooperative wealth, and democratic civilization. As Hill explained, 'this philanthropic project to the preservation to the world for its free use, forever, of the majestic scenery of the Falls of Niagara – nature's grandest gift to man … is worthy [of] the advanced thought, the grand liberality, and the true spirit of the nineteenth century'.[68] Also present were three previous New York governors; Dorsheimer, as head of the Niagara Park Reservation Commission; and Canadian park project supporters Lieut-Gov. Robinson and Attorney-General Oliver Mowat. American ability to get the job done and provide spectacular leisure facilities was displayed to an international audience as successful democracy in action in contrast to cumbersome and ineffective Provincial, Dominion, and Imperial bureaucracies.

Canadian identities were still largely subsumed in imperial identities, and Canada used both American and British models of national identity-building.[69] Support for scenic and representational interests came at elite levels and from railway companies wanting to attract settlers and tourists to the western territories. In 1884, the Canada Southern Railroad provided a prompt for action by destroying the wooded hillside above the Horseshoe Falls with landfill. Wanting to build a bridge across the gorge between Clifton House and the Falls, the Company felled trees and dumped earth. When it refused to pay clearing and removal costs, government was obliged to act or be seen failing the international public interest at this beauty spot. Mowat established a Niagara Falls Park Commission and in 1885 the Niagara Falls Park Act – An Act for the Preservation of the Natural

---

67   *New York Times* (Feb. 15, 1884), 2.
68   *New York Times* (July 16, 1885), 5.
69   Barbara J. Messamore, 'British Imperialists and Canadian "Nationalists,"' in Colin M. Coates, ed., *Imperial Canada, 1867-1917* (Edinburgh: University of Edinburgh Centre of Canadian Studies, 1997), 87-89, 97-98.

Scenery about Niagara Falls passed into legislation. Promoted as an 1887 Jubilee celebration the Queen Victoria Niagara Falls Park, opened on Victoria Day, 1888, acquired an imperial rather than a national or provincial identity. *The New York Times* refrained from mentioning the international idea but regretted the 'somewhat ponderous' name and hoped that the imperial appellation would soon be dropped.[70] A bond covered the expenses of land expropriation supplemented by tolls and franchises. Tolls both ameliorated and created local friction; comparisons with New York's generosity to international visitors resulted in their removal. A spectacle of international order and cooperation had been achieved with the Falls as a monument framed between well-governed parks.[71]

iv

Instituting reserves was not simply a matter of class or sentiment but of cultural diplomacy. Although riparian rights along the Niagara River remained an open issue, the possibility of cooperation had been suggested at the waterfalls in an uncontentious context. The 'reservation' campaigns provided a vehicle for sophisticated, powerful groups to express United States-Canadian goodwill and British-United States relations at levels that were both sub-political and above politics. The scheme gained favourable publicity through wide press coverage. Campaigns to preserve the scene continued as developments in hydro-power challenged the appearance of the waterfalls themselves. Unrestricted water use by Anglo-American hydro-electric companies concerned American preservationists during the 1890s. By 1900 additional constituencies for a riparian agreement developed. Demands by American shipping and defence interests in the Great Lakes-St Lawrence system for canal construction to improve navigation raised the prospect of massive interference with boundary waters.

Tourism and national sentiment were powerful constituencies when combined with shipping and defence interests and supported at high levels. By 1900, the New York Niagara Reserves alone attracted around 750,000 visitors annually and scenic-minded newspapermen exploited friction between hydro-power, shipping, and commercial-tourist interests. Preservation activist J. Horace McFarland, a Harrisburg printer and editor of the 'Beautiful America' section of the *Ladies Home Journal*, mounted Falls preservation campaigns with William B. Howland, editor of the influential New York magazine, *The Outlook*. Support was channelled into activism via the American Civic Association, formed in 1902 and led by McFarland. Howland continued to promote Anglo-American friendship through preservation, as Thomas Otte notes elsewhere in this volume.[72]

---

70 *New York Times* (May 21, 1888), 2.

71 Seibel, 26-37; Evans, 171.

72 Evans, 206, 377-92; Dreisziger, 437-58.

American navigation and defence interests in the Great Lakes alerted government. Water rights, development and conservation, and stream flow had become political issues.[73] On President Roosevelt's proposal an International Waterways Commission to consider boundary waters was established in 1902-03. The Commission recommended that power franchises come within its remit as demands for water extraction grew from Anglo-American-Canadian power companies on both banks. The Commission's work was not a priority for the American, British, or Canadian administrations. However, President Roosevelt was a long-standing supporter of the Niagara campaign and with his backing, in May 1904, Secretary of State John Hay opened talks with Great Britain for a treaty to preserve the Falls. An International Waterways Convention (IWC) was established in November 1905 to regulate equitable division of the water-power between the United States and Canada. The American arm of the IWC desired limiting water diversions so as not to further diminish the Falls' monumental character.

The President called for federal protection of the Falls until a treaty could be negotiated. Roosevelt had doubled the size of America's national parks, and aspired to include Niagara Falls in this chain of federal reserves.[74] In February 1906 he wrote to Hay's successor, Elihu Root, 'I enclose herewith a petition for the negotiation of a treaty for the preservation of Niagara Falls. I of course am not able to say offhand whether a treaty would be the right way to go about it, but I understand that you already have under way negotiations with the British Government of this subject ... I earnestly desire that action be pressed as rapidly as possible. There is no more worthy object before the people at this time'.[75] The following month, Roosevelt urged Congress to back the IWC and do 'all in its power to preserve the great scenic wonder, the existence of which should be a matter of pride to every dweller on this continent'.[76]

Anglo-American cultural vehicles assisted cooperation. Roosevelt appealed to the then Governor General, Lord Grey, at a forestry convention in Ottowa through Gifford Pinchot, Chief of the United States Forestry Commission.[77] Joseph Choate, the recent and popular Ambassador to Britain (1899-1905), invited Grey to address the influential, trans-Atlantic Pilgrim Society in New York in April 1906.

---

73   Samuel P. Hays, *Conservation and the Gospel of Efficiency, The Progressive Conservation Movement 1890-1920* (Cambridge, Mass: Harvard University Press, 1969), 5-9.

74   Dalton, 239-48.

75   Roosevelt to Root, Feb. 14, 1906, quoted in Elting Morison, ed., *The Letters of Theodore Roosevelt* (8 vols, Cambridge, Mass: MIT Press, 1951-54), vol. 5, 154.

76   US Congress, Senate, Preservation of Niagara Falls, 59th Cong. 1st see, Mar. 27, 1906, S. Doc. 242, 1-2, quoted in Evans, 438.

77   Hays, 38-48. Pinchot's views on preservation versus conservation were politically nuanced. He developed a policy of the sustainable use of resources by industry that took into account domestic needs against protectionist 'preservation' of reserves for elite sport.

Afterwards Root explained that the preservation of Niagara Falls was a subject 'very dear' to their hearts, hinting that the American government was 'prepared' to take unilateral action in the matter.[78]

American representational preservation interests sufficiently aligned with those of national security for the Falls' protection to gain political backing. The IWC linked Niagara Falls' protection to American defence issues and Congress passed a temporary protection measure in June 1906. The so-called Burton Act limited water diversion at the Falls on the American side.[79] Negotiations for water-extraction limitations at Niagara Falls thus far were between the American government and Britain's representative in Canada, supported by American newspaper and tourist interests and an active scenic lobby. Support was by no means unanimous. Charles Macdonald, president of the American Society of Civil Engineers, told Pinchot it was an 'international crime that so much energy [be] allowed to go to waste'.[80] More serious objections came from popular Canadian demands for cheap industrial and domestic power that far outweighed scenic-representational considerations at the site.

Nevertheless, cultural diplomacy was proving its utility. The successful preservation of the monumental grandeur of Niagara Falls as an aid to international mediation on a broader raft of political matters gained recognition at the level of high politics. The IWC specified water limits deemed necessary to preserve the Falls' monumental value as 20,000 cubic feet per second on the American side and 36,000 on the Canadian. To achieve limitation an international boundary waters treaty was required.[81] The British government's replacement of Sir Henry Durand as British Ambassador to Washington in early 1907 by James Bryce, a well-known preservation activist, indicates a desire to facilitate resolution and to maintain some representational status at Niagara Falls. Peter Neary observes that, 'Bryce recognized in the projected complete settlement of Canadian-American differences a unique opportunity to build upon the Anglo-American rapprochement of the previous decade'.[82] Canadian Prime Minister Sir Wilfred Laurier agreed to support Falls preservation if, in return, he could further Canadian national interests and place Canada's relations with the United States on clearer legal foundations.

The ability of America, Britain, and Canada to preserve scenic-leisure interests at Niagara Falls against competing nationalist and advancing industrialist interests provided a celebration of progressive governance and international cooperation as a manifestation of civilization. The Boundary Waters Treaty was concluded in January 1909 during the final months of Roosevelt's presidency, 'when a

---

78   Relayed by Grey to Laurier, Apr. 4 and 14, 1906, Laurier Papers, vol. 7, 31, quoted in Dreisziger, 453-54, 449.

79   Neary, 357-80.

80   Dalton, 246; Hays, 127.

81   James Simsarian, 'The Diversion of Waters Affecting the United States and Canada,' *American Journal of International Law*, vol. 32, no. 3 (1938), 494-506.

82   Neary, 363 *et passim*.

comprehensive agreement was reached on the uses and diversions of waters along the International Boundary' assured the appearance of Niagara Falls.[83] In February, Roosevelt held a North American Conservation Conference which promised continued international cooperation. However, the United States had gone as far as it was prepared to go towards promoting such international peaceful initiatives. The outgoing president was unable to persuade Congress of the desirability of a World Conference on Conservation.[84]

V

The first example of an international preservation effort occurred at a site where visual grandeur *and* industrial progress were sufficiently powerful to dominate cultural memories. To maintain an acceptable balance between the two required cooperative action. Cooperation at Niagara Falls was followed by decades of efforts to find common Anglo-American heritage objectives and continue the process. The joint preservation of Thomas Carlyle's London home followed in 1894, and of John Harvard's maternal grandfather's Stratford home in 1907.[85] Cooperation took place through voluntaristic associations, and at professional and sub-diplomatic levels. Further, the Niagara campaign provided influential precedence.

Although it is now common to divide the protection of cultural and natural resources and indeed to see them in terms of national interests, this was not entirely the case for late nineteenth-century Anglo-American preservationists. Societies to protect both cultural and natural resources were established during the 1890s in both countries in the wake of Niagara: the Trustees of [Public] Reservations, Massachusetts (1890); the National Trust for Places of Historic Interest and Natural Beauty, London (1894-95); and the New York (later American) Scenic and Historic Preservation Society (1896). There were connections between the three, and Niagara Falls helped foster an Anglo-American epistemic community. For such preservationists the world contained cultural monuments and 'nature's pictures'.

The prospect of further cooperation bolstered the founding aims of the National Trust. Although later circumscribed by its operative Act of Parliament (1907) the Trust's early sphere of activity included the whole of Britain and Ireland, together with the United States and Canada.[86] The Trustees of Reservations was represented on the Trust's Council. In 1886 James Bryce, an overlooked but important founder-member of the Trust and an Eliot family friend, had provided Charles

---

83   Dreisziger, 457.

84   Dalton, 246.

85   Hall, 'Politics of Collecting,' 349, 355-56.

86   '*A Report By Mr C.R. Ashbee to the Council of the National Trust for Places of Historic Interest and Natural Beauty, on his visit to the United States in the Council's Behalf, October, MDCCCC, to February, MDCCCC,*' 24, National Trust Archive, Swindon.

Eliot, an associate of Olmsted's practice and cousin of Charles Eliot Norton, with a proposal for an organization to hold multiple properties. Made by (Sir) Robert Hunter (another Trust founder) in 1884, the proposal had laid the foundations for the National Trust; but it was Eliot's Trustees that established precedence in law. A short-lived branch of the National Trust was established in Washington DC.[87] As Thomas Otte shows George Washington's ancestral home, Sulgrave Manor, was among the initial properties proposed for the Trust's collection.

The Niagara Falls preservation campaign effectively demonstrated the relationship between government, technology and industry, and social and economic need in which the spectacular value of the resource was maintained. This marked a significant stage in the recognition of natural monuments and of their value for representational preservation. It was a more modern compromise and affirmed the United States' arrival as a power on the world stage. The preservation campaign attracted widespread support and had an extraordinarily broad reach. As the first successful international preservation effort, it presented the possibility of preserving sites of common interest as a representation of progressive Anglo-American civilization and rapprochement. At the same time, it demonstrated the popular utility of preservation not only as an aesthetic leisure amenity but also as a means to facilitate agreement on more contentious political issues. Appealing to sensory appreciation and the association of monumental resources with inter-national, indeed world representation it drew on much earlier traditions of diplomacy and statecraft and on interconnected ideals and structures of western civilization.

At Niagara Falls change was inevitable. Scientific and industrial developments required political action between more than one state. While the preservation effort monumentalized its target resource as a spectacle of order and peace under the guardianship of two world powers, quite clearly the United States drove the agenda. However, both Britain and Canada found advantage in agreement. Remarkably, compromise and concord were achieved without undue acrimony. Such collaborations took diplomacy beyond the legal realms of international treaties; by protecting sufficient magnificence of a natural monument the value of cooperation was materially demonstrated while at the same time showing respect for and acknowledgment of the fragility of magnificence. Although the Falls' monumental identity was irrevocably compromised and ideological hierarchies were instituted, the Niagara Falls campaign established a precedent for the value of celebrating peaceful accord through preservation, and preservation's use as a model of cultural diplomacy. It also demonstrated the civilized multiplicity of uses for a site. Arguably, it provided precedence for the concept of world heritage.

---

87   Hall, 'Politics of Collecting,' 353.

Chapter 2

Redeeming Holy Wisdom:
Britain and St. Sophia

Erik Goldstein*

The Hagia Sophia, the Church of the Holy Wisdom, was the great church of Constantinople. Built in a remarkably short period of time, 532-537, by the Emperor Justinian, it was the centre of Christian life for much of the Byzantine empire. With the fall of Constantinople to the Ottoman Turks in 1453 it became the mosque of the Aya Sophia, and as such the chief place of worship of the Sultan, who would later assume the additional title of Caliph ul-Islam in 1517, following Selim I's conquest of Egypt. In its new guise it was slowly almost lost to Western memory. It was, however, never entirely forgotten, and as European powers began to intrude into the Eastern Mediterranean world, particularly from the early nineteenth century, a fascination grew with what had once been the greatest church in Christendom. At the same time the Eastern Question, as the problem of the Ottoman Empire came to be called, was concerned with the consequences of the slow decline of a state now commonly referred to as 'the sick man of Europe'. Turkish rule, and the Turks themselves, were viewed with great negativity, and there was a great deal of anti-Turkish agitation from 1870s through the First World War. In Britain there was great concern about the treatment of the Ottoman's Christian subjects and St. Sophia became a subject of fascination. Britain was emerging as an Eastern Mediterranean power and for a few years, at the end of the First World War, it occupied Constantinople. The desire to save St. Sophia, both materially and spiritually, grew with the expansion of British power into the region throughout the nineteenth and early twentieth centuries, and would involve early preservationists and devout Christians, who were often one and the same.

* Thanks are due to Dr Thomas Otte of the University of East Anglia, Profs. Houchang Chehabi, Husain Haqqani and Melanie Hall of Boston University; Prof. George Marcopoulos of Tufts University; and Prof. Gerald Studdert-Kennedy of the University of Birmingham, as well as to the Rt. Hon. Earl of Crawford and Balcarres for access to the papers of the 27th Earl, and to the Leeds Library for allowing access to its collections. Access to and permission to quote from the papers of the St. Sophia Redemption Committee was made possible by courtesy of The Revd. A.T. John Salter, A.K.C., General Secretary of the Anglican and Eastern Churches Association.

St. Sophia only slowly re-emerged as an icon of Christendom and western civilization into the consciousness of the west, one that had they had been robbed of by the Muslim Turks. The historian Edward Gibbon decried the building's fate observing, when discussing its construction and interior decoration, that it included ' ... images of Christ, of the Virgin, of saints, and of angels, which have been defaced by Turkish fanaticism ... '[1] From the late eighteenth century the occasional intrepid Christian tourist would visit Constantinople, capital of the Ottoman empire, and make reference to the building in published accounts of their travels. One of the earliest of these, in the mid-1780s, was Elizabeth, Lady Craven who visited the building with the assistance of the British embassy.[2] For Thomas MacGill, who spent part of 1805 and 1806 in Constantinople, the building was a shrine containing a Christian relic. He recounted the story which claimed that one of the stones in the building was reputed to be one on which Mary washed the clothes of the infant Jesus and noted that, 'the few Christians who can obtain admission to this no longer church of Christ, but imperial mosque, regard this stone with great devotion, although they are not allowed to carry away the smallest particle of it.'[3]

The control of the great church by Muslims attracted poetic concern, both good and not so good. When the Greeks of the Peloponnese revolted against Turkish rule on Easter Sunday 1821, they ignited a wave of philhellenism that swept Europe and the United States.[4] Philhellenism is often associated with western fascination for an idealized ancient Greece, but it was just as much concerned with a belief in the struggle between the Christian and Muslim world, the latter of which was largely under Turkish rule. The poet, Lord Byron, who would die fighting for Greek independence against the Turks, used an image of St. Sophia to win support for the cause of Christians oppressed by the Turks, writing in *Childe Harold:*

> And whose more rife with merriment than thine,
> Oh Stamboul! once the empress of their reign?
> Though turbans now pollute Sophia's shrine,
> And Greece her very altars eyes in vain
> (Alas! her woes still pervade my strain!)[5]

---

1   Edward Gibbon, *The Decline and Fall of the Roman Empire* (6 vols, New York: Alfred Knopf, 1994), vol. 4, 212.

2   Elizabeth, Lady Craven, *A Journey through the Crimea to Constantinople* (London: G.G.J. and J. Robinson, 1789), 211-18.

3   Thomas MacGill, *Travels in Turkey, Italy, and Russia* (2 vols, London: John Murray and Archibald Constable, 1808), vol. 2, 123.

4   William St. Clair, *That Greece Might Still be Free: The Philhellenes in the War of Independence* (London: Oxford University Press, 1972).

5   Lord Byron, *Childe Harold's Pilgrimage*, in Jerome McGann, ed., *Lord Byron: The Complete Poetical Works* (7 vols, Oxford: Clarendon Press, 1979), vol. 2, 70. See

Joshua Marsden of Worcester submitted a seventeen-verse poem in 1828 to the *Imperial Magazine* which included the lines, referring to St. Sophia as a potent icon of Christian–Moslem conflict,

> Yon Temple, once sacred to worship divine,
> Justinian's glory and fame;
> Inscribed on its alter-piece, pillars and shrine,
> Shall witness God's *true* Prophet's name.[6]

It became something of a Christian coup to gain access to the building. British naval officer Adolphus Slade visited Constantinople in 1829 and in his published account of his travels noted that it had become difficult for Christians to obtain access to the building, though he and his party managed to sneak in. He commented that, ' ... it was St. Sophia, the most celebrated, and one of the most venerable of Christian temples – centre of memorable associations – and we were congratulating ourselves of being where so few Christians, during four centuries, have been ... ' able to visit.[7] Women were noticeably in the vanguard of the rescue mission. In 1835 Julia Pardoe accompanied her father, an army officer, to Constantinople. She too managed to slip into the building and recounted her excitement, ' ... this was St. Sophia! To me it seemed like a creation of enchantment ... '[8] She reported that, ' ... both the Turks and the Greeks have a firm faith in the prophecies which have been made, that St. Sophia will one day revert to the Christians ... '[9]

More specialized interest followed as St. Sophia was compared favorably with other great monuments of western architecture. Another intrepid female traveler, G.L. Dawson Damer, published an account of her travels in the region, and commented that upon visiting St. Sophia, 'We were a little disturbed by the evident ill-will with which the few Mussulmen, who still lingered at their devotions, bore our presence.'[10] She, however, was struck by the architecture, observing, ' ... the just and noble proportions of the building are, I think, quite as striking as those of any other I ever beheld, not excepting St. Peter's, (although the diameter of the former is twenty-five feet less than that of the Basilica at Rome), the great

David Roessel, *In Byron's Shadow: Modern Greece in the English & American Imagination* (Oxford: Oxford University Press, 2002).

6   Joshua Marsden, 'Thoughts on the Present Crisis of the Turkish Empire,' *Imperial Magazine*, vol. 10, no. 114 (June 1828), 553-55.

7   Adolphus Slade, *Records of Travels in Turkey, Greece, &c* (2 vols, London: Saunders and Otley, 1833), vol. 2, 232. Slade was subsequently lent to the Ottoman navy, where he was known as Mushaver Pasha, and served for seventeen years as its administrative head.

8   Miss Pardoe, *The City of the Sultan* (2 vols, London: Henry Colburn, 1837), vol. 1, 378.

9   Ibid., 392.

10   Mrs G.L. Dawson Damer, *Diary of a Tour in Greece, Turkey, Eqypt, and the Holy Land* (2 vols, London: Henry Colburn, 1841), vol. 1, 115.

room of Vicenza, Westminster Hall, York Minister, or even the Pantheon itself.'[11] As the imagery of St. Sophia grew compelling to Christians it was seen as an æsthetic ideal. In reviewing Thomas Hope's *An Historical Essay on Architecture* the *British and Foreign Review* observed that in St. Sophia 'we trace the perfect germ of Christian style.'[12]

The loss of the great church clearly troubled educated British Christian visitors to Constantinople. Aubrey de Vere published in 1850 an account of his visit, with the observation about St. Sophia that, 'The present building has suffered little external change, and may yet lift the Greek Cross into heaven, and suspend its firmament over a Christian altar, centuries after the Crescent has ceased to insult the European shores.'[13] In the early nineteenth century it was only through these occasional travellers' accounts that people in the West heard of St. Sophia.

For a period during the mid-nineteenth century a new, reform-minded Sultan, Abdülmecid, attracted many Europeans to Constantinople. This was a time that coincided with a growth in the print media. Among the Europeans were the Swiss brothers Gaspare and Giuseppe Fossati, who had established a successful architectural practice. In 1846 the Sultan asked them to undertake a full restoration of the building.[14] The Fossatis, as they confronted the commission, were in despair about the building after inspecting it. Writing that, ' ... already the vaults and the cracked domes were letting in the rain, the wind and the snow', Gaspare Fossati made architectural conservation a religious issue. He condemned, 'the negligence of the Moslem officials responsible for the conservation of the building ... '[15] In employing Christians to restore the building the Sultan also risked provoking extremist Muslim sentiment, and accordingly he timed his plans for repair work to commence with the annual pilgrimage to Mecca, encouraging the most extreme imams to participate in the *haj*.[16]

During the restoration of 1847-49 the Fossatis discovered the magnificent Byzantine mosaics that had been whitewashed when the church had become a mosque. This was the first time the building could easily be visited by foreigners and non-Muslims. The mosaics were so spectacular that Abdülmecid considered leaving uncovered the ones outside the principal prayer areas, but in the end all were again covered. They were however available in print. In 1852 the Fossatis published a portfolio of twenty-five coloured lithographs of St. Sophia.[17] As a

---

11   Ibid., 117-18.

12   *British and Foreign Review*, vol. 7, no. 13 (July 1838), 15.

13   Aubrey De Vere, *Picturesque Sketches in Greece and Turkey* (2 vols, London: Richard Bentley, 1850), vol. 2, 181.

14   On the actual restoration work conducted see Godfrey Goodwin, 'Gaspare Fossati di Morcote and his Brother Giuseppe,' *Environmental Design: Journal of the Islamic Environmental Research Centre* (1990), 122-27.

15   Lord Kinross, *Hagia Sophia* (New York: Newsweek, 1972), 116.

16   Ibid., 118.

17   Lord Lindsay, *Sketches in the History of Christian Art* (London: J. Murray, 1847).

result of this publication this largely forgotten building re-emerged into a wider – and a Christian – consciousness.[18]

These new visual images of St. Sophia had a great impact. Arguably, they ultimately spawned the Byzantine revival in architecture in the West, with a still visible impact on church architecture. This coincided with the development of the new subject area of the History of Christian Art, with such path-breaking works as Lord Lindsay's, *Sketches of the History of Christian Art* (1847).[19] In the *Seven Lamps of Architecture* (1849) and *Stones of Venice* (1851-53) John Ruskin wrote favourably of Byzantine architecture.[20] Gradually St. Sophia came to represent the pinnacle of Christian artistic achievement. Though Ruskin never visited Constantinople he would later write of Venice, ' … the abandonment of her Byzantine architecture *meant* her ruin.'[21] James Fergusson in his *History of Architecture in All Countries* wrote that ' … it remains even now an open question whether a Christian church exists anywhere, of any age, whose interior is so beautiful as that of this marvellous creation of old Byzantine art' and he concludes by calling St. Sophia, ' … the most perfect and beautiful church which has yet been erected by any Christian people.'[22] During the ensuing decades the visual knowledge of St. Sophia would continue to grow. For example at Christmas 1898 the British diplomat Cecil Spring Rice, best remembered today for his hymn 'I pledge to you my country', but also an active preservationist who had served with and admired the diplomat, preservationist and campaigners for the rights of eastern Christians, James Bryce, sent postcards from Constantinople to the children of his friend Theodore Roosevelt, noted promoter of the United States' National Parks, including one of St. Sophia, commenting that, 'I wish you could see St. Sophia – I think it the finest building I have ever been in, and its associations are overpowering.'[23]

All of this was occurring at a time of growing British official and popular concern about the nature of Turkish rule. The Marquis of Salisbury, Secretary of State for India and a future prime minister, was sent by the British government to Constantinople during 1876-77 to help forestall a possible Russian attempt to seize the city. Salisbury, however, did not want Britain to be drawn into what he

---

18   A magisterial study of the impact of St. Sophia is Robert Nelson, *Hagia Sophia, 1850-1950: Holy Wisdom Modern Monument* (Chicago: University of Chicago Press, 2004).

19   Adele Ernstrom, '"Why Should We Always Look Back?" "Christian Art" in Nineteenth-Century Historiography in Britain,' *Art History*, vol. 22, no. 3 (Sep. 1999), 421-35.

20   John Ruskin, *The Seven Lamps of Architecture* (London: Smith, Elder, and Co., 1849) and *The Stones of Venice* (London: Smith, Elder, and Co., 1851-53).

21   Ruskin, *Seven Lamps* (Sunnyside, Kent: George Allen, 1889), 96, n34.

22   James Fergusson, *A History of Architecture in All Countries* (3 vols, London: John Murray, 1867) vol. 2, 315, 321.

23   Stephen Gwynn, ed., *The Letters and Friendships of Sir Cecil Spring Rice* (2 vols, Boston: Houghton Mifflin, 1929), vol. 1, 272.

viewed as an indefensible alliance with the Ottoman empire. A devout churchgoer, he was closely and publicly associated with high Anglicanism and an aggressive defender of its doctrinal position throughout his career.[24] Salisbury wanted to see the end of Ottoman rule over the Christian populations of the Balkans. The prime minister, Lord Beaconsfield, complained to the queen that Salisbury was 'thinking more of raising the Cross on the cupola of St. Sophia, than the power of England.'[25]

One of those concerned about perceived Ottoman misrule was William Morris, a member of the Eastern Question Association, founded in 1876.[26] Morris was also the founder of the Society for the Protection of Ancient Buildings (SPAB), established the year after the Eastern Question Association in 1877. In 1878, Morris obtained a photograph which he greatly prized of the St. Sophia, writing to several friends that, ' ... the most beautiful of the buildings raised in Europe before the nineteenth century ... [was] the Church of St. Sophia at Constantinople ... '[27] When SPAB began to interest itself in buildings at risk abroad it campaigned on numerous occasions about restoration issues relating to churches. St. Sophia became one of only a handful of SPAB causes abroad.[28] Preservation of St. Sophia was related to Morris's political interests. He was also active in anti-Turkish politics and in his first public lecture Morris mentions Hagia Sophia.[29] Indeed one of his biographers speculated that Morris was drawn to anti-Turkish agitation ' ... because his whole nature revolted at the idea of the Temple of the Holy Wisdom being in the hands of the infidel.'[30] In his second public lecture he addresses the Eastern Question.[31] In 1882, in a SPAB sponsored address, Morris observed of the three buildings that he considered as marking the beginning of modern architecture,

---

24   Michael Bentley, *Lord Salisbury's World: Conservative Environments in Late Victorian Britain* (Cambridge: Cambridge University Press, 2001), 188-205.

25   Lord Beaconsfield to Queen Victoria, Apr. 17, 1877, quoted in Richard Millman, *Britain and the Eastern Question, 1875-1878* (Oxford: Clarendon Press, 1979), 274.

26   Nelson, *Hagia Sophia*, 110.

27   Morris to Aglaia Ionides Corono, Mar. 7, 1878; Morris to Thomas Wardle, Mar. 8, 1878; and Morris to Jenny Morris, Mar. 18, 1878, in Norman Kelvin, ed., *The Collected Letters of William Morris* (4 vols, Princeton: Princeton University Press, 1984), vol. 1, 453-55, 463.

28   Frank Sharp, 'Exporting the Revolution: The Work of the SPAB Outside Britain 1878-1914,' in Chris Miele, ed., *From William Morris: Building Conservation and the Arts and Crafts Cult of Authenticity, 1877-1939* (New Haven and London: Yale University Press, 2005), 187-212.

29   Nelson, *Hagia Sophia*, 108.

30   Aylmer Vance, *William Morris: His Art, His Writings, and His Public Life – A Record* (London: G. Bell, 1897), 261.

31   Eugene Lemire, ed., *The Unpublished Lectures of William Morris* (Detroit: Wayne State University Press, 1969), 292. Lecture delivered Dec. 19, 1877. The text of the lecture has not survived.

I venture to call these buildings in their present state, the first the strangest, the second the most beautiful, the third the ugliest of the buildings raised in Europe before the nineteenth century. The first of these is the Palace of Diocletian at Spalato; the second, the Church of St. Sophia at Constantinople; the third, the Church of St. Peter at Rome.[32]

In 1891 Morris proposed his disciple William Lethaby for membership of SPAB. In 1893 Lethaby, who would become a leading figure in the Arts and Crafts movement, travelled to Constantinople together with Harold Swainson to prepare the first scholarly study of the building.[33] Lethaby's father had been a lay preacher at the local Bible Christian Chapel, and he grew up in a strictly religious household which also gave him politically radical views about a fair and free society. The building had become barred to visits by non-Muslim men, though not women. The two determined architects were fortunate to meet two tall American ladies, a mother and daughter, who had succeeded in visiting the building. The ladies lent them their clothes and, wearing these garments, the men succeeded in entering the building. As the American women were in mourning they were wearing mourning veils, an accoutrement that was undoubtedly useful in hiding Lethaby's moustache. The resulting co-authored book, *The Church of Sancta Sophia: A Study in Byzantine Building* (1894) helped bring about an architectural Byzantine revival.[34] In it they wrote,

> Sancta Sophia is the most interesting building on the world's surface. Like Karnak in Egypt, or the Athenian Parthenon, it is one of the four great pinnacles of architecture, but unlike them is no ruin, nor does it belong to a past world of constructive ideas, although it precedes by seven hundred years the fourth culmination of the building art in Chartres, Amiens, or Bourges, and thus must ever stand as the supreme monument of the Christian cycle.[35]

Architectural interest about, conservation concerns for, and religious fascination with St. Sophia were now rapidly converging.

After the publication of the book, *The Builder*, the leading architectural journal, produced an article on the construction techniques used in the church, observing that 'Probably no architectural monument that has ever been erected has aroused

---

32   Vance, 262-63. This lecture was originally delivered in 1879 before the Trades Guild of Learning at the Co-operative Institute. The 1882 Lecture was published as *Lectures on Art Delivered in Support of the Society for the Protection of Ancient Buildings* (London, 1882).

33   Godfrey Rubens, *William Richard Lethaby: His Life and Work, 1857-1931* (London: The Architectural Press, 1986), 80.

34   William R. Lethaby and Harold Swainson, *The Church of Sancta Sophia: A Study in Byzantine Building* (London and New York: Macmillan & Co., 1894), v.

35   Ibid.

so much enthusiasm in ancient times, and so much interest in modern times, as the great church erected under the orders of Justinian at Byzantium.'[36] One of the first architects to be influenced by this book was J.F. Bentley who had been commissioned to build a Roman Catholic cathedral for London at Westminster. He decided to harness Byzantine style for Catholicism, in what would be the most important British Catholic architectural statement of the period. Late in 1894 Bentley travelled to the continent to find Byzantine inspiration for his new commission but failed in his efforts to visit Constantinople due to an outbreak of cholera. He travelled, however, with a copy of the Lethaby and Swainson's book and its influence is clearly discernible in Westminster Cathedral.[37]

Architect Sir Thomas Graham Jackson also read and annotated a copy of the Lethaby and Swainson book. In 1910 Sir Thomas visited Constantinople and while there the Ottoman authorities asked him to prepare a report on St. Sophia.[38] This report in turn raised concern about the condition of the building. Jackson wrote to *The Times* in 1912 about the state of the fabric and concluded with the observation, 'St. Sophia is no less sacred in the eyes of the Moslem than in our own.'[39] The following year he published *Byzantine and Romanesque Architecture.*[40] Jackson had long been interested in Byzantine architecture, and one of his most notable buildings in the style is the chapel (Anglican) at Giggleswick School in Yorkshire, which was clearly inspired by the Hagia Sophia. Indeed, he was so proud of Giggleswick chapel that in his book *Byzantine and Romanesque Architecture* the first plate, which is intended to discuss dome construction, is of neither a Byzantine nor Romanesque dome but of Giggleswick. Built 1897 to 1901 to commemorate the diamond jubilee of Queen Victoria, its interior décor is redolent with imperial imagery, in a sense providing England with its own, though small scale, imperial Hagia Sophia. The benefactor who gave the funds for the chapel was Walter Morrison, one of the founders of the Palestine Exploration Fund, who had travelled widely in the East. Although St. Sophia was becoming iconic to many Christians, and the building was very influential, it had more ecumenical adaptations in Britain including Joseph Sunlight's South Manchester Synagogue (1912-13) which, in truly ecumenical fashion, also had a tower based on that of Westminster cathedral. The *British Architect* considered it 'a very satisfactory

36   'St. Sophia, Constantinople,' *The Builder*, vol. 68, no. 2720 (Mar. 23, 1895), 213.

37   W. de l'Hôpital, *Westminster Cathedral and its Architect: Volume I, The Building of the Cathedral* (New York: Dodd, Mead and Company, n.d), 35. It was published with an introduction by Lethaby.

38   See Basil Jackson, ed., *Recollections of Thomas Graham Jackson, 1835-1924* (London: Oxford University Press, 1950), 270; John Thomas, 'Sir Thomas Graham Jackson and the Church of the Hagia Sophia,' *Architectural History* 25 (1982), 98-101.

39   *The Times*, (Sep. 23, 1912), 5F.

40   Sir Thomas Graham Jackson, *Byzantine and Romanesque Architecture* (Cambridge: Cambridge University Press, 1913).

effect of an Eastern place of worship.'[41] St. Sophia clearly had a stylistic impact on British architecture, but there were also conservation concerns and the two often go hand in hand.

Any further plans for St. Sophia came to a halt with the outbreak of the two Balkan Wars (1912-13) and the First World War (1914), by which time St. Sophia was becoming a concern in the United States as well. In 1912 when it looked likely that Bulgarian forces would take Constantinople, British and American newspapers carried stories that the Bulgars would re-sanctify St. Sophia.[42] The Ottoman Empire entered the First World War in October 1914, and not long after Britain began planning an invasion. On the eve of the invasion there was much speculation as to the consequences of a British occupation of the city. Rumours, passed through missionary channels, that St. Sophia would be handed to the Orthodox church, even reached New York.[43] In April 1915, during the ill-fated Gallipoli campaign, when at first it looked as if British forces were on the verge of capturing Constantinople, one of the propagandistic publications of the time concluded, in a wonderfully conceptually challenged statement, that with the imminent British victory, 'the cross will replace the crescent on the minaret of St. Sophia.'[44] The building continued to resonate throughout the war in Anglo-American circles. When Dr Charles Eliot, president emeritus of Harvard University, father of the noted though deceased preservationist of the same name, and a friend of James Bryce, urged his countrymen to enter the war preserving St. Sophia was in his mind as he compared the Germans and Austrians unfavourably to the Turks observing that 'Even the Turks preserved St. Sophia.'[45]

The fascination with St. Sophia was not limited to Britain and the United States. Comments about it could be found in the press of all the Allied countries. For Russia the ambition to control Constantinople went back at least two centuries. In October 1914 Britain and France signed a secret agreement with Russia detailing what Russia would receive as a result of a victorious conclusion to the war, among which was control of Constantinople, and therefore, of St. Sophia. In late 1914 Prince Trubetskoi, a liberal Russian political philosopher, began to deliver a lecture on 'The National Question, Constantinople and St. Sophia'. This was soon afterwards published in English translation as *Saint Sophia, Russia's Hope and Calling* by the Faith Press which would later publish much of the British material

41   *British Architect* (Feb. 20, 1914), 157-8. The building is now listed Grade II.

42   One example is 'Bulgars May Plant Cross on St. Sophia,' *New York Times* (Nov. 4, 1912), 3.

43   'A New Palestine if the Allies Win – St. Sophia to be Greek Catholic Church if Constantinople is Taken and Held,' *New York Times* (Mar. 22, 1915), 3.

44   *The Dardanelles: Their Story and Significance in the Great War* (London: Andrew Melrose, 1915), 160.

45   Charles W. Eliot, 'Prepare to Enter the War … ,' *New York Times* (Mar. 12, 1916), 18.

calling for the reconsecration of the church.[46] With the 1917 Russian revolution and its withdrawal from the war, the opportunity for Russia to take Constantinople passed.

In late 1918 the Ottoman government requested an armistice, and British forces occupied Constantinople. The Royal Navy anchored close by St. Sophia. Although other Allied forces were dispatched to the city as well, it had effectively passed under British control. The British government gave great thought as to the fate of the Ottoman empire, and in particular its capital. Lord Curzon, soon to become Foreign Secretary, placed before the Cabinet a plan to move the Sultan to the Asian side of the Bosphorus, observing that as the city was only forty percent Turkish and that, with luck, most of the Turks would follow him. He suggested that the city then be internationalized and concluded his report with the hope that 'Not the least of its advantages in the eyes of some is the fact that in these circumstances Justinian's great Byzantine fane of St. Sophia, which was for 900 years a Christian Church, and has only been for a little more than half that period a Mohammedan mosque, would naturally revert to its own dedication.'[47] It is worth noting that his Oxford prize essay had been about Justinian, that he was active in such preservation groups as the National Trust and the Scapa Society for the Prevention of the Disfigurement in Town and Country, and that while serving as the Viceroy of India he had established the Archaeological Survey of India.[48] He was also an admirer of James Fergusson.[49] While serving in India he told the Asiatic Society of Bengal, 'What is beautiful, what is historic, what tears the mask off the face of the past, and helps us to read its riddles and to look in its eyes – these, and not the dogmas of a combative theology, are the principle criteria to which we must look.'[50]

British interest in the building was being generated in large part by Christian sentiment, and it was only a few months earlier Britain had captured Jerusalem. The success of British arms in liberating the Christian Holy Places, the first time they had been under Christian control since the crusades, was powerfully evocative. At the time of the capture of Jerusalem the visit of the German emperor there in 1898 was still remembered, as well as the fact that when he did, ' ... the saying went round "a better man than he walked."'[51] The victorious commander General

---

46   Prince Trubetskoi, *Saint Sophia Russia's Hope and Calling* (London: Faith Press, 1916). See also Randall Poole, 'Religion, War, and Revolution: E.N. Trubetskoi's National Liberal Construction of Russian National Identity, 1912-20,' *Kritika: Explorations in Russian and Eurasian History*, vol. 7, no.2 (Spring 2006), 195-240.

47   The National Archives, London (TNA), Cabinet Office Papers, CAB 29/2/P85.

48   David Gilmour, *Curzon* (London: John Murray, 1994), 398.

49   *Lord Curzon in India, 1898-1905* (London: Macmillan, 1906), vol. 1, 184.

50   Ibid., 188.

51   Matthew Hughes, ed., *Allenby in Palestine: The Middle East Correspondence of Field Marshal Viscount Allenby, June 1917-October 1919* (Stroud: Sutton, for the Army Records Society, 2004), 93.

Allenby accordingly entered Jerusalem on foot. The historian Martin Gilbert has observed of the British capture of the Holy City, 'The capture of Jerusalem caught the imagination of the Allied world. For Catholics, the Holy Places which had been lost when Saladin conquered the city more than eight hundred years earlier were once again under Western and Christian rule. In Rome, church bells rang out in triumph, as did the bells of London's Roman Catholic cathedral.'[52] This also meant that the holiest sites of Christendom were once again under the rule of a Christian power, including the Church of the Holy Sephulcre. There were those in Britain whose imperial vision was strongly linked to their Christian faith. Arthur Hirtzel, then assistant under-secretary at the India Office, believed that, 'The Empire ... has been given to us as a means to that great end for which Christ came into the world, the redemption of the human race. That is to say, it has been given to us to make it Christian. This is to be Britain's contribution to the redemption of mankind.'[53] Bernard Lucas, a prolific exponent of a popular missionary theology, was even blunter in his statement that, 'If the Christian West does not establish an empire of Christ in the East, the non-Christian East will establish an anti-Christian empire in the West.'[54] All that now was needed was the restoration of St. Sophia to Christendom.

There was also concern about the physical condition of St. Sophia. The 1919 SPAB *Annual Report* discussed the danger posed from decay.[55] This convergence of concern about the fabric of the church and its symbolic role led to a public movement to have the building placed in Christian hands, which would in turn assure that it was properly maintained, and its mosaics would be once again revealed. Even the famous 11th edition of the *Encylopædia Britannica* (1910) in talking about Hagia Sophia and its mosaics expressed the hope that 'They may reappear in the changes the future will bring.'[56] SPAB was concerned about some aspects of the redemption movement. Thackeray Turner of SPAB wrote to Curzon worried about the possibility of a drastic restoration. SPAB also wrote to the Foreign Office about the Hagia Sophia, and while not urging reconversion it did support action to secure the structure.[57]

---

52   Martin Gilbert, *Jerusalem in the Twentieth Century* (London: Chatto & Windus, 1996), 55.

53   Sir Arthur Hirtzel, *The Church, The Empire, and The World* (London: S.P.C.K., 1919).

54   Bernard Lucas, *The Empire of Christ* (London: Macmillan, 1907).

55   'Supplement,' in Society for the Protection of Ancient Buildings, *Annual Report* (1919).

56   *Encylopædia Britannica*, 11th ed. (29 vols, New York: Encylopædia Britannica, 1910), vol. 7, 6.

57   'Supplement,' in Society for the Protection of Ancient Buildings, *Annual Report* (1919).

The movement to 'save' the building was led by the St. Sophia Redemption Committee, which was formed in January 1919.[58] At first such a label might suggest a group of cranks, but the committee and its supporters were an illustrious group of well-known public figures. It was chaired by Lord Bryce, by now a famous former ambassador to the United States, who was also involved with the National Trust, as well as an ardent campaigner for Christian rights in the East. He had been interested in the fate of St. Sophia for at least four decades. In 1878 he published *Constantinople*, in which he describes St. Sophia as 'one of the wonders of the world' and notes that it is the only Christian church that has been preserved from very early times.[59] He concludes his section on St. Sophia with an account of its fate during the fall of Constantinople to the Turks in 1453 with the observation, 'Looking round this noble monument of Christian art, and thinking of that awful scene, it was impossible not to wish for the speedy advent of a day when the fierce faith of Arabia shall be driven out, and the voice of Christian worship be heard once more beneath this sound dome.'[60] Bryce was a long-standing advocate of Christian unification. Indeed some members of the committee hoped that they could look forward to the, ' … potentiality of Intercommunion between England and the Eastern Church, which would be the almost certain prelude to complete Christian reunion were a Pan-Orthodox Council summoned to meet in a St. Sophia redeemed by British arms.'[61] Among other notable figures on the committee was Sir Henry Lunn, who after a brief missionary career made a fortune with organized tours to the eastern Mediterranean, two future foreign secretaries Samuel Hoare and Edward Wood (later Lord Halifax), and the writer G.K. Chesterton. Another member was Professor Flinders Petrie, a famous Egyptologist, who in his youth had helped his father survey the monuments of southern England; he was also deeply religious. One of the first actions of the committee was to be in contact with Lethaby 'and other Artists with a view to the promotion of a memorial by Architects and others interested in the question of the artistic and antiquarian point of view.'[62]

The people involved in this group had a long standing interest in the condition of Christians in the Ottoman empire. In their first memorial the committee stated that they wanted to safeguard the church for Christendom and to promote the security and equal treatment of the Christian population of the region, and for which purpose they supported removing the Sultan's government from Constantinople. Ronald Burrows, the Principal of Kings College, London, commented to a friend in February 1919, 'I have very much encouraged the movement as it is a sound and

---

58   Erik Goldstein, 'Holy Wisdom and British Foreign Policy, 1918-1922: the St. Sophia Redemption Agitation,' *Byzantine and Modern Greek Studies* 15 (1991), 36-64.

59   James Bryce, *Constantinople* (New York: Harpers & Brothers, 1878), 46.

60   Ibid., 53.

61   Douglas, 'The Redemption of S. Sophia,' *The Treasury* (Feb. 1919), 368.

62   'S. Sophia Redemption Committee, Report,' Jan. 1919, Templewood I:8(1), Templewood Papers, University Library, Cambridge.

useful method of gathering new supporters for the policy of turning the Turkish government out of Constantinople.'[63] His comment was in the context of an article that morning in *The Times* which reported on the activities of the committee and that they felt that restoring the church would be 'a symbol of the redemption and liberation from age-long oppression of all those Christian peoples who are still left by the armistice under the reign of cruel fear and injustice.[64]

The treatment of Christians in the footprint of the Ottoman empire would be a concern at the looming peace conference. The St. Sophia Redemption supporters decided to focus their energies not on the wider issue of the rights of the Christian populations in the East, but on the far more evocative, emotional, and tangible symbol of the Church of the Holy Wisdom. The movement was launched at a meeting of the English Church Union held in the Great Hall of the Cannon Street Hotel. Speaking to the audience the Bishop of Oxford and longtime campaigner for Eastern Christians, Charles Gore, stated that, 'The fall of St. Sophia was a judgement on the divisions of Christendom. St. Sophia was not a Mahomedan shrine. Constantinople was not a Mecca or a Medina: it was a symbol of a Mahomedan, a Turkish, conquest, and Mahomedanism had got to recognize that it going to be reversed.'[65] A further meeting was planned in February, where the following motion would be put the vote, 'THAT this meeting declares that justice requires that the great Church of St. Sophia built by a Christian architect for a Christian Emperor and people, and taken from them by a foreign conqueror in 1453, should be restored to those who still form the large majority of the inhabitants of Constantinople, and further urges that immediate steps be taken by the Allies to prevent the imminent collapse of this priceless monument of antiquity.'[66]

The committee were active lobbyists, none more so than Bryce. On one occasion he approached a member of Lloyd George's government, the historian H.A.L. Fisher, who later wrote to his wife, ' ... Bryce buttonholed me about S. Sophia which he says is tumbling to bits. I'll try to get L[loyd].G[eorge]. to save it tomorrow.'[67] There was also a flood of propaganda, notable among which was the small book produced by the Rev. J.A. Douglas, *The Redemption of St. Sophia* complete with illustrations of how St. Sophia would look as a church once again, which admittedly was one of the things that caused SPAB to become agitated about the movement.[68] Nevertheless, within three weeks a second edition was

---

63   Burrows to Murray, Feb. 4, 1919, Burrows Papers, King's College, London, KAP/BUR /333c.

64   As quoted in *The Times* (Feb. 4, 1919), 11B.

65   'The Church of St. Sophia,' *The Times* (Jan. 24, 1919), 7G. See also G.L. Prestige, *The Life of Charles Gore: A Great Englishman* (London: Heinemann, 1935), 149, 449-52.

66   G.C. Dobbs (SSRC) to Hoare, Feb. 17, 1919, Templewood I:8(8).

67   H.A.L. Fisher to wife, n.d. but postmarked Feb. 10, 1919, HAL Fisher 206, HAL Fisher Papers, Bodleian Library, Oxford.

68   J.A. Douglas, *The Redemption of St. Sophia* (London: Faith Press, 1919). The illustration came in later editions.

needed. Other publications included such titles as *Give Back Saint Sophia* and *The Great Church of Holy Wisdom: A Plea For Its Restoration, That Night in May ... Lament on the capture of Saint Sophia by the Turk. Translated from the mode.*[69]

In March a petition was sent to the Prime Minister by the members of the St. Sophia Redemption Committee. In requesting that the great church be restored to Christendom, the petitioners also argued that, ' ... we submit that the claims of art should not be set aside. The great Church is now in a serious condition, urgently needing structural repairs, and must inevitably collapse if it is not placed in capable hands.'[70] The signatories included Thomas Graham Jackson, the archaeologist Arthur Evans, and the director of the British Museum Sir Frederick Kenyon. A separate, similar, petition was organized at Oxford, and also sent to the prime minister with a copy to the American delegation at the peace conference, signed by a number of the luminaries of the university. The one hundred and twenty-three signatories to the Oxford petition included the heads of eight of the colleges, and such famous figures as C.W.C. Oman, then the Chichele Professor of Modern History, Gilbert Murray, then Regius Professor of Greek and remembered as a notable internationalist, and A.L. Smith, the master of Balliol College.[71]

Converting the Sultan-Caliph's principal mosque into a church would not only be a powerful statement, but would also in some ways have an effect on how the past was viewed. Part of this movement was clearly about controlling the imagery of the past and of establishing identity and controlling the historical narrative. A number of factors led to the movement's failure, including ecumenical discord over which Christian denomination would control the church and the fear of a backlash, by some in government, among Britain's large Muslim population in India.

British interest continued nonetheless. One instance was a report on its condition embodied in a letter to *The Times* by the architect Sidney Toy in 1920.[72] This spurred Rev. Walter Crick also to write and express his concern that the dome might collapse, ' ... a catastrophe which the whole world would, I believe, willingly united to avert.' He the presaged the concept of a world heritage site, albeit with a Christian dimension. He acknowledged that the Turkish authorities would not accept reconsecration as a church, but he proposed that the Allies raise the funds required for repairs and that in return, 'It might then be offered to the

---

69   Athelstan Riley, *Give Back Saint Sophia: a report of the speeches delivered at the opening meeting of the crusade for the redemption of Saint Sophia, held at the Cannon Street hotel, on Jan. 23rd, 1919* (London: Faith Press, 1919); H. von Loufenberg, *That Night in May ... Lament on the capture of Saint Sophia by the Turk. Translated from the modern Greek of G. Drosini by G.R. Woodward, air by H. von Loufenberg, harmonies by Dr C. Wood, etc.* (London: Faith Press, 1919).

70   Templewood I:8(35).

71   Ibid. I:8(57).

72   Sidney Toy, letter to *The Times* (June 26, 1920), 12D. Toy in 1950 would oversee the restoration of the cathedral at Nassua, Bahamas.

League of Nations to be maintained as a world's thanks offering for peace and as a permanent memorial of all those who fought and died to vindicate Christian ethics against the recrudescence of old-world theory that might is right.'[73]

The possibility remained open in the minds of some that all was not yet lost. In April 1920 the Patriarch of Constantinople at Easter services ended his address with the hope that next year's service would be held in St. Sophia, a hope which brought applause from the congregants.[74] The previous month the *New York Times* expressed its disappointment that the Sultan was going to be allowed to keep Constantinople, observing that, ' … it may be well to remember that there is in Constantinople a holy place which was the foremost church of Christendom at its foundation, and which disputed that honor with St. Peter's for nearly a thousand years thereafter. St. Sophia has been a mosque for almost five centuries, but it was a church for twice that length of time before; and to more than a third of Christians of the world it is a holy place outranking all others.'[75] This echoed arguments made in England and provides evidence of an Anglo-American shared view, at least by some. An article appeared in December 1923 in the *New York Times* on preserved houses in Northamptonshire with links to famous American families. These included the ancestral Washington home, Sulgrave Manor, which Bryce and Curzon had been associated with trying to preserve since before the war (discussed by Thomas Otte in Chapter 5). The author of the *New York Times* article, in a remarkable leap of imagination, linked the protection of important Anglo-American buildings and the need to retake St. Sophia concluding that, 'The County of Northamptonshire is thus full of places sacred to Americans. And sacred places are often a serious responsibility. It was for sacred places that the crusades were proclaimed. And there may yet be a devastating conflict between the Cross and the Crescent for the Mosque or Church of St. Sophia in Constantinople.'[76]

This sense of abandonment of this great cause is redolent in a number of publications of the period, perhaps best exemplified by a seven stanza poem which appeared in the journal *The Christian East* around Easter 1921,

VI

With Thy Cross thrice-over blazoned on our banner's conquering face,
We have left Thy Cross down-trampled in Thine ancient dwelling place.
Lest Thy gifts to us be forfeit – lest our Eastern empire sway –
We will sign away Thine honour, and Thy glorious trust betray.

---

73   Walter Crick, letter to *The Times* (June 29, 1920), 20D.

74   Adm. Mark Bristol Diary, Apr. 11, 1920, Bristol Papers, Library of Congress, Washington.

75   'Very Strong Measures,' *New York Times* (Mar. 6, 1920), 10.

76   P.W. Wilson, 'British Cradle of American Presidents,' *New York Times* (Dec. 2, 1923), Sunday Magazine, 4-5.

VII
With Thy Cross thrice-over blazoned on our banner o'er the earth –
With Thy Cross at heart rejected as a thing of lesser worth –
Shall we dare to stand hereafter and bid others put their trust
In a sign we hold so lightly that we leave it in the dust?[77]

That issue of the magazine came with a large-sized print of the drawing of how St. Sophia would look restored as a church, suitable for framing, with the caption 'St. Sophia Restored – A Dream of What Might Have Been'.

British interest in St. Sophia did not wane with their evacuation of Constantinople in 1922. In May 1924 the architect Sir Edwin Lutyens raised the parlous condition of St. Sophia. This led Lord Crawford and Balcarres, the grandson of Lord Lindsay who had written on Christian art, who had introduced in 1900 an Ancient Monuments Protection Act in parliament and was an active follower of religious affairs, to consider the idea of tying a loan to the Turkish government to stipulations over the condition of St. Sophia. Lord Crawford suggested that the Turks give, ' … us the opportunity of clearing the frescoes and mosaics, temporarily perhaps, but sufficiently long to enable students to visit them, and for reproductions to be made. There is an untold wealth of Byzantine Art concealed behind the whitewash and stucco, presumably all of it in pristine

Figure 2.1　　St. Sophia Restored – A Dream of What Might Be. *The Christian East*, **April 1921, supplement**

77　Queenie Scott-Harper, 'The Betrayal of Saint Sophia,' *The Christian East*, vol. 2, no. 1 (Apr. 1921), 14.

condition.'[78] In the end, the responsibility for undertaking needed preservation work on the building fell to Americans.

The building continued in use as a mosque, but its physical condition was of concern to the new Turkish government. All the attention St. Sophia had attracted may have led the leader of the new Turkish republic, the secularist Mustafa Kemal Atatürk, to look to the building's welfare. He often turned to Americans for advice and assistance in his modernization of the country, as they were politically more palatable as the United States had not declared war on the Ottoman Empire. In 1926 Milton Meadry, president of the American Institute of Architects, announced that the executive committee intended, when the climate was right, to approach the Turkish government with a proposal for the restoration of the fabric of St. Sophia.[79] Prof. Emerson Swift of Columbia University had already been allowed by the Turkish authorities to study the building.[80] Atatürk arranged for Prof. Thomas Whitemore of Tufts College in Massachusetts to assist with the uncovering of the mosaics.[81] In 1935 the building's history entered a surprising new phase when Kemal Atatürk closed the mosque and turned it into a museum. Worship of any sort was prohibited. In 1983 Turkey ratified the World Heritage Convention and in 1985 the historic district of Istanbul, including St. Sophia, was added to the list of World Heritage Sites. It remains, however, a potent religious symbol, for example in 1967 when a furore arose when Pope Paul VI knelt and prayed in the building.

The Hagia Sophia's history provides an example of how the late nineteenth century began to manufacture monuments, which in turn attracted the attention of the nascent preservation movement which desired to have monuments to preserve. As some of the earlier writers of the modern era who mention St. Sophia show, its role in Christian history had not been entirely forgotten, and visiting what had been Justinian's great church had the ability both to inspire awe at its wonder and despair as its current role. Such experiences though were limited to a few travellers and their readers, or those who read closely their Gibbon or Byron. The Fossati brothers' lithographs initiated a new and much wider fascination with the building.

The advent of lithography, and later photography, and then widely accessible reproductions, further spread knowledge of the building and inevitably of its earlier Christian identity. In the evolution of the preservation movement lithography and

---

78 Lord Crawford and Balcarres to Ronald Lindsay (brother and British Representative, Constantinople), May 6, 1924, Crawford 97/10, Papers of the 27th Earl of Crawford and Balcarres, National Library of Scotland, Edinburgh. The subject came up at a dinner at the Royal Academy Club. The guests, besides Crawford and Lutyens, were Edmund Dulac (artist), Sir William Goscombe John (sculptor), and Charles Sims (portraitist).

79 'Sancta Sophia: American Proposal for Restoration,' *The Times* (Nov. 30, 1926), 14D.

80 He would later author *Hagia Sophia* (New York: Columbia University Press, 1950) and a year later *Roman Sources of Christian Art* (New York, 1951).

81 Thomas Whitemore, *The Mosaics of St. Sophia at Istanbul: Preliminary Report on the Year's Work, 1931-1932* (Paris: Byzantine Institute of America, 1933-1952).

photography were important devices in popularizing many sites that later attracted movements for their preservation. Morris explicitly prized his photograph of St. Sophia, which went on to become a remarkably distant and unlikely interest of SPAB. Visual reproduction made St. Sophia an iconic monument throughout the Western, and largely Christian, world making its salvation a question for groups not only in Britain but as far away as the North America. The early preservationists themselves were engaged in numerous activities besides preservation. Two of the greatest campaigners for historic preservation of the period, Morris and Bryce, were also deeply concerned with Christian issues in the Eastern Mediterranean world. This combination of concerns led to some important convergences of interest and explains some of the attention given to the preservation of specific monuments.

The period 1870-1920, which witnessed the birth of the international heritage movement, also saw the Hagia Sophia emerge as an international monument, one that transcended identity with one state. The discussions at the Paris Peace Conference on the Turkish treaty had included provisions for the antiquities located in the Ottoman successor states. When the issue of St. Sophia arose, this led Arnold Toynbee, then employed at the Foreign Office, to make the remarkable suggestion that 'Would it not be possible, while leaving the religious status quo in S. Sophia, to give it the status of an "international monument" from the archaeological point of view.'[82] This was one of the earliest proposals of the concept of an international heritage site, presaging the development of UNESCO's World Heritage Sites scheme after the next world war. In many ways the concept of World Heritage Sites has its inception in this period with the invention of organized historic preservation, and with the concern for the future of such structures as the Hagia Sophia. There were many motivators for preservation in this period, the admiration for older buildings, the perpetuation of knowledge of ancient events, the affection for places of traditional ceremony, and not least the preservationists' views on religion.

---

82   Minute by Toynbee, Mar. 6, 1919, TNA FO 608/82/342/4/2/3507.

Chapter 3

Early Preservation Efforts in Sri Lanka: William H. Gregory at Anuradhapura and Kandy*

Anne M. Blackburn

Introduction

Heightened attention to the preservation of ancient monuments in Sri Lanka began in the middle 1800s, gaining significant momentum during the 1870s, a formative period for antiquarian and orientalist activities on the island.[1] In this chapter I focus on the preservation of ancient Buddhist monuments during the governorship of William (later Sir William) H. Gregory, Governor of Ceylon from 1872 until 1877. The history of preservation work undertaken during Gregory's tenure offers a telling example of what Michael Hunter has called 'the element of contingency' in the history of heritage and preservation.[2] Efforts to preserve ancient Buddhist sites on the island during this time were indeed contingent upon rather varied circumstances. To understand the early history of the preservation

*An early version of this chapter was read as a paper at a conference on 'The International Origins of the Heritage Movement' held at Boston University under the direction of Melanie Hall. Early research for this chapter was undertaken in part for a Cornell graduate seminar, 'The Classical in Colonial Asia.' I would like to acknowledge the stimulating environment created by the seminar students, and express my special thanks to Lawrence Chua and Jonathan Young. I am grateful to the Manuscript, Archives, and Rare Books Library at Emory University for permission to reproduce quotations from the Gregory Family Papers, and to Olin Library at Cornell University for preparing the images reproduced here. My thanks also to the Southeast Asia Visions Digital Collection, Cornell University Library, which provided the map included here.

1   In most of my work on the Buddhist histories of Sri Lanka prior to independence I refer to the island as Lanka to underscore distance from the later nation-state and nationalist appropriations of the island's past. However, since this chapter is intended for a broader readership, including those working on Britain and British Empire, I use the terms 'Ceylon' and 'Sri Lanka' interchangeably. I have also omitted all diacritical remarks for proper names in Sinhala and Pali languages, even those present in some of the original letters to which I refer.

2   Michael Hunter, ed., *Preserving the Past: The Rise of Heritage in Modern Britain* (Phoenix Mill: Alan Sutton Publishing, 1996), 12.

of such sites in Ceylon, we must attend simultaneously to several important processes underway in the colonies and their metropolitan centers. The growth of antiquarian, orientalist, and archaeological interests in Britain and on the continent played an obviously important role, as did a deepening awareness on the part of colonial administrators that archaeology and preservation were an appropriate part of the competitive display of civilization and empire. Efforts to edit and translate medieval Sri Lankan historical narratives for use in the construction of Indian historical chronology intersected with these currents, as well as with Sri Lankan and Southeast Asian Buddhist interest in the devotional reclamation of central pilgrimage sites. Moreover, as the case of Governor Gregory reveals clearly, the history of preservation efforts depended in part on the aesthetic sympathies of individuals. Such sympathies shaped the recognition of certain sites as classical and beautiful and, thus, worthy of preservation, according to a canon of taste forged in the eighteenth- and nineteenth-century rediscoveries of ancient Greece and Rome.

Antiquarianism and Archaeology in Colonial South Asia

Nineteenth-century colonial interest in preserving historical sites of South Asia was, of course, linked to broader developments in Britain and on the continent. Timothy Champion has noted that, in Britain, '[t]he early 1840s were years of acute concern for "national monuments" and for archaeology in general, shown most clearly in the fashion for founding archaeological societies.'[3] In fact, archaeology was still in formation as a discipline, and as a discourse, during the latter half of the nineteenth century.[4] This emergent discipline was constituted by the activities undertaken in Britain, Europe and the colonies, and by reflections on them in specialized and popular publications. Upinder Singh observes that the first systematic scholarly interest in ancient Indian sites and monuments dates to the mid-eighteenth century. This was part of broader expressions of orientalism and antiquarianism including the establishment of the Asiatic Society of Bengal in 1784.[5]

By the middle of the nineteenth century, however, clearer distinctions between forms of antiquarian activity emerged in colonial India as elsewhere. The Delhi Archaeological Society began in 1847, running until 1854, while Alexander Cunningham (later Director General of the Archaeological Survey of India) undertook archaeological studies of Kashmiri ruins in 1848 with

---

3   Timothy Champion, 'Protecting the Monuments: Archaeological Legislation from the 1882 Act to PPG 16,' in Hunter, *Preserving the Past*, 39.

4   Upinder Singh, *The Discovery of Ancient India: Early Archaeologists and the Beginnings of Archaeology* (Delhi: Permanent Black, 2004), 1.

5   Ibid., 7.

particular attention to the architectural conventions they revealed.[6] Around the same time, James Fergusson, the owner of an indigo factory in Bengal, began to publish extensively on Indian architecture, analyzing buildings in part with reference to racial and religious classification.[7] The first Indian archaeological survey occurred in 1861, although systematic attention to the remains of Indian sites and monuments began a decade later, in 1871, with the establishment of the Archaeological Survey of India as one of an array of surveys initiated by the Colonial Office.[8] Moreover, the beginning of the Archaeological Survey of India marked the start of a more concerted policy on the preservation, as well as the investigation, of historic sites. As Singh observes, 'For the first time in the history of British rule in India, government policy emphasized the connection between archaeological investigation and conservation and assumed responsibility for both.'[9] Increased attention to the preservation of historical monuments stemmed to some extent from pressure mounted in the metropole, including from Henry Cole, Director of the South Kensington Museum.[10] Such pressure related partly to the crystallization of discursive associations, in Britain and Europe, between the work of historical preservation, the character of enlightened government, and the manifestation of cultural attainment. Such associations led Cunningham, for instance, to seek excavation funds for India in the name of enlightened rule:

'It will not be to our credit as an enlightened ruling power,' he argued, 'if we continue to allow such fields of excavation … to remain without more examination than they have hitherto received.'[11]

According to the Viceroy of India, in 1870:

… the duty of investigating, describing, and protecting the ancient monuments of a country is one that is recognized and acted on by every civilized nation in the world. India has done less in this direction than almost any other nation. … I think the time is come when a great and enlightened Government can no longer neglect contributing to the archaeological literature of the world …[12]

------

6 Ibid., 16, 15. On Cunningham, see Tapati Guha-Thakurta, *Monuments, Objects, Histories: Institutions of Art in Colonial and Postcolonial India* (New York: Columbia University Press, 2004), 27-42.

7 Guha-Thakurta, *Monuments*, 6-27.

8 Singh, *Discovery of Ancient India*, xv, 25.

9 Ibid., 81. In the case of epigraphy, conservation could take several forms including the reproduction of inscriptions and preparation of published copies. Cunningham proposed the publication of the *Corpus Inscriptionum Indicarum* in 1875, and the first volume, on Asokan inscriptions, was published in 1877. See idem, 125.

10 Singh, *Discovery of Ancient India*, 187.

11 Quoted in Guha-Thakurta, *Monuments*, 30.

12 Cited in Singh, *Discovery of Ancient India*, 81-82.

However, the work of conservation was not undertaken by the Archaeological Survey of India but delegated to Public Works Departments, with variable local results. Singh observes that 'there was a big gap between directives and legislation on the one hand and effective implementation on the other' when it came to the preservation of archaeological remains.[13] Although initially a proponent of district authority over the preservation of ancient buildings not actively used for ritual purposes, Cunningham was later to argue that preservation should be brought within the purview of the Archaeological Survey. However, under his leadership the emphasis remained a matter of surveying, rather than excavating, ancient sites.[14] 'Surveying' meant different things to different archaeologists. Cunningham's northern and central Indian surveys proceeded according to a rather rich conception, including the investigation of inscriptions, place names, and local histories as well as matters of style. The southern and western surveys initiated in the later 1870s and 1880s, however, expressed more narrowly architectural expectations.[15]

I dwell at some length on the Indian case because experiments undertaken in India, and metropolitan decisions made with respect to India, affected approaches to archaeology and preservation in nearby Ceylon. Although that island was separately administered during this period as a Crown Colony, and many aspects of colonial rule differed greatly, colonial administrative approaches to Ceylon were sometimes shaped by Indian experiences.[16] Moreover, in the sphere of orientalist and antiquarian activity, the two colonies were closely linked in both discourse and practice through the work of scholars who explored the history of South Asia with an eye to the connections between the island and the mainland. These connections are made very clear by nineteenth-century correspondence between Colombo's Queen's House and the Colonial Office. This includes copies of letters and documents sent to and from the India Office, as well as epistolary references to scholars located elsewhere in England and Europe. During Governor Gregory's tenure, for instance, he exchanged regular letters with the India Office Library and the Secretary of State on matters related to monuments and inscriptions. Through Robert Childers of the India Office Library, Gregory had contact with orientalist scholars including Max Müller.[17]

As archaeological interest began to develop in Britain and India during the 1840s, small sums of money were pledged by the Ceylon government to begin

---

13   Ibid., 188.

14   Ibid., 87, 196, 216.

15   Ibid., 195.

16   See also Bethia N. Bell and Heather M. Bell, *H.C.P. Bell: Archaeologist of Ceylon and the Maldives* (Clywd: Archetype Publications, 1993), 26.

17   See The National Archives, Colonial Office (hereafter, CO) 54, containing correspondence from 1873-1880. R.W. Ievers, *Manual of the North Central Province, Ceylon* (Colombo: Government Printer, 1899), 234.

clearing the jungle that had encroached on the archaeological remains of the island's ancient capitals at Anuradhapura and Pollonaruva.[18] However, there was no significant support for archaeological work, nor a plan for investigation and preservation, until the late 1860s.

> In 1868 Mr Russell, Government Agent, Jaffna, wrote to Government and called attention to the official papers in the *Fort St. George Gazette* relating to the arrangements made for conserving the ancient architectural structures or their remains and other works of art in India, and suggested that similar steps might be adopted here with a like object at Anuradhapura.[19]

In the same year Governor Hercules Robinson established an Archaeological Commission for Ceylon, with a special charge to attend to the island's epigraphic heritage.[20] Although no systematic excavation occurred, the Colonial Secretary's office provided funds to Government Agents who supervised clearing operations and the photography of ruins. Robinson's interest in these matters was sufficiently great to move a relatively junior civil servant with some skill in local literary languages and the island's religious history, T.W. Rhys Davids (later famous for his studies of Pali and Buddhism), from the island's south to Anuradhapura where he took up the post of Assistant Government Agent.[21]

---

18   Bell and Bell, *Discovery of Ancient India*, 35.

19   Ievers, *Manual*, 215.

20   Bell and Bell, *Discovery of Ancient India*, 30.

21   Ananda Wickremeratne, *The Genesis of An Orientalist: Thomas William Rhys Davids and Buddhism in Sri Lanka* (Delhi: Motilal Banarsidas, 1984), 101, 104. See also Ievers, *Manual,* 219. Subsequently, a correspondence developed between Governor Gregory's office and T.W. Rhys Davids about the work of preservation. According to a letter dated May 13, 1872, Gregory had requested advice from Rhys Davids about 'the preservation of the antiquities of the island.'
Rhys Davids wrote that he
> would beg to suggest that these two cities [Anuradhapura and Pollonaruva] should be thoroughly explored first, and the results of that exploration published as soon as possible before the attention which Buddhism and Comparative Philology are at present exciting at home, has dwindled away. All perishable statues inscriptions &c might be removed to a place of safety, the ruins be thoroughly cleared from the trees which are destroying them & copies of all the plans and photographs or casts of those statues not removed might be placed in the New Museum. The ruins are interesting to different scholars philologically, historically & philologically it would be advisable for the Museum to take photographs or squeezes & for the report to be published *facsimiles in lithographe* of the Pali Elu and Sinhala *inscriptions* with the texts transliterated, & with *translations and notes ... Historically,* to render the report of any abiding value, such notices of these buildings as can be found in the native books might be published, in order to shew their dates,

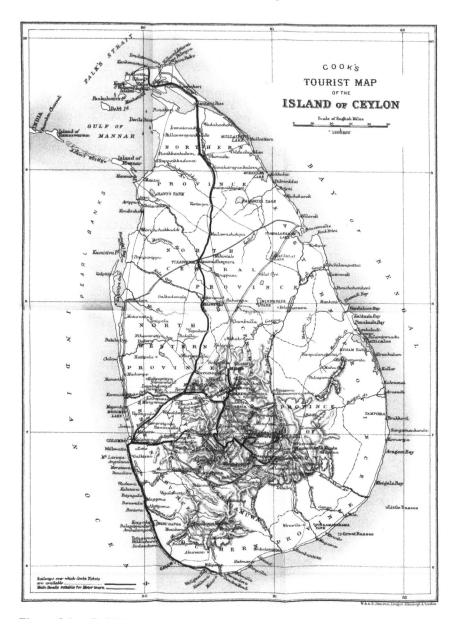

Figure 3.1   British-period map of Ceylon

Source: *India, Burma, Ceylon and South Africa: Information for Travellers and Residents.*
1909. London: T. Cook

Attention to the island's historical sites and monuments received a great fillip with the 1872 arrival of Governor Gregory to Colombo. Gregory, a former Anglo-Irish Parliamentarian, had long been an ardent supporter of the visual arts and museum-based efforts to educate the tastes of England's growing electorate.[22] He chaired the 1860 Parliamentary Committee on the British Museum and was elected a Trustee of the National Gallery in 1867, joining the Council of the Arundel Society the following year. Gregory's enthusiasm for ancient sites and classical monuments thus greatly pre-dated his arrival in Ceylon. Soon after reaching the island, Gregory authorized the most sustained preservation work to date under British rule, focused particularly at the ancient sites of Pollonaruva and Anuradhapura.[23] At Anuradhapura, for instance, Assistant Government Agent Rhys Davids was authorized to clear the areas around Buddhist relic monuments, to remove trees and roots likely to ruin existing structures, to clear the areas around other structures, to copy and report on inscriptions, and to prepare a detailed plan of the most important structures and their relationship to one another. Signaling his own interest in antiquarian and preservationist pursuits, as well as the influence of scholarly trends abroad, the new Governor Gregory emphasized the importance of the preservation of inscriptions in his 1872 address to the Legislative Council.

> These inscriptions, varying in character and dialect, will be of deep interest to the philologist, and throw light on the ancient usages, religious customs, and early history of Ceylon. It is unnecessary for me to inform you that the study of Oriental literature, language, and history is being pursued with ardour throughout Europe; but I may mention that I have received constant communications from persons of literary eminence, urging me to carry out speedily and effectually the work which I have indicated.[24]

Gregory took his own somewhat contradictory stand on restoration and conservation: the stones of failing structures could be restored if expense allowed, but all inscriptions and stone carvings of striking interest were to be removed

---

the reasons for, & the persons by whom, they were built. For this purpose the services of a pandit being required ...
Rhys Davids to the Governor's Office, May 13, 1872, Gregory Family Papers Box 25, Manuscript Archives and Rare Book Library [hereafter, MARBL], Emory University.

22   Brian Jenkins, *Sir William Gregory of Coole: The Biography of an Anglo-Irishman* (Gerrards Cross: Colin Smythe, 1986).

23   Wickremeratne, *The Genesis of an Orientalist*, 107.

24   Writing to his mother of the intentions behind an early trip to Anuradhapura, Gregory wrote: 'Lastly I wished to indulge my antiquarian tastes by a visit to this region & to see what they would afford to our new museum. And I have been amply rewarded. These ruins are full of interest. Palaces & temples & inscriptions are scattered every where. There is much that is grotesque and bad in taste but much that is very beautiful. I never was better in my life & believe the climate to be perfectly good ... ,' Gregory to the elder Mrs Gregory, July 21, 1872, Gregory Family Papers Box 31, MARBL, Emory University.

to the Government Agent's compound if possible. Moreover, Government Agent Twynham was specifically asked to report on all large objects that might be moved to Colombo, where the Governor had ambitious plans to build a museum.[25] After considerable correspondence with the Colonial Office, Gregory secured the assistance of Captain Hogg of the Royal Engineers to photograph sites and inscriptions at Anuradhapura and Pollonaruwa. He also sought a careful study of Sri Lankan inscriptions by Goldschmidt, a philologist hired after consultation with scholars in England.[26] Robert Childers wrote from London to the Governor:[27]

> You will see from the correspondence that a first rate photographer, a soldier of the RE now on duty in New Zealand in connection with the Transit of Venus expedition is under orders to go to Ceylon when the transit work is done and take photographs for Dr Goldschmidt. This arrangement leaves nothing to be desired, and lessens all risks attendant on the engagement of a private photographer in London.

> My conviction is that Dr. Goldschmidt will be thoroughly successful and his work lead to important results. I hope the discoveries made by General Cunningham at Bharhut India may be the precursor of similar discoveries in Ceylon ...[28]

Governor Gregory also approached India-based Alexander Cunningham directly, in order to ascertain methods appropriate to the intensifying preservation work occurring in Ceylon.

> I am desirous of obtaining advice and assistance from you in a matter of common interest to both of us namely in the reproduction of the many and various-[? unclear] rock inscriptions in Ceylon. In the last four months I have been employing the services of Capt. Hogg RE in photographing the inscriptions in the neighborhood of the old city of Anaradhapura [sic] & we have succeeded in getting a considerable number.

---

25   Wickremeratne, *The Genesis of an Orientalist*, 107-108 cites Sri Lankan National Archives 41.128, and an August 12, 1872 letter from the Government Agent of the Northern Province to the Anuradhapura Assistant Government Agent, Rhys Davids. On tensions between preservation and the work of museums in India, see Guha-Thakurta, *Monuments*, 63-70.

26   B. Bastiampillai, *The Administration of Sir William Gregory, Governor of Ceylon 1872-1877* (Dehiwala: Tisara Prakasakayo, 1968), 150-51.

27   See further: CO 54/594, CO54/495-C-Ceylon:/14493/74, CO 54/497, CO 54/499-Ceylon 1875: Offices & Individuals. See also Ievers, *Manual*, 234.

28   Childers to Gregory, December 8, 1874, Gregory Family Papers Box 25, MARBL, Emory University.

I want to carry on the photographing of all the legible inscriptions but there are many of great length which though partially destroyed may be of great interest. There are also others the situation of which is such, see No 2 that they cannot be properly dealt with by photography. I should be obliged if you would inform me how you work out inscriptions in similar cases in India. …

The work is novel to me but I am anxious to complete it during my tenure of Govr. As probably my successor may consider it as involving needless expense. I do not, and am fortified in that opinion by Mr Minayeff [a Russian scholar of Pali] who considers the inscriptions to be of extraordinary value to the philologist, whatever they may be to the historian.[29]

Figure 3.2   View of the Thuparama Dagoba (Buddhist relic monument) in Anuradhapura

Source: James G. Smither. *Architectural Remains, Anuradhapura, Ceylon*. 1894. Colombo: Government Printer

---

29   Gregory to Cunningham, incomplete date 1874, Gregory Family Papers Box 31, ibid. Cunningham replied with detailed instruction on the preparation of paper impressions before concluding: 'I trust that the above account will be sufficiently clear: but if not I shall be most happy to furnish information to any of the officers employed in Ceylon in the work of copying the inscriptions, who may think it worth while to refer to me on the subject,' Cunningham to Gregory, December 29, 1874, Gregory Family Papers Box 25, ibid.

The Governor's work in archaeology and preservation was encouraged by broader trends to which I have already referred and, crucially, by the colony's strong financial surplus during his tenure. Gregory pursued his interests in a growth period for the island's plantation economy, before the budget rigors induced by the eventual effects of the coffee blight that had begun in 1869.[30] The initiatives established by Gregory at Anuradhapura and Pollonaruva were continued by his immediate successors, Governors Longden, Gordon, and Havelock, with particular enthusiasm during Gordon's tenure. Following his departure from Ceylon, Sir William Gregory regularly lobbied the Colonial Office, and his successors in Colombo, for the continued clearance and surveying of the island's ancient cities, epigraphic investigation, and the preservation of literary works and manuscripts.[31]

The Reclamation of Anuradhapura

William Gregory focused particularly on the royal cities of Anuradhapura and Kandy in his work on historic preservation. Anuradhapura was an early, and enduring, fascination for Gregory, the first Governor in fifteen years to actually tour its district.[32] Making repeated trips to Anuradhapura, including one of dark importance to him, during which his first wife fell deathly ill, Gregory considered the old city and its environs properly central to any celebration of the island's history. It is revealing that Gregory initiated a survey of Anuradhapura in 1873 very soon after his arrival. According to Bastiampillai: 'Detailed plans and sections, indicating architectural measurements, and drawings of the ancient moundlike religious structures enshrining Buddhist relics, the dagabas, were made. Lithographical copies were sent to the British Museum for the reference of scholars.'[33] Anuradhapura and its north-central region occupied a central place in the Governor's paternalist vision for the colony. Formerly a site of complex and successful irrigation and, therefore, of rich agricultural production, the area had become more arid and less fertile, its elaborate system of tanks dysfunctional after the movement of the island's political centers south and west from roughly the thirteenth century.

Gregory's approach to Ceylon was substantially shaped by his experiences as landholder in Ireland. A member of an Anglo-Irish administrative family, active in Ireland for several generations, Gregory had strong views on agriculture, landholder-tenant relations, and labor. These he pursued actively in Parliament before his departure for Sri Lanka. Indeed, Gregory may have been somewhat

---

30   Bastiampillai, *Administration of Sir William Gregory*, 6-11.

31   Ibid., 154-55. See further correspondence between Gregory and Gordon in the British Library's Stanmore Papers, 49207. See also Gregory to Carnarvon, CO 54/11.

32   Jenkins, *Sir William Gregory*, 225.

33   Bastiampillai, *Administration of Sir William Gregory*, 152. See also Jenkins, *Sir William Gregory*, 232.

relieved to preoccupy himself with the problems of landholding and production at a distance from the nearly intractable difficulties faced in London and Dublin.[34] Jenkins describes Gregory's attitude to the responsibilities of Ceylon as one of 'benevolent paternalism,' a term that coheres well with the records of Gregory's letters to friends, family, and London colleagues.[35] He had strong views about Ceylonese economy and infrastructure, as about much else, and was particularly disposed to employ them as a result of his strong liking for the island and its 'native' population, as well as his conviction that hard work was salvation in the face of his bereavement.

> I am so entirely broken hearted & desolate that I do not know if I shall have strength to go on, but I have resolved on my course. As a matter of duty I will do my work honestly and unflinchingly. … It would be wrong for me to flinch, & never will I do wrong knowingly. Besides I feel that work, strenuous continued work is my only relief from suffering, and the little pleasure that I can now take in life will be from the success of my measures for the good of those I govern.[36]

To Gregory, the restoration of irrigation systems and the development of the region's agriculture appeared valuable to the island's economic development. However, he seems also to have felt a sharp abhorrence of the isolation, illness, and poverty he perceived in the Anuradhapura region. Improvements to this north-central region were, early in his tenure, envisioned as part of his intended legacy. In a letter composed to his mother early during the Ceylon tenure, Gregory wrote of the Anuradhapura district:

> The 1st point was to look carefully into the requirements of this vast & fine central district at one time the granary of Ceylon irrigated by innumerable tanks & producing rice sufficient for an enormous population. I regret to say a sad sight met my eyes. The villages are still there & so are the tanks, there are 1571 villages and 1600 tanks but *not one tank is in repair.* The consequence is that land formerly producing crops is now waste & bare or has returned into jungle infested with wild beasts. The population has dwindled away & is steeped in poverty … I am immediately going to send a Commission to report on a system of thorough irrigation of the district, and I trust before I leave the island to see broad expanses of green rice fields waving over what is now brush wood, & to replace penury with plenty. It is now 15 years since a Governor has visited Anuradhapura. I do not in the least blame my predecessors, they were doing

---

34 On Gregory's involvement in Irish questions, see Jenkins. On Ireland in relation to India see David Cannadine, *Ornamentalism: How the British Saw Their Empire* (New York: Oxford University Press, 2001), 15.

35 Jenkins, *Sir William Gregory*, 231.

36 Gregory to the elder Mrs Gregory, June 20, 1873, Gregory Family Papers Box 31, MARBL, Emory University.

great & good work elsewhere, & it is what I have seen done elsewhere that has opened my eyes to the necessity of immediate action in that district.[37]

In 1873 Gregory created a new province, the North Central Province, in order to administer the Anuradhapura region more closely.[38] He was, eventually, exceedingly pleased with the results.

> I am at the moment at the old city of Anaradhapura [sic] once the capital of the Kandyan kings[39] the district round was about 1,000 years ago the granary of Ceylon all owing to the magnificent system of irrigation maintained by the native monarchs. The district had fallen into decay. When I came to Ceylon there were not 60,000 people in it. The rest had perished from starvation & from a dreadful disease called 'parangi' ... I made a series of magnificent roads ... I then set to work to restore the tanks & their feeders, & passed an ordinance whereby the villagers received & had put in gratuitously a masonry sluice into each of their village tanks on condition of their all turning out & repairing their wretched broken down bunds ... Thanks to the energy & tact of the official Mr Dickson to whom I entrusted the new Province, the place succeeded, the villagers were infused with confidence & have worked bravely at their lands some hundreds of which have been & are being supplied with sluice ... The produce of the last five years is five times greater than of the previous five.[40]

Anuradhapura and the Mahavamsa

Intensifying scholarly interest in the Pali language 'chronicle' titled the *Mahavamsa, (Great Chronicle/Lineage)*, composed in Sri Lanka early in the sixth century, also influenced the concentration on Anuradhapura visible in the preservation projects of the 1870s. The *Mahavamsa* is a chronicle of royal and Buddhist monastic activities, composed in Sri Lanka in several installments. Its early chapters treat the arrival of Buddhism to the island and the construction of Buddhist devotional sites and other elaborate urban structures at Anuradhapura from the third century BCE into the first centuries CE. This text gained attention in the middle years of the nineteenth century in part because its detailed chronology promised to resolve confusions about the historical chronology of kingdoms in mainland South Asia. Moreover, scholars working in the rather new fields of Buddhist studies and comparative religion, as well as in historical philology, came to see the text as one that would shed considerable light on the history of Buddhism

---

37   Gregory to the elder Mrs Gregory, July 21, 1872, ibid.
38   Jenkins, *Sir William Gregory*, 238.
39   On his identification of Anuradhapura with Kandy, see further below.
40   Gregory to Layard, February 22, 1884, Gregory Family Papers Box 32, MARBL, Emory University.

in South Asia, Buddhist relations to Hinduism, and the development of Indo-European languages. A first and partial translation of the *Mahavamsa* into English (1837) preceded Gregory's arrival in Ceylon. This was deemed insufficient by him. During his tenure, he supported the preparation of a Pali language edition of the entire text, and its translation from Pali into Sinhala. These plans were made with some understanding of metropolitan orientalist needs and wishes, as is evident from his correspondence with Robert Childers about work on the text. It also reflected Gregory's interest in the ancient literature of the island, manifested in support for the study of 'Oriental' languages and literatures, the preservation of manuscripts and the creation of 'Oriental' libraries in Kandy and Colombo as well as the Colombo Museum.[41] The Governor, whose education at Harrow and Oxford was formatively classicist in character,[42] possessed (as did many of his caste and class) a certain affinity for and appreciation of ancient literatures.

Support for work on the *Mahavamsa* became a central focus of Gregory's efforts to illuminate the island's cultural heritage, and formed part of the legacy he wished to identify in his farewell address of 1877 made to the Legislative Council of Ceylon.

> We have entrusted the revision of the second portion of the Mahawansa, left untouched by Turnour, to two learned Pali scholars – Sumangala, the High Priest of Adam's Peak, and the Pandit Batuwantudawe. This is now completed, together with the variants of the text and also a translation into Sinhalese. The translation into English has been undertaken by De Zoysa Mudaliyar. The revision of the Tika, which is the very important commentary on this interesting account of the history of Ceylon, and the translation into Sinhalese of the first 36 chapters of the Mahawansa have been entrusted to the same editors. I have written to Siam to endeavour to procure a correct copy of the Dipawansa, a history of Ceylon of earlier date than the Mahawansa, but have failed to obtain it. I do not, however, despair of its being acquired, and I trust it will be edited in the same manner, and that continued encouragement will be given to learned natives to cultivate and make known the ancient literature of Ceylon.

Childers' correspondence with Gregory reveals the close relationship between textual scholars of the *Mahavamsa* and scholars of Ceylonese archaeology during the 1870s.

> Especially glad am I to hear that a survey of Anuradhapura and the "Appian Way" is actually in progress. I cannot tell you with what excitement I am looking forward to an authentic map of that glorious classic ground. ...

---

41   See also Anne M. Blackburn, *Locations of Buddhism: Colonialism and Modernity in Sri Lanka* (Chicago: University of Chicago Press, 2010), esp. ch. 3-4.

42   Jenkins, *Sir William Gregory*, 28-44.

I am greatly interested in your account of the recent discoveries at the Ruanweli Tope (Mahathupa). … I have always considered the account of the building of the Mahathupa one of the most interesting things in the Mahavansa. I hope you will get at the interior of the relic shrine (the bell shaped structure). It must have been of the most surpassing beauty, see pp 149 and 182 [of the *Mahavansa*]. … When we have a minute and exhaustive account of the present state of the building, after it has been thoroughly excavated, I have no doubt we shall be able to identify many things described in Mahavansa. The large decoration mentioned in the already edited portion of Mahavansa is I think the famous diamond ring placed on top of the dome to ward off lightning.[43]

Archaeology and preservation in Ceylon developed along with studies of the *Mahavamsa* in a reinforcing process, as scholars began to compare sites noted in the text with those identifiable by inscriptions found on ruins from the site, and as access to the text incited further interest in Sri Lankan archaeology at home on the island and abroad.

Preservation, Devotion, and Disestablishment

Archaeological and preservation efforts on the island were greatly abetted by broader orientalist and antiquarian developments, and by the growing sense that a civilized empire was one prepared to manifest concern for the cultural heritage of its colonies. However, they were sometimes hampered by debates about colonial involvement with the religious institutions and practices of those they colonized. Initial British commitments to maintain the responsibilities of local rulers *vis-à-vis* Buddhism on the island gave way over time to greater degrees of disestablishment under pressure from Christian missionary activity at home and in the colonies. With respect to Anuradhapura specifically, in 1869 the colonial administration assigned control over the central pilgrimage sites of Anuradhapura to the Atmasthana Committee of monks and lay persons charged with maintenance of the site.[44] Colonial efforts to clear the jungle around the ancient cities of Anuradhapura and Pollonaruva, and the archaeological surveys that took place, added pressure to the disestablishment debate since sites increasingly accessible to scholarly investigation were also more accessible to Buddhist pilgrims and Buddhist authorities. Some Christians found room for criticism in the administration's involvement with spaces clearly of live religious important to local Buddhist residents, and Gregory himself was sometimes charged with undue sympathy for

---

43   October 30, 1873, Childers to Gregory, Gregory Family Papers Box 25 (p. 229). MARBL, Emory University.

44   Bell and Bell, *Discovery of Ancient India,* 106-109. See also Elizabeth Nissan, 'The Sacred City of Anuradhapura : Aspects of Sinhalese Buddhism and Nationhood' (PhD diss., London School of Economics and Political Science, 1985).

the religion and culture of local Buddhists.[45] Replying to his mother, the Governor wrote:

> I am amused at your reference to my alleged partiality towards the Buddhists. I care very little for them though their religion is by far the purest in the world after the Christian, but I do care for the old remains of ancient Kandyan glory and I have worked hard to restore them & have I flatter myself made Kandy a very different town from what it was when I came. I send you a newspaper marked containing an account of a meeting of the Kandyan aristocracy – all the great chiefs & heads of families were present to inaugurate an endeavour to revive the study of the ancient literature of the island.[46]

Gregory's own relationship to Buddhist institutions and religious practice was complex, and influenced by his experiences with Irish Catholicism. By the time Gregory reached Ceylon, he had moved toward greater support of Irish Catholic institutions, recognizing 'the conservative and restraining role the Catholic Church might play, if only it was treated with respect and generosity.'[47] A draft letter composed by Gregory to the Secretary of State for the Colonies in 1872, and later sent as an enclosure with an 1876 letter (as matters related to the administration of Buddhist temples drew more attention from colonial authorities), reveals some of the Governor's thinking on the relationship between the administration of Buddhist institutions, social stability, and civilization.

> It is said that if the Buddhists see their Temples and sacred places falling into ruin by the neglect and fraudulent conduct of their priests they will conceive a

---

45   See also L.A. Wickremeratne, 'The Rulers and the Ruled in British Ceylon: A Study of the Function of Petitions in Colonial Government,' *Modern Ceylon Studies*, vol. 1, no. 2 (1970), 213-32.

46   Gregory to the elder Mrs Gregory, December 8, 1873, Gregory Family Papers Box 31, MARBL, Emory University.

47   Jenkins, *Sir William Gregory*, 176. Jenkins noted the connections between Gregory's Irish experience and his approach to the Governorship of Ceylon: '... his Irish background and experience moulded his general approach to his duties as Governor ...' 231. In his autobiography (edited by his second wife, Lady Gregory), Gregory notes, with respect to his work on Ireland: 'I was reproached with being indifferent to religion, as I seemed to draw no distinctions, and to treat all alike. To dogmatic religion, I replied, I was comparatively indifferent, but to religion in general terms I considered the State was bound in policy to afford every possible encouragement and assistance. I never could look with aversion or displeasure on a religion not my own. I regarded it as having its particular merits, its special adaptation to its followers, and, instead of searching to explore its demerits, I was quite satisfied to see many dwelling in it safely and content. ... I have ever looked on the Roman Catholic Church as the great central fortress of law and order, resisting wild theories and destructive innovations ...' in Augusta Gregory, ed., *Mr Gregory's Letter-box, 1813-1830* (London: Smith, Elder, & Co., 1898), 167-68.

contempt for their religion and become Christians. Apart from the immorality of this argument (if it be our duty to intervene) I deny on the most competent authority the correctness of this assertion – Devil dancing not Christianity is the alternative. This was the old superstition which was superceded by Buddhism but it still lingers among the people, and strange to say I have heard of cases of well educated Sinhalese reverting to these barbarous ceremonies just as cultivated moorland has ever a tendency to revert to heath. Although it is highly desirable that Buddhists should become Christians, still it is not desirable that they should exchange a singularly pure religion for either entire disbelief in any religion whatsoever, or for the barbarous rites of devil dancing.[48]

The religious implications of preservation projects were made more complex by local Buddhist involvement in the preservation of Anuradhapura monuments. As early as 1841, a Buddhist monk collected funds to restore a major relic monument, the Thuparama, though without any attempt to restore the structure to its original style and proportions.[49] In the mid-1870s, another Buddhist monk, Naranwita Sumanasara, undertook to rebuild another relic monument particularly precious to Buddhist devotees, the Ruvanweli Dagaba. Governor Gregory requested the Assistant Government Agent of Anuradhapura, H. Hay Cameron (also the son of a family friend) for information on the monk's activities. Cameron replied:

> Yes! There is a pilgrim priest here, one Naranwita Unnanse who proposes to restore the Ruwanwelli Dagoba. He is trying to raise a subscription amongst the people, and has already begun to collect the loose bricks preparatory to commencing his work. The Modern Sinhalese attempts at restoration are so decidedly unsuccessful that I do not think it would be desirable to permit him to meddle with the Shrine, at all events, until he should have submitted his plans for the approval of the Maha Nayaka [Chief Priest of the monk's monastic order] who would of course be directed higher in his decision. The man himself does not show any inclination to go in for clearing the jungle, his sole ambition is the restoration of the Dagoba. I shall be glad to know your wishes as to this point – for many of the old buildings have already been disfigured by such tinkerings.[50]

To this Buddhist effort Governor Gregory contributed financial support, personally, rather than in his official capacity, to support 'the conservation of a great national monument', but the resulting effort was unable to stand heavy rains.[51] Cameron,

---

48   CO 54/503.
49   Bell and Bell, *Discovery of Ancient India*, 35.
50   Cameron to Gregory, September 10, 1872, Gregory Family Papers Box 25, MARBL, Emory University.
51   Bell and Bell, *Discovery of Ancient India*, 35-36. 'Could not something be done to conserve the ruins at Anuradhapura? Captain Hogg tells me the Buddhist priests have excavated with too much enthusiasm at the Ruwanwali Tope, that the whole thing is likely

working closely with the Governor, continued attempts to mediate between colonial sponsorship of preservation and the interests of local Buddhists.

> I should have dispatched my reply, Sir, to your letter earlier, but wished to be able to give you definite answer with reference to the removal of the two monoliths from the Thupa Rama Dagoba, which is the one indicated by your description, I have sounded the feelings of the Chief Priest and the Kandyan chiefs, and find that tho' they would not consent, and indeed are forbidden by their religion, to part with any thing belonging to their sacred buildings, *freely and for nothing*, yet for so great a consideration as the restoration of the remaining columns and the replacing of their capitals they would yield a ready consent. I do not think it is too much to give; for the work is worth the doing for its own sake, is it not Sir?[52]

Conflicts between Buddhist restoration programs and the interests of non-Buddhist colonial officers and scholars were by no means restricted to Ceylon, as shown by the striking case of Bodh Gaya in eastern India during the same period. In 1877, the King of Burma sent three Burmese to repair the Maha Bodhi temple, a major Indian pilgrimage site for Asian Buddhists. They received permission from the Hindu priest in charge of the site and began their work without the awareness of the colonial government.[53] Eventually apprised of their efforts, the Bengal Government sent an archaeologist-orientalist, Rajendralal Mitra, to report on matters at Bodh Gaya. Mitra's account reveals a gap between two notions of heritage preservation circulating in late nineteenth-century Buddhist Asia:

> The Burmese gentlemen are undoubtedly very pious and enthusiastic in the cause of their religion, but they are working on no systematic or traditional plan. They are ignorant of the true history of their faith and perfectly innocent of all knowledge of architecture and the requirements of archaeology and history and the mischief they have done by their mis-directed zeal has been serious.[54]

In time, colonial administrators in India and Ceylon began to experiment with ways of drawing on Buddhist financial support for restoration and preservation according to metropolitan scholarly expectations for the restoration of authentic style and proportion. At Bodh Gaya Alexander Cunningham eventually proposed that the Burmese King be prevented from contributing to costs associated with the ancient sites, but encouraged to contribute to new buildings, such as residences

---

to come down with a run some day. This would be a grievous pity,' Childers to Gregory, December 8, 1874, Gregory Family Papers Box 25, MARBL, Emory University.

52   Cameron to Gregory, February 5, 1873, ibid.
53   Singh, *Discovery of Ancient India*, 220.
54   As quoted in ibid., 220. See also idem, 291.

for Buddhist monks.[55] A decade later, the Ceylonese government was prepared to use the funds donated by a Siamese Buddhist prince for the excavation of the Mirisaveti Dagaba at Anuradhapura using prison labor, but insisted that the work be administered by the colonial administration.[56] As the Government Agent at Anuradhapura later remarked:

> In 1888 I had some unofficial communication with Sir Arthur Gordon [Governor of Ceylon], who informed me that a Siamese prince, during his stay in Colombo, had placed in the Governor's hands a sum of money which he desired should be expended upon some Buddhist monument of historical interest and Sir Arthur Gordon suggested that the money might be devoted to the restoration of the Mirisawetiya Dagaba ... & rebuilding the upper portion which had fallen away. I understood that the Siamese prince accepted the suggestion, & that Sir Arthur Gordon offered to allow prison labor to assist in the work as an act of courtesy & friendly assistance to the King of Siam.[57]

The private correspondence between the Anuradhapura Government Agent and Governor Gordon in the middle 1880s reveals considerable latitude in the collaborative experiments undertaken to continue excavation and preservation work despite budget shortfalls on the island. Government Agent Ievers proposed to high-ranking local (Ceylonese) administrators from the Anuradhapura region that they support further efforts to clear historic sites. They were offered the opportunity as an act of merit-making in accordance with Buddhist devotional dispositions. This appears also to have been an attempt (though ultimately unsuccessful) to defuse Buddhist criticism of excavation projects that might intrude on the relic chambers of relic monuments.[58]

> ... speaking about these works [to the Rate Mahataya Hikewewe], suggested his undertaking one at which he expressed much pleasure and offered to do [? unclear] Dagoba, not only in compliance with the wish expressed, but as a religious act. Consequently we settled that he should undertake the clearing of Abayagiri which would not be so difficult as it was cleared some years ago.

---

55 Ibid., 221-22.
56 Bell and Bell, *Discovery of Ancient India*, 38.
57 Letter from R.W. Ievers, Government Agent, Anuradhapura, November 20, 1890, Sessional Papers, 56, 1890.
58 Ievers to Gordon, 17.1.85, 5.4.87, 28.4.87, Stanmore Papers, 49208; Bell and Bell, *Discovery of Ancient India*, 39-41. With reference to controversial excavation of a dagaba, or relic monument, see Ievers to Gordon May 4, 1887: '... I think I can manage it without raising an outcry.—But I shall wait until you come here and I can better explain,' Stanmore Papers, 49208. This almost certainly refers to Ievers' plan to enter the interior relic chamber of the Abhayagiri relic monument in order to search for ancient texts. See Nissan, 'The Sacred City of Anuradhapura', 259-60; Ievers, *Manual*, 237.

I have written to Galugaswewa R.M. informing him of this, and offering him the clearing of the spire and platform of Iryatawana Rama…

I shall see Ratwatte R.M. at the end of the month and if he cares to participate in the good work we can utilize his men.[59]

Aesthetic Sympathy and Preservation: 'Beauty' and 'Taste'

Thomas Metcalf suggests that in India the British approached historical Indian monuments in aesthetic terms, while emphasizing the commemorative aspect of their own memorial construction.[60] Although this distinction may not do full justice to the architectural choices of the Raj, Metcalf's comment on the aesthetic dimension of colonial encounter is astute, and helps to frame a discussion of the role of aesthetic sympathy in the preservation of Buddhist sites in nineteenth-century Ceylon.[61] I have already indicated a number of reasons for the centrality of Anuradhapura to early archaeological and preservation work in Sri Lanka. In addition to those already mentioned, however, it is important to attend to the ways in which Governor Gregory's aesthetic appreciation for architectural styles to be found at Anuradhapura stimulated his interest in work at that location. Gregory approached a variety of cultural forms – including architecture, sculpture, and literature – from a particular understanding of what constituted 'beauty' and 'taste'. This understanding interacted in striking ways with the evaluations he made of South Asian social groups, and the manner in which he ranked religions according to hierarchies of refinement, and purity. Such hierarchies are apparent, for example, in his distinctions made between Buddhists and Hindus on the island of Ceylon. In an 1877 Address to the Legislative Council, speaking of his efforts to control pilgrimage to the Sri Lankan site of Kataragama, which Gregory understood as a Hindu rather than Buddhist location, he argued that:

---

59   Private letter from Ievers to Gordon, January 17, 1885, Stanmore Papers, 49208.

60   Antonella Pelizzari, ed., *Traces of India: Photography, Architecture, and the Politics of Representation, 1850-1900* (Montréal: Canadian Centre for Architecture and New Haven: Yale University Press, 2003), 243.

61   The distinction may hold for memorial projects understood rather narrowly but, as Metcalf's own 1989 striking work on the architecture of the period suggests, major public building projects undertaken by British colonizers drew on metropolitan understandings of style in which aesthetic judgments and understandings of political symbolism were closely linked. See also Philip J. Ayres, *Classical Culture and the Idea of Rome in Eighteenth-Century England* (New York: Cambridge University Press, 1997); Richard Jenkyns, *The Victorians and Ancient Greece* (Cambridge: Harvard University Press, 1980); and Frank Salmon, *Building on Ruins: The Rediscovery of Rome and English Architecture* (Aldershot: Ashgate, 2000).

[i]t is idle to allege the slightest similarity between these pilgrimages and those of the Buddhists and the Roman Catholics to their respective holy places. The Buddhists are of the most respectable classes, and visit certain sacred shrines, influenced solely by religious feelings; they are cleanly in their habits and abstemious. They do not make their passage through the country a scene of revelry.

Unlike some of his predecessors and successors, William Gregory was greatly drawn to the visual surroundings of the island, to Buddhism (as he understood it), and to the way of life (again, as he understood it) of what he took to be the island's oldest aristocracy, the Kandyan elites.[62] His appreciation of the island's physical beauty seems to have been immediate, and enduring.

I am perfectly charmed with this island – its beauty is beyond any thing that even one with great imagination can conceive – The Central Province is full of the grandest mountain scenery, peaks & [? unclear] & waterfalls, but it is never bare or austere, & the vegetation is glorious.[63]

Writing nearly a decade after his departure from Ceylon, to then Governor Gordon, he noted:

I hope I may be strong enough to pay Yapahoo [a historic city towards the west of the Kandyan region] a visit one of these days. I think the view from the portal of the palace one of the most beautiful in Ceylon and that is a bold word. How extraordinary is the general grimness of Indian scenery in comparison with the charm of that of Ceylon.[64]

The visual attention with which Gregory encountered the physical landscape of the island extended to his responses to other visual forms, as he reflected on the degree to which the island's persons, painting, sculpture, and architecture might be considered beautiful, tasteful or works of art. Strikingly, he was fully prepared to accord high marks to local patrons for works of visual grace. Writing of his own architectural restoration undertaken in the area of the eighteenth-century Kandyan royal palace and Temple of the Tooth (*Dalada Maligava*), Gregory remarked:

---

62   Gregory interacted frequently, on social as well as administrative terms, with high-caste Kandyan families from Sabaragamuva and the Central Province. He accepted the leadership of certain high-ranking Kandyans with respect to Buddhist matters in Anuradhapura and the North-Central Province, and viewed his Kandyan contemporaries as descendants of the distant rulers of Anuradhapura. See Gregory to Layard, February 22, 1884, Gregory Family Papers Box 32, MARBL, Emory University. On the ascription of connection between Kandy and Anuradhapura, see Nissan, 135.

63   Gregory to Layard, June, date incomplete but almost certainly 1872, Gregory Family Papers Box 31, MARBL, Emory University.

64   November 10, 1886, Stanmore Papers, 49207.

The town of Kandy has been greatly beautified by the restoration of the works of the Kandyan kings, which had been mutilated and destroyed by the barbarity of English officials. The late King of Kandy, though a ruthless and bloody tyrant, *was nevertheless a man of consummate taste, and in all restorations I must say that that I have gone upon the lines laid down by him and by the guidance of old drawings* (1877 Address to the Legislative Council, emphasis added).

I have already done a famous bit of work at Kandy having completed in the old Kandyan style a wall left unfinished by the late king of Kandy which surrounds a Temple and I have cleared all the grand old wall of the great Temple of the rubbish which had accumulated round it ... I am also restoring some Kandyan work which our ruthless predecessors had pulled down, but had fortunately left enough of to see a pattern. In about a year and a half Kandy will be worth a visit, and be far more lovely than it is now which it is hard to imagine.[65]

Much at Anuradhapura and Kandy satisfied his eyes, as did the architecture of Rajasthan and some to be found in Gujarat. He recoiled from many works of what he considered a 'Hindu' style, happy neither in southern India nor before some of the remains at Pollonaruva in Ceylon. Not long after his arrival in Sri Lanka, the Governor wrote with characteristic detail to his mother:

I returned last week from an excursion about 80 miles to the north of this to the ancient cities of Anuradhapura & Pollanaruva. The first was the capital of Ceylon before the Christian Era and from the old accounts & from the existing remains must have been of enormous size. They say it was 16 square miles in extent. The other city was in its grandeur between 1100 & 1200 of our Era, and there is a marked difference between the architecture of the two. The one seems of a pure native style, the other is leavened with the Hindooism of the invaders from India & is in my opinion very inferior.[66]

Further, later writing to the Secretary of State for the Colonies in 1874 about the Ruvanweli Dagaba for which he had given personal support to the restoration work of the monk Naranwita, Gregory remarked tellingly on this ancient relic monument:

This design so entirely different from the florid Hindoo work of a later period, is the work of the 2nd Century B.C. and will be interesting to architects from its beauty and to archaeologists from the high civilization which it denotes to have prevailed in this Island at such an early epoch ...[67]

---

65   Gregory to Layard, November 25, 1872, Gregory Family Papers Box 31, MARBL, Emory University.

66   Gregory to the elder Mrs Gregory, July 21, 1872, ibid.

67   CO 54/492, Gregory to Kimberley 15.1.74. Explaining his personal contribution to the Buddhist monk's restoration of the Ruvanweli Dagaba, Gregory noted not just its

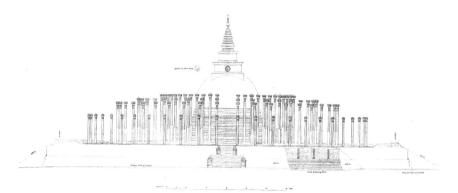

Figure 3.3   Architectural drawing of the Thuparama Dagoba Restoration, Anuradhapura

Source: James G. Smither. *Architectural Remains, Anuradhapura, Ceylon.* 1894. Colombo: Government Printer

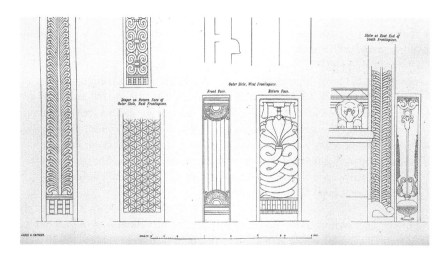

Figure 3.4   Architectural drawing of decorative detail on the Ruwanwali Dagoba, Anuradhapura

Source: James G. Smither. *Architectural Remains, Anuradhapura, Ceylon.* 1894. Colombo: Government Printer

Figure 3.5   Architectural drawing of proposed restoration of the Abhayagiri Dagoba complex, Anuradhapura

Source: S.M. Burrows. Report on Archaeological Work in Anuradhapura and Pollonnaruwa. 1886. Colombo: Government Printer

William Gregory's preference for an effect of architectural lightness and relatively unencumbered surfaces is repeatedly apparent in the lengthy and forthright correspondence maintained with his dearest friend Henry Layard.

> I hate this [Madras] Presidency, the scenery except on the western coast is mostly uninteresting, the climate always detestably hot, mosquitoes swarming & aggressive (we had hardly any in upper India) and this southern Hindoo architecture heavy, grotesque, ungainly disgusts me though of course is worth seeing. It is very different from the graceful beauty of the Hindoo architecture in Rajpootana.[68]

> As for Oodipur [Udaipur] it is the loveliest spot I have ever seen a kind of paradise amid the austerity of Indian scenery. It is enclosed by hills & on the side of a long but not very wide lake. The city is on one side terminated by the huge palace of the Maharana an enormous pile 100 ft high almost rising out of

---

status as a 'national monument' but the fact that it was 'characteristic of the best period of Sinhalese art and identified with one of the most powerful and enlightened kings who ever ruled this country,' SLNA 6/41/1309A, cited in Bell and Bell, *H.C.P. Bell: Archaeologist of Ceylon and the Maldives*, 35.

68   Gregory to Layard, February 19, 1886, Gregory Family Papers Box 32, MARBL, Emory University.

the water and its great expanse is broken by pretty little balconies and stone canopies. The other side is the line of temples & tombs, & private palace built with white marble and in the most varied fashion a series of domes, & canopies & pillared porches & marble steps & in the lake are two islands covered with palaces of the Maharana also in marble. The whole is so light, so brilliant & so fantastic though never grotesque, that it really seems to have been called at once into existence by some powerful and very tasteful 'Jinni.'[69]

At Aurangzabad we had great festivities the Mahommedan Governor gave us ... after that we returned to Bombay which I think more odious than ever, & its huge ungainly buildings were even more loathsome than before after seeing the lightness and fineness of the old Mahommedan structures. Since then I have seen new buildings not Mahommedan both here & at Baroda which are beautiful, some of them with domes others flat roofed but all in accordance with the character & beauty of the country ... We have come on to Ahmedabad well worth a visit. The marble work of the dome of the tombs is enchanting, so delicate, so light & so varied...[70]

Gregory's predilection for Buddhist and 'Mahommedan' architecture, and his dismissal of many Hindu buildings, was not uncommon in his day.[71] Indeed, it often closely resembled (and was almost certainly informed by) the classifications developed by James Fergusson in a series of publications initiated in the 1840s, and extending over several decades.[72] With their emphasis on early Buddhist purity and later Hindu degeneracy, Fergusson's architectural classifications stood in a position of mutual reinforcement to many orientalist studies of South Asian literature and religion. Early studies of Hinduism and Buddhism developed in close proximity to historical philology and a rising interest in the Indo-European roots of European culture, leading to an orientalist emphasis on the ancient origins of South Asian religions. The most esteemed form of Buddhism was often that of ancient origins prior to ritual 'accretions.'[73] Discoveries of Indo-Greek connections in the Buddhist

---

69   Gregory to Layard, January 18, 1886, ibid.

70   Gregory to Layard, January 5, 1886, ibid.

71   On British responses to Islamic architecture, see Thomas Metcalf, 'Monuments and Memorials,' in Pelizzari, ed., *Traces of India*, 35-49.

72   Guha-Thakurta, *Monuments,* 16-18. See also Thomas Metcalf, *An Imperial Vision: Indian Architecture and Britain's Raj* (New Delhi: Oxford University Press, 2002), 50.

73   On the colonial period reception and 'invention' of Buddhism and its relationship to Hinduism, and the place of religion in colonial apparatus, see, for instance, Philip Almond, *The British Discovery of Buddhism* (Cambridge: Cambridge University Press, 1988); Bernard S. Cohn, *Colonialism and Its Forms of Knowledge: The British in India* (Princeton: Princeton University Press, 1996); Richard King, *Orientalism and Religion: Post-colonial Theory, India and the Mystic East* (London: Routledge, 1999); and Donald S. Lopez, ed., *Curators of the Buddha: The Study of Buddhism Under Colonialism* (Chicago: University of Chicago Press, 1995).

artistic production of northwest India intensified positive orientalist associations between Buddhism and a pure and ancient culture. 'Just as the excellence of Hellenic art was associated with the greatness of the civilization of Periclean Athens, the superior qualities the British perceived in Buddhist art came to hinge on a broader construct of the purity of Buddhist religion and culture in India.'[74]

Nineteenth-century understandings of race and comparative religion developed in potent interconnection with architectural and archaeological studies of southern Asia. By Gregory's time, there could be no aesthetic response to South Asian forms isolated from such discursive trends. It is crucial to recognize the central role played by visual experience and judgment in the ascription of historical value by colonial administrators and scholars to particular sites and structures in the colonies. Such visual judgment was historically conditioned in the ways just mentioned but, of course, to the viewer it was simply *natural*. Moreover, the apparent naturalness of aesthetic evaluation increased the power of the judgments made.[75] Some colonial administrators and observers of southern Asia, of whom William Gregory provides a striking example, received their impressions of these regions and their cultures primarily through a stylistic and aesthetic orientation, rather than from a religious or philosophical standpoint. Gregory had no strong religious affiliation of his own; his conscious evaluations of religions were typically utilitarian. Indeed, he sought the disestablishment of the Church of England in Ceylon. To 'art' and 'beauty,' however, Gregory responded strongly. An aesthetic sympathy for particular places and persons profoundly shaped his responses as Governor of Ceylon. A certain emotional-intellectual comfort generated by the experience of shared 'taste' influenced his conduct of social relations across racial lines.[76] Moreover, Gregory's aesthetic judgment influenced his identification of some colonial sites as particularly worthy of appreciation and preservation. In turn, this led to the preservation and creation of some local historical memories rather than others. Finding many of the ancient sites at Anuradhapura beautiful and aesthetically intelligible; he was inspired to render them stable and more visible. Subsequently, those sites engaged the attentions of generations of archaeologists and historians. This helped to shape dangerously potent local Sri Lankan narratives in the post-independence years that privileged the ancient world of Anuradhapura

---

74   Guha-Thakhurta, *Monuments*, 36.

75   Note Edward Said's early remarks on the importance of 'taste' where he glosses 'power cultural' with reference to 'orthodoxies and canons of taste, texts, values,' in Edward Said, *Orientalism* (New York: Vintage Books, 1979), 12.

76   David Cannadine's remarks are apposite here: 'To be sure, the Enlightenment brought about a new, collective way of looking at peoples, races and colours, based on distance and separation and otherness. But it did not subvert the earlier, individualistic, analogical way of thinking, based on the observation of status similarities and the cultivation of affinities, that projected domestically originated perceptions of the social order overseas,' in Cannadine, *Ornamentalism*, 8. Further, ' ... there were at least two visions of empire that were essentially (and elaborately) hierarchical: one centred on colour, the other on class,' ibid., 9.

as a Sinhala-Buddhist heartland, and claimed for multi-religious and multi-ethnic Sri Lanka a more exclusively Sinhala-Buddhist destiny essential to majority religious and political identity on the island.[77]

When we look at the structures in Anuradhapura and Kandy that received particular accolades from William Gregory in terms of their tastefulness and artistic merit, we see his marked preference for clean lines and basic geometric forms, evocative of neo-classical styles. This is not surprising, since late-eighteenth- and nineteenth-century British and Northern European encounters with Greco-Roman contours – in person, through architectural drawings, and through the resulting neo-classical architectural forms in northern Europe – had created something akin to what Michael Baxendall has called a 'period eye' for men of William Gregory's class, education, and artistic inclinations.[78]

His preferences in the sphere of Sri Lankan architecture reveal that Gregory *re-cognized* – with pleasure and sympathy – elements associated with the classical forms of the Mediterranean world within the visual field of southern Asia. This may be understood as part of what David Cannadine has termed 'the domestication of the exotic.'[79] Although his contemporaries typically expressed a strong preference for the products of the *ancient* South Asian past, Gregory's aesthetic sympathies led him to a somewhat more historically capacious embrace of southern Asian sites and buildings. For William Gregory, the most compelling forms could be chronologically ancient, but need not be: certain qualities of light and grace were to be celebrated and preserved whether they occurred in ancient or medieval Asian forms. In any case, they were reassuringly familiar, part of the understanding of

---

77   See, for instance, Steven Kemper, *The Presence of the Past: Chronicles, Politics, and Culture in Sinhala Life* (Ithaca: Cornell University Press, 1991); Nissan; and Stanley Tambiah, *Buddhism Betrayed?* (Chicago: Chicago University Press, 1994). An evocative comparative case from Buddhist Southeast Asia is Ankor Wat. See Penny Edwards, *Cambodge: The Cultivation of a Nation, 1860-1945* (Honolulu: University of Hawaii Press, 2007) on French colonial involvement in the preservation of Angkor Wat, and the subsequent place of that site in Cambodian nationalist politics, including the destructive utopianism of the Khmer Rouge.

78   Michael Baxandall, *Painting and Experience in Fifteenth Century Italy* (2nd edition, Oxford: Oxford University Press, 1988), 40, and further: ' … some of the mental equipment a man orders his visual experience with is variable, and much of this variable equipment is culturally relative, in the sense of being determined by the society which has influenced his experience. Among these variables are categories with which he classifies his visual stimuli, the knowledge he will use to supplement what his immediate vision gives him, and the attitude he will adopt to the kind of artificial object seen.' Thomas Metcalf notes the centrality of this aesthetic orientation to *building* projects undertaken by the Raj. 'The use of classical forms to express the spirit of empire was, for the late-Victorian Englishman, at once obvious and appropriate, for classical styles, with their reminders of Greece and Rome, were the architectural medium through which Europeans always apprehended empire …,' Metcalf, *An Imperial Vision*, 177-78.

79   Cannadine, *Orientalism*, xix.

'civilized taste' and a 'civilized past' still arguably rather new to Victorian England and Britain. In his classic work, Michael Baxandall remarked that [m]uch of what we call 'taste' lies in this, the conformity between discriminations demanded by a painting and skills of discrimination possessed by the beholder ... If a painting ... rewards our virtuosity with a sense of worthwhile insights about [its] organization, we tend to enjoy it: it is to our taste.[80]

The buildings and monuments to which William Gregory responded most positively were those he was best equipped to recognize as beautiful and successful. That is, they were those edifices standing in evocative proximity to the physical contours of a classical Mediterranean past as it was understood in mid-nineteenth-century London and Dublin. Gregory's letters on matters of art and aesthetics suggest that to him the Buddhist sites of Anuradhapura were primarily beautiful, and secondarily of historical importance. This beauty – which he associated with a purity he perceived in Buddhism – intensified his attachment to Sri Lanka and to its local elite, deepening his long-standing interest in the work of museum-based education and historical preservation.

---

80 Baxandall, *Painting and Experience*, 34, 109.

Chapter 4

Conflict and Neglect: Between Ruin and Preservation at the Church of the Nativity, Bethlehem

Raymond Cohen

Introduction

The Church of the Nativity, Bethlehem, revered by untold generations of pilgrims as the birthplace of Jesus Christ, has had a tumultuous history.[1] The original basilica, in the form of a columned Roman temple, was built by the Emperor Constantine in 330-33, following his adoption of Christianity as the religion of the empire and the visit of his mother Queen Helena to Jerusalem and nearby Bethlehem in 325. Those two places became the jewels in the crown of the Byzantine Empire, the settings of the key events in the life of Christ.

Excavations carried out in the 1930s showed that Constantine's structure had burnt down, possibly in 529 when the Samaritans rebelled against Byzantine rule. Extensive mosaic floors and the foundations of walls under a thick layer of ash mixed with charred wood and broken tiles, gave witness to a devastating conflagration. A splendid new cruciform church, erected by the Emperor Justinian (527-569), followed the original design of atrium and nave with four colonnades. At the west end of the basilica a porch or narthex was added, while the east end, over the underground Grotto of the Nativity, consisted of a spacious cruciform transept and chancel with attached apses. The walls were lined with marble or mosaics, surmounted by a timber roof. Maintenance of the roof and mosaics, so susceptible to decay, has been a recurrent problem from that day to this.

In later years, Justinian's church repeatedly skirted destruction. In 614 Persian armies swept through the Holy Land. They sacked its cities but somehow spared the Nativity Church. Then in 638 the Byzantine Empire lost Palestine to the Caliph 'Umar ibn al-Khattab at the head of a conquering Muslim army. The Caliph promised the Patriarch Sophronius to protect the church and, with occasional lapses, subsequent Muslim rulers mostly kept that promise. Even the fanatical Egyptian Caliph al-Hakim, who destroyed the Church of the Holy Sepulchre in 1004, did not touch the Church of the Nativity. In 1099 the Crusaders conquered

---

1   The following section is based on R.W. Hamilton, *The Church of the Nativity Bethlehem* (Jerusalem: Government of Palestine, Department of Antiquities, 1947), 15-29.

the Holy Land, indiscriminately massacring the population of Jerusalem, whether Greeks, Jews or Muslims. They established a Latin Kingdom, remaining in Jerusalem and Bethlehem until their defeat in 1187 by Saladin and the restoration of Muslim rule over Palestine.

('Catholics' as we now tend to call them, members of the Western Church under the authority of the Pope, the Bishop of Rome, are still known as 'Latins' in the Jerusalem area, following the original language of their liturgy. 'Orthodox' as we now tend to call them, originally the Church of the Roman Empire in the East based on Byzantium or Constantinople, now Istanbul, are still known as 'Greeks' in the Jerusalem area, following the original language of *their* liturgy. Most of the parishioners of both communities have long been Arabic-speakers.)

The Crusader period was crucial for the Nativity Church. Before the arrival of the Crusaders the Royal Church of Byzantium possessed *praedominium* or primacy at the Holy Places, as it had had since the pilgrimage of Queen Helena. The Patriarchate of Jerusalem was in its hands; other communities, including the Latins, Armenians, Syrian Orthodox and Copts, had their own altars but never questioned their own subordinate status. Under the Crusader Kings of Jerusalem the Greek Patriarch was sent packing. Latins now had primacy at the Holy Places and control of the Patriarchate. Under the Muslim Sharia Law that governed Palestine after 638 construction and repair of a structure constituted proofs of ownership. Moreover, the one who roofed a building *ipso facto* owned that building. It was therefore as proprietors that the Latins rebuilt the Church of the Holy Sepulchre and opulently restored and refurbished the Church of the Nativity. They installed a new roof of cedar beams, sheeted with lead, and redecorated the walls of the nave and apses with portraits of Christ's ancestors, scenes from his life, and representations of Church councils.[2]

When the Crusaders departed, the Byzantine Church regained its primacy but never forgot or forgave its painful expulsion from the Jerusalem Patriarchate, the depredations of the Crusaders, or the theological schism between the Churches that rendered their conflict so bitter and unrelenting.[3] In contrast, the Roman Catholic Church looked back on the Crusader period as a golden age, considering the Orthodox Church to be heretics and interlopers. Ever since then Latins and Greeks have been locked in a struggle for primacy at the Holy Places, a surrogate battle for the Christian birthright. For a time one might gain the upper hand, only for that supremacy to be overturned at a later date by imperial preference in some new international constellation. This fierce competition continued for the 400

---

2   E.T. Richmond, 'Introduction,' in William Harvey, *Structural Survey of the Church of the Nativity, Bethlehem* (London: Oxford University Press, 1935), viii-x.

3   The schism is conventionally dated to 1054, when leaders of the two Churches excommunicated each other. In 1204 a Crusader army sacked Constantinople. A last attempt at reconciliation at the Council of Florence (1439) was repudiated by Byzantine opinion. In 1453 Constantinople fell to the Ottoman Turks, little help forthcoming from the Catholic West.

years of Ottoman Turkish rule from 1516 until 1917 when the British arrived. It has been contained but never abandoned, and still occasionally flares up.

Our story here is of the record of preservation at the Church of the Nativity from the mid-nineteenth to mid-twentieth century, under the Ottoman and British Empires. It is largely a study in frustration and dilapidation. This is more to do with law and politics than architecture. According to the Turks, the Holy Places, including the Church of the Nativity, were *waqf*, religious property held in perpetuity by a religious foundation as a sacred trust. Unlike *mulk* or private property, ownership was inalienable and could not be transferred, whether by sale or bequest. However, rights of possession and usage could be graciously bestowed by the sovereign, acting as trustee, upon whichever Christian community was favoured at the time.

The Turks frequently exercised their sovereign prerogative, occasionally transferring various rights by imperial decree or *firman* from one Church to another as they saw fit. On the whole, they preferred the Greek Orthodox community because they were Turkish citizens. Sometimes, though, French diplomatic influence (because of France's 1536 treaty with Turkey and status as protector of Catholics in the Ottoman Empire) was effective in obtaining advantage for the Latins. As a Muslim power the Ottoman Empire assiduously maintained the Holy Places of Islam but was frankly indifferent to the fate of the Christian Holy Places, except as a source of income. As far as it was concerned their preservation was the problem of the major resident communities – Greek Orthodox, Roman Catholic and Armenian Orthodox – not theirs. The authorities would issue permits to conduct specified repairs in return for exorbitant presents and fees but would not take the initiative.

As long as a single community enjoyed undisputed primacy in the church, thanks to the fickle grace of the Sultan, that community's concern to maintain the property outweighed the sovereign's passivity. Once the Ottoman Empire went into decline and the Sultan lost the effective power to determine possession at the Christian shrines, the system broke down. Given the logic of Muslim property law, Greeks and Latins were rarely able to agree on repairs, let alone settle disputes. Arguing over every hook and nail, they feared that cooperation in making a repair would constitute recognition of the other's right to the item being repaired. In particular, rivalry for possession of the roof over the church proved gravely detrimental. Thus the very religious ardour that hallowed the site also impeded the communities' collaboration. Each unswervingly viewed its claim to be the True Church to entitle it to the exclusive right to the birthplace of Christ. This unfortunate state of interchurch rivalry was frozen on the eve of the Crimean War (1853-56) and inherited by the British when they took over Palestine in 1917. It impeded British officials at the Holy Places for the entire Mandate.

Turkish Rule

Long before the arrival of the Turks in 1516, a pattern of official indifference and venality had set in at the Church of the Nativity. During the Mamluk period (1260-1516), repairs could only be carried out by the special authorization of the Sultan based in Cairo and this was only grudgingly granted. Permission to repair the damage to mosaics and paintings caused by rain leaking in was not forthcoming and nothing was done to stop vandalism and looting. Still, in 1480 permission was obtained to carry out repairs to the roof and, in a combined operation, Philip Duke of Burgundy supplied the craftsmen, wood, iron and tools; King Edward IV of England the lead; and the Doge of Venice the ships.

Under the new Turkish dispensation marble was openly removed from the church for the decoration of Muslim buildings, including the Haram es-Sharif in Jerusalem. By the mid-seventeenth century little of the marble facing of the walls remained and the roof was again rotting away as the protective lead sheathing had been stripped off for the manufacture of bullets. Eventually, the Greek Orthodox Patriarchate, in the ascendant at that point, acquired the necessary firmans for an extensive programme of repairs. By 1671 a new roof and icon-screen or iconostasis had been installed at a cost of 100,000 silver écus, donated by a Greek of Constantinople named Manolakes Kastorianos, 20,000 écus going on presents and fees. A special road had to be built for forty oxen to drag timber for the roof

Figure 4.1   Church of Nativity, Central Nave

Source: John Carne, *Syria, The Holy Land, Asia Minor, &c. Illustrated*, vol. 3, London: Fisher, Son, & Co., 1838, 83. Engraved by William Radclyffe after Thomas Allom

on wagons up through the Jerusalem Hills to Bethlehem. Thanks to the repairs, the fabric of the edifice was preserved for another 160 years.[4]

In 1842, following an earthquake some years before which had caused serious damage to the Nativity Church, the Greek Orthodox Patriarchate obtained an imperial firman authorizing much-needed repairs. They were able to restore the carpentry throughout the basilica and entirely replace the lead of the roof. The walls, from which the marble had been removed in previous centuries, were plastered over, framing the surviving mosaics in white and covering up Latin inscriptions. The Greeks also repaved the nave and chancel or choir, separating them with a new twenty-foot high wall. The purpose of this controversial partition (demolished in 1918) was to shut off the east end of the church for worship at a time when the nave was used as a market. Much the worse for subsequent decay, this is more or less the church we see today.[5]

While the restoration enabled the Greeks to decisively stamp their primacy on the main body of the church in Bethlehem, the Latins were left with the underground Grotto of the Nativity.[6] At the Church of the Holy Sepulchre in Jerusalem the situation was similar. In 1847 Pope Pius IX resolved to regain the Catholic Church's lost position in the Holy Land, appointing a Latin Patriarch for the first time since 1187. Rivalry between Greeks and Latins, supported by their respective great power patrons, Russia and France, greatly intensified. In October 1847 the silver star attached to the altar in the grotto, bearing the Latin inscription HIC DE VIRGINE MARIA JESVS CHRISTVS NATVS EST was stolen by Greek monks in order to remove a prominent symbol of Catholic presence. Before Christmas there was a violent altercation between Greek and Latin monks.

This set the scene for the drawn-out international crisis that ended with the Crimean War of 1853-56 between Britain, France, and Turkey on the one side, and Russia on the other. Louis-Napoléon Bonaparte's France, exercising its traditional role, first intervened on the side of the Latins to press for the restoration of rights lost in the eighteenth century. Ahead of Christmas 1852 the Sultan of Turkey, Abdul Mejid I, granted concessions at the Holy Places to the Latins and replaced the silver star with its Latin inscription at his own expense. In reaction, Tsar Nicholas I of Russia intervened on behalf of the Greeks, threateningly deploying two armies along the Danube. In an attempt to stave off a Russian invasion the Sultan

---

4   Hamilton, *Church of the Nativity*, 34-36.

5   H. Vincent and F.-M. Abel, *Bethléem: Le Sanctuaire de la Nativité* (Paris: J. Gabalda, 1914), 203.

6   They also had limited rights of passage in the main basilica. Through the influence of the Emperor Napoleon III they later acquired the Altar of the Manger. Details of the Status Quo in the Church of the Nativity are found in a 1929 British document known as the 'Cust Report': L.G.A. Cust, facs. ed., *The Status Quo in the Holy Places* (Jerusalem: Ariel, 1980). A revised and updated guide to the Status Quo in 'The Basilica of the Nativity at Bethlehem,' undated but clearly dating to 1941 or later can be found among miscellaneous papers in Israel State Archives (henceforth ISA) 3058/17-פ.

Figure 4.2   Church of the Nativity, Grotto

Source: Henry Stebbing, *The Christian in Palestine* (London: George Virtue, 1847), 3. Artist, William Henry Bartlett

issued firmans in 1852 and 1853 confirming the status quo at the Holy Places. The rights of possession and usage as they existed at that time were removed from the purview of the courts and were never to be altered. In effect, this preserved Greeks primacy at the key shrines of the Holy Sepulchre and the Nativity in perpetuity.[7]

Note that, contrary to received truth, this arrangement was imposed on the Churches, and was not an 'agreement' that they voluntarily accepted. Indeed, all the communities were convinced that they were victims of an injustice, deprived of altars rightfully theirs. This is why they never ceased to question its validity.

With hindsight, historians tend to date the emergence of the Status Quo (capitalized) as the inviolable legal regime at the Holy Places to the 1852 and 1853 firmans. The Crimean War, it is argued, which broke out soon after, demonstrated the danger of local disputes at the Holy Places escalating into general war. To avoid this, the status quo (not yet capitalized) was elevated into a kind of 'ceasefire in place,' enforced by the great powers, freezing and hence stabilizing the situation on the ground. In this way local ecclesiastical rivalries would be insulated from

---

7   On the events at the Church of the Nativity preceding the Crimean War see 'Correspondence Respecting the Rights and Privileges of the Latin and Greek Churches in Turkey,' *House of Commons Parliamentary Papers (HCPP)* vol. 71 (London, 1854). For a contemporary autobiographical account see James Finn, *Stirring Times or Records from Jerusalem Consular Chronicles of 1853 to 1856* (London: C. Kegan Paul & Co., 1878).

international politics. As we shall see, the status quo was in fact only permanently solidified and legitimized a quarter of a century later.

Meanwhile, neglect and dereliction had already begun to take hold in the Church of the Nativity, contested rights of possession precluding measures of preservation. Writing in 1864, an American visitor gloomily described the disrepair of the 'old edifice':

> In the interior the framework [of the roof] is visible, the flags of the pavement are broken, and the roof is sadly out of repair. Of the beautiful pictures and inscriptions which once adorned the walls nothing remains. The reason of this neglect is that this church is the common property of Latins, Greeks, and Armenians. One will not allow the other to undertake any improvement; and thus the oldest and most interesting church in the world, after the Church of the Holy Sepulchre, is allowed to fall into decay. The nave is used as a bazaar for the sale of Bethlehem trinkets.[8]

Nor did the 1853 status quo eliminate competition from the basilica. In the affair of the grotto tapestry, 1869-74, the Greeks threw down a major challenge to the 1853 status quo. For many years, Latin rights in the Grotto of the Nativity had been bitterly contested by the Greeks. In order to reaffirm those rights, the Latins had sought to replace a tattered tapestry that covered the vault and walls of the crypt with new wall-hangings. They feared that the disintegration of the worn tapestry, bearing Latin inscriptions and emblems, would efface the visible proof of their possession of the shrine. For their part, the Greeks inflexibly opposed this request. In 1869 a fire in the crypt destroyed Latin furnishings, including the tapestry.

The fire produced a tremendous outcry. Greeks and Latins traded accusations of arson and brought tough young monks into Bethlehem to defend their rights. While Muslim gendarmes guarded the grotto, France and Turkey entered into negotiations to replace the tapestry.[9] The outbreak of the Franco-Prussian War in 1870, however, brought about a suspension of the talks. Taking advantage of France's defeat, the Greeks went ahead and installed their own tapestry in the crypt in February 1872. This breach of the 1853 status quo was intended to enforce the Greek claim to the shrine but was repudiated by the French and Turkish Governments. They resumed their talks and agreed that wall-hangings, identical to those destroyed in the fire, would be installed at French expense. On March 15, 1873 the new tapestry was hung.

The Greeks now resorted to violence. On the night of April 25-26, 1873, Greek monks, backed by a detachment of Greek marines on pilgrimage to the Holy Places, entered the grotto. While the local gendarmes turned a blind eye,

---

8   Albert Rhodes, *Jerusalem As It Is* (London: J. Maxwell, 1865), 129.
9   'L'Affaire de L'Église de Bethléem,' *Le Temps* (May 6, 1873).

the Greeks wrecked the tapestry and all other Latin furnishings, injuring five Franciscan friars in the assault.[10]

The outcome of the episode ultimately validated rather than undermined the 1853 arrangements. Following the brawl, and the considerable damage done to Catholic property, a mixed Franco-Turkish commission was set up at French insistence with full powers to assess the damage and reestablish the status quo. The commission arrived at two decisions. First, the Greeks were obliged to pay damages of 500 Turkish pounds to the Latins. Second, a new fireproof asbestos tapestry would be installed, the personal donation of Marshall Patrice MacMahon, the President of France. To remove all ambiguity about possession, the tapestry was decorated with four of the principal scenes relating to the infancy of Jesus with explanatory Latin texts, French *fleurs-de-lis*, the crossed arms of the Franciscan Order, and red Terra Sancta or Jerusalem crosses. At the installation ceremony in 1874 the tapestry was hung by Latin workers in the presence of the acting Turkish governor, the acting French consul, the chastened Greek Orthodox Bishop of Bethlehem, and the dragoman of the Greek Orthodox Patriarchate.[11]

Even so, Turkish justice did not come cheaply. The archive of the Apostolic Delegate to Jerusalem and Palestine contains a list of the payments made to Turkish officials. This included the sum of 2,000 gold francs paid to the *mahasibigy*, the accounts manager of the Pasha (the honorary title carried by the governor); 800 gold francs to the *cadi* or Muslim judge's Sharia court; 300 gold francs to the *mufti* or Muslim legal authority; 150 gold francs to the *mudir* or local governor of Bethlehem; 190 gold francs to the members of the Waqf religious trust; 100 gold francs to officers and *zabtie* gendarmes; 60 gold francs to the janissaries guarding the French consulate; 50 gold francs to the governor's servants, and so on. Altogether 4,560 gold francs, about £55,000 in modern values, were spent oiling the wheels of Ottoman administration in Jerusalem and Bethlehem.[12]

Payment of *douceurs* or bribes was normal practice in the Ottoman Empire, inherent in a society where officials were chronically underpaid. The list of beneficiaries maps out the network of Ottoman civil, military and legal authorities in the Jerusalem area with which the Churches had to deal on issues of preservation. The reference to Muslim jurists and institutions should come as no surprise because these were the people responsible for making legal decisions about Holy Places, whether Muslim, Christian, or Jewish. This included all matters touching on the repair of religious property. Significantly, the list does not include anybody who actually knew anything about preservation. Until the departure of the Ottoman

---

10   Vincent and Abel, *Bethléem*, 204. A more detailed account of the incident is given in a memorandum submitted by Fr. L. Hugues Vincent to the British authorities dated July 28, 1932, ISA 24/4-מ.

11   Apostolic Delegate to High Commissioner, May 21, 1932, ISA 24/4-מ. See also attached letter, Patriarch Bracco to Cardinal Franchi, Sept. 16, 1874.

12   Ibid.

Empire in 1917 no civil body dealt with the subject in the Jerusalem area.[13] Progressive antiquities laws were enacted in Istanbul in 1874 and 1884, but these dealt with excavations and artifacts, not with the preservation of ancient buildings. Nor did the Ottoman Government have the people to ensure their enforcement.[14]

The lesson of the affair of the grotto tapestry was obvious: that the status quo could no longer be discretionary, but had to become binding under international law. The opportunity for this decisive step, the 'apotheosis' of the Status Quo at the Holy Places, occurred at the Congress of Berlin following the Turkish defeat in the Russo-Turkish War of 1877-78. At the insistence of France, Article 62 of the 1878 Treaty of Berlin formally incorporated the Status Quo into public international law.[15] As intended, this ensured that the 'ceasefire in place' at the Holy Places would be maintained indefinitely. Apart from the odd murder and riot, peace was largely preserved at the Church of the Nativity for the remainder of Ottoman rule.[16] But the Status Quo also condemned the Church of the Nativity to generations of disrepair.

In the grotto, unresolved disputes meant that the vault and walls behind the tapestry remained blackened with soot from the 1869 fire until they were thoroughly cleaned and disinfected at the initiative of the British district commissioner ahead of Christmas 1944.[17] Under the terms of the Status Quo the sovereign was permitted to carry out essential minor repairs at the Holy Places, were the communities unable to agree. On several occasions the Turkish authorities replaced nails missing from the silver star. Yet as the Ottoman Empire grew feebler, it became ever more reluctant to decide or to fix anything, however trivial, for fear of upsetting the great powers.

One anecdote will serve to illustrate the point: Early in 1908, during the changing of the guard, a Turkish soldier accidentally knocked down an Armenian picture in the grotto, breaking the glass. The Turkish officer authorized the Armenian dragoman to sweep up the broken glass but the Greeks and Latins protested that this was a violation of the Status Quo. At this, monks from the three major communities descended *en masse* on the grotto. To defuse the situation, the Turkish mudir of Bethlehem rushed to the scene with a strong contingent of

---

13   Shimon Gibson, 'British Archaeological Institutions in Mandatory Palestine, 1917-1948,' *Palestine Exploration Quarterly* 131 (1999), 128.

14   George Young, *Corps de droit ottoman* (7 vols, Oxford: Clarendon Press, 1905), vol. 2, 389-94.

15   'The rights possessed by France are expressly reserved, and it is well understood that no alterations can be made in the *status quo* in the Holy Places.'

16   In October 1893 an Italian monk was shot dead and two others wounded when they barred entry to a Croat who insisted on bringing two Russian pilgrims into the church during celebration of the Latin Mass ['A Monk Killed in Bethlehem,' *New York Times* (Oct. 29, 1893); 'Turquie d'Asie,' *Le Journal des Débats* (Nov. 3, 1893)]. In 1905 the Greek Orthodox celebration of their Christmas Mass was interrupted by a brawl between Greek and Latin monks ['Clergy Fight at Bethlehem,' *New York Times* (Jan. 10, 1905)].

17   The episode is documented in ISA 30/44-ב.

Figure 4.3   Church of the Nativity, Grotto Altar

Source: H. Vincent and F.M. Abel O.P., *Bethléem: Le Sanctuaire de la Nativité* (Paris: J. Gabalda, 1914), plate XVII. The photograph is credited to the American Colony

troops. He was soon followed by the French, Greek, Italian and Russian consuls. The next day the governor proposed that the Armenians replace the glass of the picture themselves but the Greeks and Latins protested this proposal, too. Over the next nine months of deliberations a mutually agreed outcome proved unattainable. Before Christmas, with the Armenian Patriarch threatening to boycott the traditional seasonal ceremonies unless the question was settled, the mudir replaced the glass himself.[18]

The British Mandate

General Edmund Allenby's capture of Jerusalem from Turkey in December 1917 at the head of British Empire forces replaced Muslim by Christian rule after 700 years. His humble entry on foot through the Jaffa Gate – breached by the

---

18   'The Profanation of the Holy Sepulchre,' *New York Times* (Oct. 24, 1909). 'Do it yourself' became the authorities' traditional solution to minor repairs upon which the communities could not agree.

Ottoman Government in 1898 so that Kaiser Wilhelm II could ride into the city in triumph on his white horse – heralded a transformation in the government approach to preservation. First and foremost, British officials, the products of Victorian culture, possessed a *sensibility of preservation*, that is, a consciousness of the value and importance of safeguarding ancient buildings in their unspoiled setting. In addition, steeped in the Bible, they had an enthusiasm for the Christian Holy Places that had simply never before existed in governing circles in Palestine. Finally, they felt themselves to be under a moral obligation to the Christian world to safeguard its historical heritage for future generations.

It is worth recalling that the Palestine Exploration Fund was set up in London as early as 1865 in order to survey and investigate the Holy Land, a part of the world that fascinated the Victorians. Its maps and reports laid the foundations for all subsequent archeological research on the area. In a similar spirit, the first comprehensive, scientific survey of the art and architecture of the Church of the Nativity was published on behalf of the Byzantine Research Fund and the British School at Athens in 1910.[19] It was based on a careful investigation carried out between December 1908 and March 1909 by William Harvey, then a Gold Medallist of the Royal Academy of Arts. The drawings and photographs that Harvey made in Bethlehem are an invaluable record of the condition of the church, particularly the mosaics, towards the end of the Turkish period.[20]

Harvey found the mosaics in a fragmentary form. He observed that:

> The fragments are uninjured in themselves, and remain to all appearance in their original state. Restoration does not seem to have been attempted and could only be attempted even now if the work were authorized by the Turkish Government and the artist surrounded by Turkish soldiers to protect him from the fanatical rivalry of the sacristans.[21]

During the British Mandate of Palestine the 'fanatical rivalry' of the communities did not change but the commitment of the government to preservation most certainly did. On their arrival in Jerusalem the British military authorities were confronted by a daunting challenge, the result of centuries of neglect. The city's historic buildings and medieval *suqs* were in a ruinous state, roofs caving in, walls collapsing, filth and dilapidation everywhere. The citadel's surroundings

---

19   R. Weir Schultz, ed., *The Church of the Nativity at Bethlehem* (London: B.T. Batsford, 1910).

20   During the First World War Harvey worked in the Ancient Monuments Branch of HM Office of Works and in 1925 published a textbook on conservation. In the 1930s he conducted a second survey of the structural condition of the Nativity Church. See William Harvey, *The Preservation of St. Paul's Cathedral and Other Famous Buildings* (London: The Architectural Press, 1925); William Harvey, *Structural Survey of the Church of the Nativity Bethlehem* (London: Oxford University Press, 1935).

21   Schultz, *Church of the Nativity*, 33.

were disfigured by refuse tips and cesspits, once magnificent gates appropriated for such enterprises as baking dung cakes, houses and shops built up against the walls, corrugated iron and tin can used as building materials, ancient stonework pillaged for private use.[22]

As early as the spring of 1918 the military (from 1920, civil) governor of Jerusalem, Colonel Ronald Storrs, a classicist and aesthete, threw himself into the task of rehabilitating the city and its surroundings. How was this to be done in the absence of funds, organization and manpower? On April 8, 1918 he issued the following public notice and declaration of intent:

> No person shall demolish, erect, alter, or repair the structure of any building in the City of Jerusalem or its environs within a radius of 2,500 metres from the Damascus Gate (Bab el-Amud) until he has obtained a written permit from the Military Governor. Any person contravening the orders contained in this proclamation, or any term or terms contained in a licence issued to him under this proclamation, will be liable upon conviction to a fine not exceeding £E.200.[23]

In order to actively advance the cause of preservation Storrs took as his inspiration the work of the National Trust, founded in Britain in 1894-95 as an independent charity to protect open spaces and historical buildings under threat. He conceived the idea of a Pro-Jerusalem Society along the same lines, supported by the great and good of the city irrespective of race or creed, able to raise funds from private well-wishers. Among its goals would be to provide and maintain parks and open spaces, establish cultural institutions such as museums and art galleries, protect and preserve local antiquities, and encourage arts, handicrafts and appropriate industry. The society began its activities in the spring of 1918 and was inaugurated in September of that year.

For the post of civic advisor and secretary Storrs turned to the architect and designer Charles Ashbee, who was then in Cairo. A disciple of William Morris and a central figure in the Arts and Crafts movement, Ashbee proved an enterprising and indefatigable organizer. He wrote the report that launched the society and was dedicated to reviving traditional local crafts, such as hand blown Hebron glass and weaving. Storrs also enlisted Ernest Richmond, another experienced architect, who had worked before the First World War on the preservation of ancient buildings in the Egyptian Ministry of Public Works. He was employed as consulting architect to the Waqf on the magnificent seventh-century Dome of the Rock. To re-cover the octagon, which had been losing its royal blue and turquoise arabesque ceramic tiles for many years, Storrs brought in an Armenian expert, David Ohanessian,

---

22 A very full visual and textual record of the condition of Jerusalem at the time can be found in C.R. Ashbee, ed., *Jerusalem 1918-1920* (London: John Murray, 1921).

23 Ibid., 77.

who set up a potteries industry in the Old City with Armenian craftsmen from Kutáhia in Turkey.[24]

Two other important figures in the cause of preservation should be mentioned: One was Harry Luke, the assistant governor of Jerusalem, a classical scholar who became an expert on the Holy Places and the local Christian communities. He was of great assistance within the civil administration, which took over from the military in 1920. The other was the Dominican archeologist and Biblical scholar Père Louis-Hugues Vincent of the École Biblique. Together with Père Félix-Marie Abel he had written definitive works on Jerusalem and Bethlehem and was to provide learned advice to the government for many years.

The society's first big project in Jerusalem was cleaning and repairing the citadel with its famous Mamluk tower, a massive job. Later the Ottoman city walls were cleared of accretions and dangerous spots repaired, and an uninterrupted walk around the ramparts completed. The road in front of the Jaffa Gate was opened up and the walls surrounded with a belt of open land. The vaults, roofing and walls of the Mamluk masterpiece Suq al-Qattanin were restored, looms installed, and seventy weavers put to work. By the end of his term as civil governor in 1925 Storrs and the Pro-Jerusalem Society, whose role was taken over by the government, had indeed placed the task of preserving Old Jerusalem on a sound footing (and, though this is another story, building a modern planned city outside the walls).

From the start, Bethlehem, which he cherished as of 'surpassing merit,' proved an irresistible attraction to Colonel Storrs. On Christmas Eve 1917 he celebrated Mass in the Latin Church of St. Catherine's with about forty British officers, then walked in procession through the Church of the Nativity and down into the crypt for an evocative service.[25] He commented in his diary on 'the beautiful Greek basilica,' but was appalled by the 'hideous rubble wall' built in 1842 by the Greeks, cutting the nave off from the chancel or choir and blocking the view of the iconostasis and eastern apse. He determined to remove it, giving the job to Ashbee and Richmond.

The Greek Patriarchate agreed at first to the government carrying out the demolition, then asked to do it themselves, which immediately provoked Latin and Armenian protests. Impatient of the complications of the Status Quo (with its implicit link between repair and possession), Storrs disregarded the opposition and pressed ahead, using his own contractor. The Greeks were deeply unhappy, because they considered the wall to be theirs and feared a loss of rights. However, in the circumstances of 1918 they had no choice but to comply, saving face by wielding the first pickaxe. Storrs was content, arguing that if the wall 'fell not

---

24 Ronald Storrs, *Orientations* (London: Ivor Nicholson & Watson, 1937), 364-68.

25 Ibid., 330-31. Catholics are not permitted to conduct Mass in the main body of the Church of the Nativity and do so in their adjacent Church of St. Catherine's.

under the edict of a military despot, it would stand for ever.'[26] However, the Greeks never forgot this breach of their rights and became incorrigibly suspicious of government intentions.

In July 1920 civil government was established in Palestine and within a short time an Antiquities Ordinance was promulgated with the aim of protecting antiquities in the country. The ordinance set up a Department of Antiquities under a director responsible for administering the law, to be assisted by an international advisory board consisting of representatives of the major archeological schools in Jerusalem (including the Palestine Exploration Fund).[27] 'Antiquity' meant any construction or artifact produced earlier than the year 1700. Like the 1884 Ottoman Antiquities Law, the 1920 ordinance dealt extensively with archeological excavations and finds. A significant innovation was the special attention paid to preserving historical monuments. Since the department had no independent means to carry out repairs, it was authorized to hand over the care and conservation of historical sites to approved societies. Explicit mention was made of the role of the Pro-Jerusalem Society in caring for certain medieval buildings in Jerusalem.[28]

For our purposes, the key provision of the law was the exclusionary clause giving ecclesiastical or other religious bodies sole responsibility for sites in their possession or use. This included the Christian Holy Places. Here the Department of Antiquities had the duty of approving and inspecting repairs and additions. It could warn and encourage the communities about the state of their shrines but was given very limited powers of initiative or implementation.

There was a very good reason for this. Immediately after the British occupation of Palestine a number of disputes had arisen between the communities at the Holy Places. For example, Storrs had not been military governor a month before a brawl broke out between Greeks and Armenians in the Church of the Nativity over liturgical rights.[29] After long and careful consideration the British Government concluded that it had no interest in entangling itself in such religious disputes and

---

26   Ibid., 352-53; Waggott telegram, Sept. 17, 1918, in Jane Priestland, ed., *Records of Jerusalem, 1917-1971* (9 vols, Cambridge: Archive Editions, 2002), vol. 1, 234-36. For a politically incorrect account see C.R. Ashbee, *A Palestine Notebook* (New York: Doubleday, 1923), 14-15.

27   The first director was Professor John Garstang of the University of Liverpool, who helped draft the ordinance. See John Garstang, 'Eighteen Months' Work of the Department of Antiquities for Palestine July, 1920-December, 1921,' *Palestine Exploration Fund Quarterly Statement* (1922), 57-62.

28   Norman Bentwich, 'The Antiquities Law of Palestine,' *Journal of Comparative Legislation and International Law* 3rd series, no. 6 (1924), 251-54. A 1929 revised antiquities ordinance tightened up the law by adding substantial detail but this did not affect the limited power of the government to preserve church property.

29   Storrs, *Orientations*, 473.

that a commission should be appointed with the approval of the League of Nations to determine the rights and claims of the parties.[30]

As a result, the Mandate for Palestine, formally confirmed on July 24, 1922, contained two articles on the Holy Places that in the event were to prove incompatible. Article 13 conferred on Britain the responsibility for 'preserving existing rights' at the Holy Places, while Article 14 made provision for a special commission 'to study, define and determine' those rights and claims. Once it proved impossible to appoint such a commission, because the Catholic nations could not agree which of them should sit on it, the Palestine Government found itself saddled with enforcing rights that remained contested. In a historic 1924 order in council, moreover, the high commissioner, Herbert Samuel, decided that the local courts should not settle disputes relating to the Status Quo (since these had international ramifications). Rather, these should be referred to the government which would make an administrative ruling about current practice, noting that 'no change in the Status Quo was held to have occurred.'

To further complicate matters, every year British officials had to report on developments at the Holy Places to the Permanent Mandates Commission of the League of Nations. Here they were questioned by representatives of governments – Greece or Italy, say – supporting one side or the other in some ongoing dispute. In the circumstances, for all their commitment to preservation the authorities found themselves in a position not totally unlike that of the Turkish Government before the First World War. As the chief secretary of the Palestine Government, Colonel Stewart Symes, told the commission in 1926:

> The duty of the Administration, by the terms of the Mandate, was solely to preserve the status quo with regard to the Holy Places, and no provision had been made for changing that status quo, Difficulties as they arose were settled by negotiation with the parties concerned.[31]

Ironically, in one respect the British Government had even less room for maneuver than its Ottoman predecessor. Deeming the Holy Places to be waqf, the inalienable property of a religious trust, the Turks held themselves as sovereign to be the trustee of last resort (with what in Common Law would be considered rights of ownership) fully entitled in principle to take whatever action was necessary to preserve the fabric of the shrine. In opposition, the British Colonial Office ruled that the Mandate of the League of Nations did not legally confer on the Palestine Government 'any responsibility for, or rights to carry out repairs of non-Moslem religious buildings.' Nor was it in a position, it also ruled, to charge the three major

---

30 'Holy places etcetera in Palestine: origins of Article 13 of the Palestine Mandate,' Jan. 27, 1940, The National Archives, London (henceforth TNA) FO 371/23249.

31 League of Nations Permanent Mandates Commission, Minutes of the Ninth Session, 22nd meeting, June 22, 1926, C.405.M.144.

communities with responsibility for repairs in the Church of the Nativity.[32] The authorities then followed the post-1853 Ottoman practice of making only urgent minor repairs at their own expense.

During the Mandate, Britain did its best to do its duty at the Church of the Nativity. It could chalk up a number of successes: the edifice was thoroughly surveyed and remains of the original Roman basilica, including floor mosaics, discovered; Marshall MacMahon's tapestry was repaired, the grotto was cleaned and restored; column paintings were revealed. Unfortunately, because of the legal constraints on the sovereign, referred to above, the most serious structural defect in the entire church, the dilapidated roof, which was far from watertight, could only ever be temporarily patched up and not replaced. Neither George V nor George VI were ever able to emulate Edward IV.

With the onset of the rainy season in 1926 the roof was found to be leaking. As in the 1918 case of the demolished wall, the Armenians and Latins claimed the right to share the cost of the repair with the Greeks, on the grounds that they were co-proprietors of the basilica. Rebutting this claim, the Greeks insisted that they were sole owners and should bear the entire expense. Since the matter was of the utmost urgency, given that in the Jerusalem area 550 mm of rain falls on average in the rainy season of November to March – a heavier average monthly rainfall than Manchester – the authorities felt that they had no choice but to carry out temporary repairs themselves right away, suspending in the meantime the matter of payment.[33]

Time and again, for the rest of the Mandate, the leaky roof returned to haunt the Palestine Government. It was the curse of the Church of the Nativity. The rains of winter 1930-31 proved that the decay of the roof, and especially the lead sheathing, was spreading. In consequence, the narthex was flooded 'as if there was no roof at all,' while rainwater streamed down the walls of the building, doing irreparable damage to their priceless mosaics.[34] On inspection, the director of public works concluded that the condition of the roof was so bad that it 'may have to be entirely replaced.'[35] This diagnosis was confirmed by the architect and preservation expert William Harvey in a comprehensive survey of the entire church carried out in 1934. All the timbers of the main roof and aisle roofs were worm-eaten and some affected by dry-rot. The lead sheeting was in a very bad condition and in some places entirely rotted away. His recommendation was that the roof be entirely replaced.[36]

---

32   'Memorandum: Repair of the holy places in Palestine,' June 17, 1933; Minute, June 16, 1933, TNA CO 733/247/14.

33   Cust, *Status Quo*, 39-40.

34   Archbishop Evdoros to deputy district commissioner, March 10, 1931, ISA 4308/19- מ.

35   Campbell to Moody, Dec. 14, 1931, ibid.

36   Harvey, *Structural Survey*, 11-12, 14-15.

For this to happen, though, there would either have to be agreement between the communities on sharing the cost or else the government would have to pay. Unfortunately, all attempts during the Mandate to solve the dispute over ownership proved fruitless. For its part, the government, mainly on legal grounds, but also because of grave financial and political objections, declined to foot the bill for a new roof. In the absence of an answer to the problem of payment, the best that could be done was to plug holes in the roof with tarred felt. So in 1931, 1936, 1944 and 1947 this was the temporary and utterly inadequate expedient chosen. In 1944 the mosaics were again flooded after heavy rain. On the eve of the British departure from Palestine the district engineer yet again lamented the 'very poor condition' of the roof.[37]

Conclusion

Sadly, this story has no happy end. For all its good intentions, the Palestine Government under the Mandate failed to solve the critical problem of the Nativity roof. It never even attempted to preserve, let alone restore, the medieval mosaics in the basilica. And where the British Empire failed, the governments of Jordan and Israel in later years were unlikely to succeed. An examination of the church in 1997 for UNESCO found that as a result of the chronic dilapidation of the roof 'the majority' of the wall mosaics and paintings had been 'damaged beyond repair.' Extensive damage had also been done to the walls at some points.[38]

There is no question that the British authorities made an historic contribution to the preservation of antiquities in Palestine. Ancient buildings were saved that would otherwise have been lost. Theirs was an enlightened and progressive trusteeship that laid the foundations of scientific archeology and architectural preservation throughout the country. Where buildings in religious use were concerned, it was another matter. Restoration of the two single most important Christian edifices in the Holy Land, the Churches of the Holy Sepulchre and the Nativity, was fatally compromised by the legal constraints imposed by the Mandate of the League of Nations and the dead hand of the Status Quo.

Britain's moment in Jerusalem and Bethlehem began with bright optimism when Colonel Storrs, inspired by the ideals of the National Trust, demolished the 'loathsome wall' of the Nativity Church, as Charles Ashbee called it. Ironically, this romantic act proved to be the first and last glorious fling of the British authorities in the church. It aroused the everlasting suspicion of the powerful Greek Orthodox Patriarchate and so did nothing to advance cooperation in the long run. Ever after the Palestine Government found itself impeded by Greek obstructionism.

---

37   Newton to Director of Public Works, Jan. 31, 1944; Concannon to District commissioner, Mar. 31, 1947, ISA 28/1-מ.

38   F. Wenzel, S. Szaktilla, and H. Pliett, 'The Roof of the Church of Nativity in Bethlehem' (Paris: UNESCO, 2000), 7.

As a result, decay not preservation became the rule. The last thing that the Palestine Government could be accused of was indifference to the fate of the basilica. Unlike the Ottoman authorities British officials were emotionally deeply committed to saving the Holy Places. Unfortunately, they found themselves hamstrung by factors beyond their control. We are left with an object lesson in the helplessness of governments and professional preservationists when confronted by the determined hindrance of contentious religious forces.

Chapter 5

'The Shrine at Sulgrave':
The Preservation of the Washington
Ancestral Home as an 'English Mount
Vernon' and Transatlantic Relations

T.G. Otte*

[T]he union of the Empire rests on sentiment and not on calculation of apparent material interest.

W.M. Ramsay[1]

The concept of heritage is invested with multiple and complex meanings. Although it draws on history as an academic discipline, concerned with understanding the past, it transcends the confines of careful, evidence-based reasoning. It is rather an exercise in historical imagination, a re-imagining of a particular version of the past in the present. And it does so in a manner suited to the needs of the present and the anticipated future.[2]

That relationship between the present and the imagined past is central also to preservation. Indeed, it is at the root of the dialectic that characterizes all attempts to preserve the past and its relics. In its efforts to stem the progress of decay, to replace what has been lost, and to keep what remains, preservation creates an illusion of permanence and perpetuity. In preserving the remnants of the past, it transforms them by reinterpreting and recreating them in light of present concerns. Rather than containing the pressures of change, preservation is itself part of the process of change. In selecting some relics of the past as worthy of preservation, it privileges particular versions of the past and reinforces these through tangible

* I am grateful for their advice and encouragement to Melanie Hall, Erik Goldstein and Joanna Otte, and to Martin Sirot-Smith, former Curator at Sulgrave Manor, for allowing me access to the archives there.

1 Sir William Mitchell Ramsay, *The Imperial Peace: An Ideal in European History* (Oxford: Clarendon Press, 1913), 28.

2 The *locus classicus* is David Lowenthal, *The Past is a Foreign Country* (Cambridge: Cambridge University Press, 1985). See also the essays in Michael Hunter, ed., *Preserving the Past: The Rise of Heritage in Modern Britain* (Stroud: Alan Sutton, 1996); and the more polemical Robert Hewison, *The Heritage Industry: Britain in a Climate of Decline* (London: Methuen London, 1987).

monuments. The history of preservation, then, ought properly to be located in the broader political and socio-cultural contemporary contexts.[3] Tangible monuments of the past ought to be understood as part of any given period's political iconography. Indeed, unlike other forms of appreciating the past, the preservation of ancient buildings or historical monuments is an appeal to the senses, mobilizing hopes and providing a sense of belonging and unity.[4]

Preservation in Britain was, to a large degree, the product of late-Victorian campaigns to protect landscapes and buildings from the ravages wrought by the twin evils of industrialization and sprawling conurbations.[5] By international comparison, British efforts at preservation and restoration lagged somewhat behind. In typically British fashion, they were the product of antiquarian impulses, private initiatives and usually less than wholehearted government support. Yet, following the passing of the Ancient Monuments Protection Act in 1882 and its two subsequent amendments, and the foundation, in 1895, of the National Trust, the movement gathered pace.[6]

The high value accorded to the preservation of historical monuments and artefacts was not merely a byproduct of the contemporary, intellectual dominance of history, with all the trimmings of a distinct and increasingly well-organized academic discipline, endowed university chairs, dedicated scholarly journals, and specialist archives. It also reflected broader intellectual currents. It was an integral part of the secularization of the European mind in the nineteenth century. As the hold of the established Churches and their doctrines declined, history began to occupy a prominent place in Western life and thought.[7] Indeed, from the middle of century onwards, parallel to the secularization of moral and intellectual thought

---

3   For an instructive example see Charles Dellheim, *The Face of the Past: The Preservation of the Medieval Inheritance in Victorian England* (Cambridge: Cambridge University Press, 1982).

4   See the observations in George L. Mosse, *The Nationalization of the Masses: Political Symbolism and Mass Movements in Germany from the Napoleonic Wars Through the Third Reich* (New York: H. Fertig, 1975).

5   For useful discussions of some of the origins see Nikolaus Pevsner, 'Scrape and Anti-Scrape,' and Jane Fawcett, 'A Restoration Tragedy: Cathedrals in the Eighteenth and Nineteenth Centuries,' in Jane Fawcett, ed., *The Future of the Past: Attitudes to Conservation, 1174-1974* (London, 1975), 35-54, 75-116; and Peter Mandler, *The Fall and Rise of the Stately Home* (New Haven and London: Yale University Press, 1997), 22-68.

6   The Ancient Monuments Protection Acts 1882 and 1900, 45 & 46, Vict., c. 73 and 63 & 64 Vict., c. 34. For some discussion of the Trust see Merlin Waterson, *The National Trust: The First Hundred Years* (London: BBC Books, 1994), 14-59; Jennifer Jenkins and Patrick James, *From Acorn to Oak Tree: The Growth of the National Trust, 1895-1994* (London: Macmillan, 1994), 1-42; Graham Murphy, *Founders of the National Trust* (London: National Trust: 2002), 99-125.

7   For this transformation see Owen Chadwick, *The Secularization of the European Mind in the Nineteenth Century* (Cambridge: Cambridge University Press, 1995), 189-228. For some discussion of the growth of history as an academic discipline see John P. Kenyon,

there was a form of sacralization of the past and its relics. The presence of the past permeated many aspects of public life. Most public spaces were dominated by evidence of conscious 'national self-historicization'[8]: historic buildings were preserved, new ones built in 'historic' styles, and monuments extolling past achievements erected in prominent positions.

At one level, the campaign, in the years before the First World War, to preserve Sulgrave Manor, one of the ancestral homes of the Washington family, fits into this pattern of 'self-historicization'. Yet, in other respects, it offers an important variation on this theme. Whilst preservation has tended to be treated as part of the 'invented traditions' of individual nations,[9] the efforts to save and restore Sulgrave were international in character, and largely, but not wholly, independent of national political contexts. Their history provides a useful case study of the multifaceted and complex history of preservation.

Figure 5.1   The British Home of the Washingtons. Sulgrave Manor (Northants) and Its Associations

Source: Issued by the British-American Peace Centenary Committee, 1912, London

---

The History Men: The Historical Profession in England since the Renaissance (Pittsburgh: University of Pittsburgh Press, 1984), 144-99.

8   Joseph Leerssen, *National Thought in Europe: A Cultural History* (Amsterdam: Amsterdam University Press, 2006), 189-90.

9   The *locus classicus* is Eric Hobsbawm and Terence Ranger, eds, *The Invention of Tradition* (Cambridge: Cambridge University Press, 1982). See also Roland Quinault, 'Westminster and the Victorian Constitution,' in *Transactions of the Royal Historical Society*, 6th series, ii (1992), 79-104.

i

The village of Sulgrave lies on a ridge in the undulating pasture land of the South Westernmost tip of Northamptonshire, some seven miles North East of Banbury. Built in the warm gingerbread Marlstone of that part of the county, its houses are threaded along two near-parallel streets, with an enclosed space in its middle and dominated at either end by the manor house and the church with the adjacent remnants of Norman earthwork defences.[10] By the middle of the nineteenth century, the manor house had fallen into considerable decay. The Western half of the building had been pulled down sometime before 1780, and since then a succession of tenant farmers had done little to improve what remained. Indeed, one visitor in the early 1880s described it as 'little more than a quaint and interesting ruin.'[11] Buried in the rural isolation of the Midlands, it was not on account of its artistic or architectural merits that Sulgrave Manor attracted public attention in the years before 1914, but because of its historical associations. Of the former it had none; it had never been more than a modest gentry dwelling. The latter rested on its Washingtonian ancestral connections.

Although originally of North-country descent, in Tudor times a branch of the Washington family had established itself in Northamptonshire under Lawrence Washington, a prosperous wool-stapler and one-time Mayor of Northampton. His wealth and his judicious connections more especially with the Spencers, the local magnates, enabled him to take advantage of the dissolution of the monasteries by purchasing the Manor of Sulgrave from the Crown in 1539. The estate remained in the hands of his direct descendants for the next 120 years until, in 1659, it was sold out of the family.[12]

For much of the nineteenth century, Sulgrave's connection with the great-grandfather of the first President of the United States had been little more than an antiquarian curiosity. As official relations between the two countries improved from the 1870s onwards, however, interest in Washington's ancestral home grew. The Washington family pew and Lawrence's memorial brass in Sulgrave church attracted some attention. Of greater interest still was Lawrence's coat of arms, carved in the spandrels of the moulded stone arch of the entrance porch to the manor house, with its two bars and three mullets, the original 'stars and stripes'.[13]

---

10   Nikolaus Pevsner and Bridget Cherry, *The Buildings of England: Northamptonshire* (London: Yale University Press, 1973), 20, 421.

11   E.W. Tuffley, 'The Origin of the Stars and Stripes,' *St. Nicholas* (Nov. 1, 1883), 69.

12   For a history of the manor and the house see H. Clifford Smith, *Sulgrave Manor and the Washingtons: A History and Guide to the Home of George Washington's Ancestors* (New York: Macmillan, 1933).

13   Tuffley, 67-72; Sir H. Dryden, 'The Washington House at Sulgrave,' *Northamptonshire Notes and Queries* (Apr. 1885); C. Edwards, 'The English Home of the Washingtons,' *Atalanta* (Jan. 1, 1898), 213-18.

As so often in the history of preservation, it was the spectre of dismantling by foreign elements that provided a first stimulus for efforts to save Sulgrave. In 1902, the local *Banbury Guardian* reported that the manor house had been purchased by a wealthy, but anonymous, American, and that it was to be taken down 'brick by brick and stone by stone' to be re-assembled as part of the 1904 World Fair in St. Louis, Missouri.[14] The 'Sulgrave scare' proved to be without foundation. But it was sufficient to give rise to a scheme for the National Trust to acquire and preserve the place.[15] Ultimately, this first scheme for the preservation of Sulgrave came to nothing, and it was not until 1910/11 that more determined efforts were made to save it.

As in 1902, it was an external stimulus that galvanized preservation efforts. This time, however, it was not the threatened intervention of quasi-demonic foreign plutocrats that made Sulgrave the target of a determined, well-organized and broadly based preservationist campaign. It was rather the result of a more benign cultural, but no less powerful, influence. Historians have commented on the role of centennial commemorations for gauging contemporary attitudes. Centenary celebrations helped to revive, perpetuate and reinforce public perceptions of past events and people in collective memory. They were fairly infrequent before 1850, but increased in size and numbers by the end of the long nineteenth century, so much so that a 'cult of the centenary' had become established throughout the wider Western world.[16] Such events were ephemeral affairs. Some of them, however, left a lasting legacy. And one such legacy was the preservation of Sulgrave Manor.

The growth of centennial commemorations was rooted in the increased interest of the Victorian and Edwardian public in the past. This was no detached interest, but rather reflected very much contemporary concerns. It was scarcely surprising, then, that the 'Great Rapprochement' in the relations between Britain and America should find public expression in a centenary celebration.[17] By 1910, attention

---

14   'Washington Genealogy,' *The New York Times* (Sept. 14, 1902); Smith, *Sulgrave Manor*, 80. For an interesting parallel, the 1910 'Tattershall scare,' see Mandler, *Fall and Rise*, 184-89.

15   Melanie Hall, 'The Politics of Collecting: The Early Aspirations of the National Trust, 1883-1913,' in *Transactions of the Royal Historical Society*, 6th series, xiii (2003), 356. For the Trust's early interest in manor houses see also G.M. Young, 'The Manor House,' in James Lees-Milne, ed., *The National Trust: A Record of the First Fifty Years* (London: B.T. Batsford, 1945), 55-60.

16   Eric Hobsbawm, 'Mass-Producing Traditions: Europe, 1870-1914,' in Hobsbawm and Ranger, *Invention of Tradition*, 281. For a detailed discussion of the growth of this phenomenon see Roland Quinault, 'The Cult of the Centenary, c. 1784-1914,' *Historical Research*, vol. 76, no. 171 (1998), 303-23. For case studies see also Iain Pears, 'The Gentleman and the Hero: Wellington and Napoleon in the Nineteenth Century,' in Roy Porter, ed., *Myths of the English* (Cambridge: Polity Press, 1992), 216-36; and Andrew Lambert, *Nelson: Britannia's God of War* (London: Faber, 2004), 339-62.

17   For a discussion of Anglo-American relations see Bradford Perkins, *The Great Rapprochement: England and the United States, 1895-1914* (London: Gollancz, 1969); and

had come to focus on the forthcoming anniversary of the 1812-14 war, the last Anglo-American armed conflict. It was a squalid little war, fought over American insistence on maritime rights and British security concerns along the Canadian frontier, and exacerbated by British tactlessness and American bellicosity.[18] But it had ended in a draw, and this made it all the easier to celebrate its centenary. Whilst the outbreak of the war did not suggest itself as an obvious object of celebrations, its conclusion did. The 1814 peace treaty of Ghent left unresolved a number of issues that had caused the war, yet it laid the foundations of the future rapprochement between Britain and her former colonies. Indeed, the very fact that the peace treaty had not provided for a comprehensive final settlement eventually led to a series of bilateral agreements on specific points of contention.[19] In the minds of successive British foreign secretaries and American secretaries of state, arbitration had become the preferred tool of settling disputes between the two nations.

In turn, the central role of arbitration in nineteenth-century Anglo-American relations helps to explain the importance that more especially international lawyers attached to the forthcoming centenary. In 1910, at the annual Lake Mohonk conference of international lawyers and arbitration activists, a resolution was passed to form a committee for the purpose of organizing nationwide celebrations of the peace centenary. The committee, in fact, never met, but it was absorbed, in the spring of 1911, in the 'National Committee for the Celebration of the One Hundredth Anniversary of Peace among English-Speaking Peoples', under the chairmanship of John A. Stewart, a prominent New York Republican politician.[20]

The committee enjoyed the backing of powerful social and political figures from across the political spectrum. The former US President Theodore Roosevelt and the philanthropist-steel magnate Andrew Carnegie acted as its public figureheads.[21] Among its honorary chairmen were Roosevelt's Secretary of State,

---

Kathleen Burk, *Old World, New World: The Story of Britain and America* (London: Little, Brown, 2007), 411-36.

18   Bradford Perkins, *Castlereagh and Adams: England and the United States, 1812-1823* (Berkeley: University of California Press, 1964); C.P. Stacey, 'The Defence of Upper Canada,' in Morris Zaslow, ed., *The Defended Border: Upper Canada and the War of 1812* (Toronto: Macmillan, 1964), 11-20.

19   For a contemporary assessment see Henry E. Richards, *The Progress of International Law and Arbitration: An Inaugural Lecture delivered at the University of Oxford* (Oxford: The Clarendon Press, 1911), 13-14.

20   Pope to Perley, Nov. 4, 1911, Borden Mss, National Archive of Canada/Archive Nationale du Canada, Ottawa, MG26H (1a), vol. 17. For the Lake Mohonk conferences see also C. Reid, 'American Internationalism: Peace Advocacy and International Relations, 1895-1916' (PhD thesis, Boston College, 2005).

21   Carnegie was a prominent advocate of international arbitration, offering a 'pile [of money]' for the 'holy work,' Carnegie to Bryce, Mar. 25 and Apr. 16, 1911, Bryce Mss, Bodleian Library, Oxford, MS Bryce USA 4. See also Peter Krass, *Carnegie* (Hoboken: J. Wiley & Sons, 2002), 486-99.

Elihu Root, Grover Cleveland's vice-president Adlai E. Stevenson, the former presidential contender and future Secretary of State William Jennings Bryan as well as Joseph Hodges Choate, until 1905 US ambassador to Britain. Although a private initiative, the prominent role of such public figures lent the committee a quasi-official character. Indeed, its sister organizations in Britain and Canada were subsequently to adopt a similar format.

Like other centenary commemorations of the period, the tentative programme drawn up by the committee in early 1911 was 'to a large degree popular and educational in its character, rather than formal and official.'[22] The committee chose a deliberately historical approach, with emphasis on scholarly lectures, historical exhibitions, and other educational activities, such as issuing school textbooks. Its programme, however, also had political content in that it advocated an unlimited Anglo-American arbitration treaty, a not uncontentious issue in US politics at that moment. Among its more tangible, popular elements were celebrations in Ghent and London, to be held on Christmas eve 1914, the exact anniversary of the peace treaty, 'and [to be] observed throughout all English-speaking lands.' There were also plans for an international park along the Niagara frontier with a free memorial bridge across the river. Finally, the programme envisaged the purchase of Sulgrave Manor by popular subscription in America and Britain 'as a visible monument to the cordial relations existing between the two great branches of the English-speaking peoples.'[23] The project was 'original to our Movement, and ... gave it both colour and impetus.'[24] Its focus on the Washington connection, moreover, reflected the growing cult of saving places associated with the first president, which had emerged in the United States in the aftermath of the civil war.

The committee made only slow progress in 1911, its enthusiasm apparently exhausted after the publication of its prospectus. Early in the year, Stewart, very much the moving spirit behind the committee's work, made contact with James (later Viscount) Bryce, Britain's ambassador at Washington, with a view to involving him in the centennial preparations. The ambassador was an obvious choice. A distinguished and respected late Victorian scholar-turned-politician, he had served in several Liberal cabinets, though had never quite reached the top flight. In many ways, his appointment, in 1907, as ambassador to the United States marked the culmination of his public career. Bryce was very much a public ambassador. The publication, in 1888, of his *American Commonwealth* had already established him on both sides of the Atlantic as the foremost British authority on the American

---

22   *One Hundredth Anniversary of Peace among the English-Speaking People, 1814-1914: A General Prospectus of Plan and Purpose of the National Committee for Its Celebration* (New York, s.a. 1911), 6 (copy in Borden Mss, vol. 26).

23   Ibid., 3-4. See also *Penny Illustrated News* (Aug. 12, 1911), though this placed Sulgrave in Yorkshire!

24   Perris, 'Milestones of Anglo-American Friendship' (unpublished TS), Perris Mss, Stanford University Library, fo. 24. Perris was secretary to the British Peace Centenary Committee.

republic. It also established the intellectual origins of the later twentieth-century Anglo-American 'special relationship'.[25] As the then Permanent Under-secretary of the Foreign Office later noted somewhat maliciously, Bryce 'had the quality of liking to make long and rather dull speeches on commonplace subjects ... popular with the American masses.'[26] But it meant that, throughout his six years at Washington, the indefatigable ambassador travelled the length and breadth of the United States, addressing universities, bar associations, civic dignitaries, and ecclesiastical conferences. Throughout that time, too, he encouraged Anglo-American preservation efforts. No one, observed Prime Minister H.H. Asquith at Bryce's retirement in 1913, had 'contributed more largely to create and foster this temper [of cordiality] between the two great kindred peoples.'[27] No ambassador also had advocated the cause of arbitration more assiduously than Bryce.[28]

Whatever Bryce's genuine enthusiasm for Anglo-American unity, he was nevertheless a cautious diplomatic operator. The approach by the US committee lacked precision, he decided; there had been no further communications from that quarter; and the focus of the proposed celebrations ought to be broadened to include Canada.[29] This latter idea chimed in with Stewart's own thinking. In the spring of 1911, he had suggested to the influential French-Canadian businessman and Senator Raoul Dandurand that Canada should be involved in the centenary movement. Nothing definite was decided at the time, but the participation of the United States' northern neighbour was now firmly on the agenda. This was to change the dynamics of the centenary preparations as well as the tone and content of the planned commemorative events in 1914.[30]

On the Canadian side, the matter was taken by Sir Joseph Pope, the capable and experienced Under-secretary of the embryonic Department of External Affairs, and George H. Perley, a millionaire lumber baron, former chief whip and trusted lieutenant of the leader of the Canadian Conservatives, Robert Laird Borden, in whose newly formed government he was a minister without portfolio. For his part, Pope stressed the imperial nature of the 1814-1914 event, and insisted that the

---

25   For studies of Bryce's American experiences and connections see Edmund Ions, *James Bryce and American Democracy, 1870-1922* (London: Macmillan, 1968); and Hugh Tulloch, *James Bryce's American Commonwealth: The Anglo-American Background* (Woodbridge: Boydell Press, 1988).

26   Charles, Lord Hardinge of Penshurst, *Old Diplomacy* (London: J. Murray, 1947), 132. For a more balanced judgement see Keith Robbins, 'History and Politics: The Career of James Bryce,' *Journal of Contemporary History*, vol. 7, no. 3-4 (1972), 37-52.

27   As quoted in Bryce obituary, *The Times* (Jan. 23, 1922).

28   See for instance his address at the 1909 Lake Mohonk conference, 'Allegiance to Humanity,' in James Bryce, *University and Historical Addresses: Delivered during a Residence in the United States as Ambassador of Great Britain* (London: Macmillan, 1913), 249-63.

29   Bryce to Pope, Oct. 31, 1911, and Pope to Perley, Nov. 4, 1911, Borden Mss, vol. 17. For Bryce's views on Anglo-Saxon unity see Tulloch, 44-48.

30   Pope to Perley, Nov. 4, 1911 (second letter of that date), Borden Mss, vol. 17.

imperial government in London be consulted.[31] This was not only constitutionally proper, it also reflected Canada's somewhat ambiguous semi-autonomy in her external affairs. The current state of Canada's relations with her southern neighbour, moreover, posed further practical problems. For years, US-Canadian relations had been bedevilled by the difficulty of arriving at a satisfactory, reciprocal trading arrangement. In early 1911, a comprehensive agreement had been reached. But this was then sabotaged by a group of businessmen, spearheaded by the Toronto financier Sir Edmond Walker, who perceived their vested interests in transport and commerce to be threatened. Their anti-reciprocity manifesto argued that the commercial agreement would 'weaken the ties which bind Canada to the Empire ... and make it more difficult to avert political union with the United States.'[32] The campaign defeated the agreement, and paved the way for Borden's victory at the polls in September 1911. But it had also highlighted the strength of imperial sentiments in Canada as much as the continued volatility in US–Canadian relations.[33]

It was to Walker, Dandurand and the influential lawyer-politician Sir Alexandre Lacoste that Pope and Perley turned for advice on the planned peace celebrations. They all agreed that, '[i]n view of recent events, ... [a joint US-Canadian meeting] could not fail to be embarrassing all round, and the interchange of compliments on such occasions somewhat hollow and insincere.' At the same time, it seemed impolitic to 'turn the cold shoulder to them in this business.' Prominent public figures should, therefore, be approached to form a Canadian centenary committee in case matters progressed further. The committee was to be of a strictly private and unofficial nature so as not to bind the government in Ottawa. Finally, it was agreed that Britain 'should properly take an active part' in the planning and preparations for the centennial commemorations.[34] Even so, on 4 June 1912, the Canadian Peace Centenary Association was established, with Walker as its president.[35]

The centenary initiative now gathered pace. In Ontario, scene of Borden's recent triumph at the polls, talk of centenary celebrations was 'exciting a certain amount of interest out in the Niagara Peninsula', and it seemed sensible, on electoral grounds, to harness such popular sentiments.[36] South of the border, the American

---

31 Pope to Perley, Nov. 4, 1911, ibid. For Pope's role in the creation of the Department of External Affairs see J. Eayrs, 'The Origin of Canada's Department of External Affairs,' in Hugh L. Keenleyside, ed., *The Growth of Canadian Policies in External Affairs* (Westport: Greenwood Press for Duke University, Commonwealth-Studies Center, 1960), 16-18.

32 As quoted in Robert C. Brown and Ramsay Cook, *Canada, 1896-1921: A Nation Transformed* (Toronto: Mclelland and Stewart, 1974), 181.

33 For a detailed discussion see Robert C. Brown, *Robert Laird Borden: A Biography* (2 vols, Toronto: Macmillan, 1975-77), vol. 1, 173-92.

34 Quotes from Pope to Perley (confidential), Nov. 10, 1911, and Perley to Borden, Nov. 21, 1911, Borden Mss, vol. 17.

35 Hamilton to Perley, Mar. 24, 1913, Borden Mss, vol. 26.

36 Pope to Perley (private), Nov. 22, 1911, Borden Mss, vol. 17. Pope's informant was William Manly German, an anti-reciprocity Liberal, who represented an Ontario riding

committee was considerably enlarged. The mayors of New York and other big cities joined at the end of 1911, and a separate committee was established in New York State to organize the centenary celebrations along the Niagara frontier.[37] By early 1912, bills of appropriation, sponsored by the American committee, were before Congress and the New York state legislature for an aggregate appropriation of $8 million in support of the projected 1914 centenary events.[38]

At the same time, and once again in response to representations by the US centenary movement, a British Peace Centenary Committee was set up in December 1911 along similar lines as its sister organization in America. As was the case with the latter, its members were well-known Society or political figures from the fringes of Westminster. This made it all the more important to maintain the pretence of the committee's entirely unofficial nature. There was to be 'no reference to politics'; and the committee's sole focus was to be on 'celebrating in the most suitable way the one hundred years of peace and [on] promoting good feeling between the United States and the British Empire, without allowing it to be used in any way to further controversies or other ideas – such as universal disarmament – no matter what their intrinsic merits might be.'[39] The committee's Secretary, Harry Shaw Perris, nevertheless provided a link to the nascent peace and arbitration movement in Britain. A former Unitarian minister and author of an anti-military tract, he had been secretary to the National Peace Council until 1909 and was well-connected in pacifist circles, not least also through his brother, the journalist George Herbert Perris, a leading light in the International Arbitration and Peace Association.[40] Much of the organizational work of the committee, however, was carried out by its honorary secretary, Robert Donald, and Harry E. Brittain, the chairman of its Dominions sub-committee. The former was the distinguished editor of the Liberal-leaning *Daily Chronicle*, and enjoyed excellent contacts among members of the Asquith government.[41] Brittain, a Unionist-supporting gadfly journalist, was actively involved in promoting the causes of the British Empire and of Anglo-American unity. He was one of the founders of the Pilgrims Club, had played a minor role in the Tariff Reform League, and

---

in the Canadian House of Commons.

37   Stewart to Borden, Dec. 4, 1911, Borden Mss, vol. 26; 'Peace Centenary Celebrations,' *The Times* (Nov. 21, 1911).

38   Stewart to Borden, Jan. 12 and 31, 1912, Borden Mss, vol. 17.

39   Quotes from Stewart to Borden, Dec. 14, 1911, and Perley to Borden, Feb. 5, 1912, ibid.

40   41 Intriguingly, on leaving the NPC, Perris worked for the Shakespeare Memorial Committee, see Martin Ceadel, *Semi-Detached Idealists: The British Peace Movement and International Relations, 1854-1945* (Oxford: Oxford University Press, 2000), 176; and Paul Laity, *The British Peace Movement, 1870-1914* (Oxford: Clarendon Press, 2001), 133-34, 179-81.

41   Donald obituary, *The Times* (Feb. 18, 1933); Smith, *Sulgrave Manor*, 205.

had helped to organize the first Imperial Press Conference in 1909.[42] Other well-known members of the committee were the former Liberal MPs Thomas, Lord Shaw of Dunfermline, a Scotch lawyer, and Philip Stanhope, Lord Weardale. A. Shirley Benn, *quondam* British vice-consul at Mobile, Alabama, and now MP for Plymouth, represented the Unionist opposition. The financier and former Liberal MP Sir Weetman Pearson, recently elevated to the peerage as Lord Cowdray, who had substantial commercial interests in North and Central America, provided a link to the City of London. This link was reinforced by the two honorary treasurers, Nathan Meyer, 1st Baron Rothschild, and John Baring, 2nd Lord Revelstoke.

The driving force behind the committee's early work was its first chairman, Albert, 4th Earl Grey of Howick. A traditional Whig, Grey had parted company with his ancestral party following Gladstone's embracing of Home Rule, and since then had steadily shifted to the right.[43] Under Conservative governments he had held senior colonial posts, first in Southern Rhodesia, and then, from 1904 to 1911, as Governor-General of Canada. Grey threw himself into his new role with gusto. He habitually thought on the grandest scale,[44] and his original plans for the centennial celebrations were no exception. But they also revealed a darker undercurrent that influenced the early work of the committee. The organizing principle behind Grey's thinking was that of Anglo-Saxon racial unity. And the forthcoming anniversary was to furnish the opportunity to celebrate it. To that end, he informed Prime Minister Borden in early 1912, the committee had discussed the purchase of the Crystal Palace and its grounds. No definite plans had been formulated

> but our present contemplation is to raze the glass Palace to the ground and to erect thereon some monument, which, 500 years hence, will appeal to all men of British and American descent, and causes them to a higher opinion of the greatness and conception and of the boldness of execution of Britishers in the early days of King George V.

As Governor-General, Grey had played a key role in preserving the Plains of Abraham as a National Park. His aim then had been the 'imperializing' of the site of General Wolfe's defeat of the French in North America. Now, as chairman

---

42   Brittain obituary, *The Times* (July 10, 1974). See also Sir Harry Brittain, *Pilgrims and Pioneers* (London: Hutchinson & Co., 1946), 103-6, 197-223; Andrew S. Thompson, *Imperial Britain: The Empire in British Politics, c. 1880-1932* (London: Longman, 2000), 75-76.

43   Donald Southgate, *The Passing of the Whigs, 1832-1886* (London: Macmillan, 1965), 408-11.

44   As president of the Royal Colonial Institute, he planned offices '50% greater than the floor area of St. Paul's,' Grey of Howick to Wrench, Mar. 28, 1913, Wrench Mss, British Library, Add.Mss. 59542.

of the peace centenary committee, he drew on that experience. The 'Centennial of Peaceful Fellowship between the British and American Peoples' had to be used 'for the purpose of imperializing and on a racial and not [a] national basis, the Crystal Palace grounds ..., adapted for Race monumental purposes.' Grey envisaged the redevelopment of the grounds on a grand scale. The 200-acre site around Sydenham Hill should be 'treated on the assumption that the work would take 100 years to complete and would cost 100 million sterling.' Westminster and St. Paul's Cathedral were 'the true Valhalla of our race ... but it should not prevent us from seizing the present opportunity from securing the Crystal Palace grounds ... and holding it in trust not only for the Empire, but for the English-speaking world.'

With his grand scheme for the 'Race Imperialization of the Crystal Palace' Grey was not merely paying homage to an abstract principle. It was also a reaction to the crisis of the Edwardian state and society. The recent strike wave and the growing strength of the Labour movement, Grey feared, threatened to plunge Britain into 'a state of civil war against those forces which are endeavouring to substitute collective for private ownership.' In the incipient struggle between the forces of 'individual liberty and voluntary co-partnership between labour and capital, and those who champion the solidarity of labour and the tyranny of state socialism', Grey's scheme was meant to harness the energies of the former.[45]

Grey's letter to Borden was significant on a number of counts. In the first instance, his chairmanship provided the necessary energetic impulse for the activities of the British committee. Further, by focusing on the racial unity between the British people in the wider Empire and the American people, Grey set the tone for the committee's future deliberations and plans. In so doing, he amplified on a theme that had come to play a significant role in late-Victorian and Edwardian political discourse. 'Race' as a category was well established already in mid-Victorian thought.[46] The roots of 'Anglo-Saxonism', the notion of a racial and cultural bond between Britons and Americans, certainly extend to this period, as evidenced by the writings of Goldwin Smith.[47] By the later 1890s, it had come to imbibe Social Darwinian precepts and aspects of the evolutionary, social scientific theories of such diverse thinkers as Benjamin Kidd or Karl Pearson; and it stipulated that Anglo-Saxons were innately superior to other peoples. Above

---

45   Grey of Howick to Borden, Feb. 27, 1912, Borden Mss, vol. 6.

46   For some discussion of this see Peter Mandler, '"Race" and "Nation" in Mid-Victorian Thought,' in Stefan Collini, ed., *History, Religion and Culture: British Intellectual History, 1750-1950* (Cambridge: Cambridge University Press, 2000), 224-44.

47   Goldwin Smith, *Three English Statesmen: A Course of Lectures on the Political History of England* (London: Macmillan, 1868), 168-69; Goldwin Smith, *The United States: An Outline of Political History* (New York: Macmillan, 1983), v. See also Elisabeth Wallace, 'Goldwin Smith on England and America,' *American Historical Review*, vol. 59, no. 4 (1959), 884-94. See also Reginald Horsman, 'Origins of Racial Anglo-Saxonism in Great Britain before 1850,' *Journal of the History of Ideas*, vol. 37, no. 3 (1976), 387-410.

all, it had established itself as a pervasive intellectual force among the political elites in both Britain and North America.[48] And it was a notion which could unite 'diehards' like Grey and progressive Liberals like Bryce, in whose mind issues of racial unity were indissolubly linked with the civilizing mission of imperial trusteeship.[49]

Grey's explicit reference to Britain's domestic malaise, moreover, highlights the importance of the wider social and political context in the history of the centennial movement. There was an indubitable, inherent tension between Grey's espousal of racial unity and his staunch defence of individual liberty against the forces of collectivism. Even so, his 'race monumental' scheme was meant to emphasize an immutable element of strength amidst the turmoil and uncertainty of Britain's current internal problems. Imperial reform, based on Anglo-Saxon unity, thus, was meant to offer a panacea for all of Britain's domestic ills. To that extent, Grey's thinking reflected also the deeper malaise of the Edwardian Right, and laid bare the fissures within popular conservatism.[50] And it is in this context that the efforts of the peace centenary committee to preserve Sulgrave Manor ought to be seen.

Throughout the first half of 1912, Grey continued to pursue the Crystal Palace scheme for the 'permanent glorification of our English-speaking Brotherhood' and 'the commemoration of the racial unity of the English-speaking peoples of the world.'[51] Soon, however, practical problems in Anglo-American relations made it necessary to put matters on hold. For one thing, the forthcoming presidential elections in November 1912 and the prolonged transition period from the Taft presidency to the newly elected administration of Woodrow Wilson paralyzed the

---

48   See for instance, Theodore Roosevelt, *Biological Analogies in History: The Romanes Lecture 1910* (Oxford: Clarendon Press, 1910), 22-23. For a detailed account of this see Stuart Anderson, *Race and Rapprochement: Anglo-Saxonism and Anglo-American Relations, 1895-1904* (East Brunswick: Associated University Press, 1981), 26-61 and 73-94; also Bernard Semmel, *Imperialism and Social Reform* (London: G. Allen and Unwin, 1960), 20-24; Serge Ricard, *Théodore Roosevelt: Principes et pratique d'une politique étrangère* (Aix-Marseilles: Aix-en-Provence Publ. de l'Univ. de Provence, 1991), 244-48; and W. Mock, 'The Function of "Race" in Imperialist Ideologies: The Example of Joseph Chamberlain,' in Paul M. Kennedy and Anthony Nicholls, eds, *Nationalist and Racialist Movements in Britain and Germany before 1914* (London: Macmillan, 1981), 190-203.

49   For a discussion of Bryce's complex views on this see Tulloch, 44-48, 188-208.

50   For some of this see G.R. Searle, 'The "Revolt from the Right" in Edwardian Britain,' in Kennedy and Nicholls, 21-39; Frans Coetzee, *For Party or Country: Nationalism and the Dilemmas of Popular Conservatism in Edwardian England* (Oxford: Oxford University Press, 1990), 155-58; and E.H.H. Green, *The Crisis of Conservatism: The Politics, Economics and Ideology of the British Conservative Party, 1880-1914* (London: Routledge, 1996), 194-206.

51   Quotes from Grey of Howick to Borden, Apr. 6, 1912; and to Knollys, (confidential), Apr. 1, 1912, Borden Mss, vol. 6.

activities of the centenary committees on both sides of the Atlantic.[52] For another, uncertainty over the new president's attitude towards Britain and the emergence of a number of contentious issues in Anglo-American relations made London tread warily in the matter of joint centenary preparations. The decision, in March 1912, by the US Senate to block ratification of a general Anglo-American arbitration treaty caused consternation at the Foreign Office.[53] And so did American plans to exempt American coastal shipping from toll charges on the near-complete Panama Canal. The planned exemption flouted the provisions of the 1900 Hay–Pauncefote treaty, which had stipulated that US and British merchant vessels passing through the canal should be treated on terms of equality. For the British and Canadian governments, the contentious plans were as much a matter of legal principle as of commercial interests.[54] Certainly, the plans suggested that recent centennial rhetoric in no way reflected official US policy; manifest American interests were always likely to trump sentiments of Anglo-American amity.

The toll question was left hanging fire during the presidential elections. Both Taft and Wilson equivocated on it. But the considerable vote for the Progressive movement of the renegade Republican Theodore Roosevelt, who had campaigned for an exemption of US shipping, meant that the issue continued to cause a considerable headache for policy-makers in London, Ottawa and Washington. And this affected the work of the centenary movement.

At the end of 1912, Borden returned to the issue. No great progress had been made in the preparatory work of the US committee, he noted, and none in Britain. The Canadian Association, by contrast, was eager to move forward; and the provincial governments of Ontario and Quebec had stressed the 'importance of a worthy and notable celebration of the century of peace.' In consequence, Borden had set up a high-powered sub-committee of the Canadian Privy Council, chaired by his trusted lieutenant Perley. The premier impressed upon the imperial government the desirability of joint action by London and Ottawa. The forthcoming anniversary, he argued, 'will be especially important for a joint celebration by the two great kindred nations, which, if carried out in a moderate

---

52   Grey of Howick to Borden, June 7, 1912, Borden Mss, vol. 6; tel. Bryce to Grey (no. 11), Jan. 19, 1913, The National Archives, London (hereafter TNA), FO 371/1856/2857.

53   Tel. Bryce to Grey (no. 54), 24 Apr. 1912, *BD* viii, no. 514, Mitchell-Innes to Borden, June 28, 1912, Borden Mss, vol. 24; for some of the background see Ions, *Bryce*, 233-37; see also T.G. Otte, *The Foreign Office Mind: The Making of British Foreign Policy, 1865-1914* (Cambridge: Cambridge University Press, 2011), 368-69.

54   Borden to Grey, July 26, 1912, and reply, Aug. 1 1912, Borden Mss, vol. 231; for some further discussion see P.A.R. Calvert, 'Great Britain and the New World, 1905-1914,' in F.H. Hinsley, ed., *British Foreign Policy under Sir Edward Grey* (Cambridge: Cambridge University Press, 1977), 388-91; Walter LaFeber, *The Panama Canal: The Crisis in Historical Perspective* (New York: Oxford University Press, 1978), 52-62.

and dignified manner, should greatly assist in bringing these two countries into closer and warmer relationship.'[55]

At the same time, the American committee extended an invitation to its British sister organization to attend a joint conference in Washington in April 1913, at which a detailed programme for the centenary commemorations was to be finalized.[56] Borden's intervention and the latest initiative by the US committee shifted the focus on to the official attitude of the British government. As the centenary question affected official relations between Britain and America, the Colonial Office turned the matter over to the Foreign Office. There, Borden's communication was the first the department had heard of the Canadian Association. But it seemed 'undesirable that such a committee should act independently of Earl Grey's [British] committee', argued Gerald Spicer, the senior clerk in the Foreign Office's American department. His fellow clerk Rowland Sperling added 'that the zeal of the US committee tends to outrun its discretion and that many of their original suggestions for the celebrations have had to be discouraged owing to the controversial questions which they might raise.'[57]

London's reluctance to take the lead was reinforced by ambassador Bryce. Little interest was taken in the United States in the movement's activities, he reported. Stewart and Humphreys, the chairman of the New York state branch of the US committee, were 'both worthy and respected men, but chiefly known as Republican politicians.' Their programme was too ambitious, yet their plans not detailed enough. The fate of the appropriation bill, now pared down to US$7 million, was uncertain. Finally, the failure of the Taft administration's reciprocity and arbitration legislation was 'a visible setback to the sentiment of international friendliness among politicians here.' President-elect Wilson was likely to be sympathetic to the movement's aims, 'unless Irish politicians get at him.' Little could be done, Bryce argued, until after Wilson's inauguration in early March 1913.[58]

The ambassador fully appreciated the potential political utility of the planned celebrations. The recent reciprocity row had been a blow for US–Canadian relations. It was therefore all the more important that the centenary celebrations should take place. Earl Grey's committee should visit the United States in the spring, and the Canadian Association should take 'prompt action' as soon as a programme had been finalized.[59] At the Foreign Office, however, doubts remained.

---

55   Borden to Prince Arthur of Connaught, Dec. 11, 1912, TNA, FO 371/1856/1355; Perley to Borden (confidential), Mar. 25, 1913, Borden Mss, vol. 26.

56   Perris to Grey of Howick, Jan. 8, 1913, Borden Mss, vol. 26.

57   Quotes from mins., Spicer and Sperling, Jan. 10 and 14, 1913, on Lambert [Colonial Office] to Foreign Office, Jan. 9, 1913, TNA, FO 371/1856/1355; also Just [Assistant Under-secretary, CO] to Mallet (confidential), Feb. 12, 1913, ibid./6922.

58   Bryce to Grey (no. 14, confidential), Jan. 20, 1913, min. Spicer, Jan. 29, 1913, TNA, FO 371/1856/4180.

59   Bryce to Grey (no. 31), Feb. 10, 1913, TNA, FO 371/1856/7741.

Recent press reporting suggested that the incoming Democrat administration was less friendly towards Britain than its predecessor. This made it 'more important than ever to refrain from any governmental action till the situation is cleared.'[60]

Indeed, with the Isthmian canal tolls and the future of the arbitration treaty still hanging fire, British diplomacy remained generally passive.[61] The political situation in Washington remained 'rather mixed up', observed Alfred Mitchell-Innes, the embassy counsellor there. American public opinion was 'very sensitive on the question of our interference with what they believed to be a matter of domestic policy, even if we have a treaty right to do so.'[62] Bryce's successor, Sir Cecil Spring-Rice, took an equally gloomy view of the chances of the arbitration treaties taking the Senate hurdle. '[O]pposition of an unexpected strength [had] developed', he noted with dismay.[63]

Government endorsement of the semi-official celebrations of the peace centenary, then, seemed scarcely advisable. For the moment, of course, London and Ottawa kept tabs on the organizing committees in Britain and Canada, whilst officially remaining aloof to them.[64] While the two governments dragged their feet, however, the committees began to develop a momentum of their own that would eventually carry the governments with them. The British committee organized a large meeting at the Mansion House meeting on 18 December 1912 to launch a fund-raising campaign with the aim of raising some £50,000-60,000. At this gathering Earl Grey called for closer coordination of the efforts of the three national centenary committees. Their aim should be 'celebrations ... of such a character as will impress the popular imagination, and thus tend to strengthen still further those ties of mutual knowledge and of mutual affection and goodwill which already unite in common hopes and common aspirations the peoples of the Empire and the Republic.'[65]

The planned celebrations were now on a more modest scale. There were two strands to them, the first an ambitious educational scheme: the endowment of lectureships at the principal universities, essay prizes, and the financing of school textbooks. Public festivities were the second strand. Although these were to include celebrations in Ghent around Christmas 1914, their frame of reference remained the racial bond between the kindred peoples of Britain and America.

---

60   Min. Spicer, Feb. 20, 1913, on 'Anglo-American Peace,' *Morning Post* (Feb. 20, 1913), ibid.

61   Mins. Sperling and Spicer, Mar. 17, 1913, on tel. Bryce to Grey (no. 57), Mar. 15, 1913, TNA, FO 371/1857/12304.

62   Innes to Borden, Jan. 26, 1913, Borden Mss, vol. 24.

63   Tel. Spring-Rice to Grey (no. 123), June 6, 1913, TNA, FO 371/1857/26066.

64   Perley to Borden (confidential), Mar. 25, 1913, Borden Mss, vol. 26.

65   *Celebrations of the 100 Years of Peace among the English-Speaking Peoples: Report of Public Meeting at the Mansion House, December 18th, 1912* (London, 1912) [non-paginated brochure]; 'British-American Peace,' *The Times* (Dec. 19, 1912). All following quotes are from these as well.

The public persona of the first US president was to serve as the principal focus of these events. His 'part in a rebellion', Grey opined, would come to be seen 'as the first foundation of eventual Anglo-Saxon and Celtic consolidation.' A statue of George Washington was to be erected in either Westminster Abbey or Westminster Hall, the only remaining mediaeval part of the Houses of Parliament. Given the emphasis on Washington, the future of Sulgrave Manor acquired a new significance. Indeed, it would become the focal point of British centenary activities. Washington's ancestral home, Grey prognosticated, would become the destination of 'an agreeable pilgrimage to the increasing numbers of American visitors to England.' There was no more reference now to Grey's grand Crystal Palace design. Yet he had not abandoned all ambitions in that direction. Anglo-American sentiment, he argued, might 'require that some grand and impressive monument, symbolical of the unity of purpose that binds together the English-speaking peoples of the world, shall be erected on the best available site in London, right in the heart of the Motherland.'[66]

The committee also published a series of pamphlets and brochures on aspects of Anglo-American relations since 1814. One of these was dedicated to Sulgrave, described rather in the manner of an eager estate agent as 'a fine old house of white stone, in practically unaltered condition, with walled gardens, homestead and grass paddocks around.' Its projected purchase and preservation would furnish 'a symbol of the blood-relationship of the British and American races, ... a tribute to the memory and character of the great Patriot, and ... a rendezvous for pilgrims from both sides of the Atlantic.'[67]

The centennial activities now steadily gathered pace. In January 1913, the US committee formally renewed its invitation to a joint conference in the United States in later in the spring. The British committee duly accepted the invitation, though public confirmation of this was delayed until after Wilson's inauguration.[68]

The end of the hiatus in American politics also reinvigorated the activities of the Canadian Association. It, too, agreed to attend the New York conference in May.[69] Under Perley's chairmanship its executive committee was much enlarged

---

66   It is not entirely clear why the Crystal Palace scheme was dropped. This may have been on account of a rival peace exhibition, organized by the King Edward National Memorial Fund, of which Lord Weardale was a leading member, see *The Great 1912 'Progress of Peace' Exhibition and Festival, Crystal Palace* (s.loc., s.a.), Borden Mss, vol. 26.

67   *The British Home of the Washingtons: Sulgrave Manor and Its Associations* (London: Issued by the British-American Peace Centenary Committee, 1912). The other pamphlets were *The 100th Anniversary of Peace among English-Speaking Peoples: Meaning and Importance of the Coming Celebrations, Century of the Treaty of Ghent: December 24th, 1814 - December 24th, 1914*, and *The Peace of the Canadian Frontier: The Rush-Bagot Agreement*.

68   Carnegie et al. to Grey of Howick, Jan. 10, 1913 (copy), and Grey of Howick to Grey, Feb. 21, 1913, TNA, FO 371/1856/7149 and /8935.

69   Perley to Borden (private), Apr. 16, 1913, Borden Mss, vol. 26.

to include a number of senators, financiers and lawyers, as well as the former Liberal minister, and future Prime Minister, William Mackenzie King. The Canadians concurred with the American and British committees on the 'value of physical memorials in bringing the subject to mind and inspiring the proper emotions with regard to it.' It also mirrored the emphasis on educational activities of the sister organizations in Britain and the United States. Perley, moreover, was a staunchly Protestant Quebec representative, and this helps to explain the palpable religious overtones of the Canadian centenary plans. At the forefront of 'any worthy commemoration', the committee emphasized, had to be 'a feeling of devout thankfulness to Almighty God for His great mercy to us that two great nations ... have ... avoided war and sought peaceable solutions to their difficulties.' In consequence, there ought to be 'widespread religious exercises, extending to every place of worship in the land.'[70]

In April 1913, Woodrow Wilson consented to become the nominal head of the US committee, with William Jennings Bryan, his Secretary of State, acting as his deputy. This latest development changed the internal dynamics of the transatlantic centennial movement. Its immediate consequence was, as Spicer at the Foreign Office argued, 'that neither H[is] M[ajesty's] nor the Canadian Gov[ernmen]t can any longer remain unrepresented on the British and Canadian committees.'[71] Indeed, the names of most Cabinet ministers as well as the leaders of the Conservative and Labour parties were now added to the list of committee members.[72]

Now bearing an official imprimatur, the committees worked with dispatch. A general centennial manifesto was agreed at the international conference in New York in May 1913, attended by delegations from the three national committees as well as representatives from Newfoundland, then still a separate colony, Australia, and the city of Ghent. Guidelines were laid down for festivities of an international character, and the North American delegates agreed on suitable memorials ('Peace Arches') along the US–Canadian frontier. The 'Manifesto to the Nations' reflected more strongly now the internationalist leanings of the American centennial activists. The time had come, it stressed, 'when international rivalries and differences, though numerous and severe, may be settled without the carnage and the horror of war.' The racial element remained, but it had become entwined with internationalism. The forthcoming anniversary 'touches directly both the interests and the imagination of every land to which Great Britain's sons have gone, as well as those of every nation from which the present-day population of the United

70   Quotes from memo. Canadian Association, Feb. 11, 1913, encl. Grey of Howick to Grey, Mar. 4, 1913, TNA, FO 371/1856/10526; and Hamilton [Secretary, Canadian Association] to Perley, Mar. 24, 1913, Borden Mss, vol. 26.

71   Min. Spicer, Apr. 4, 1913, on tel. Bryce to Grey (no. 69), Apr. 4, 1913, TNA, FO 371/1856/15345; see also Bryce to Borden, May 3, 1913, Borden Mss, vol. 17.

72   Min. Spicer, May 1, 1914, and Bonham Carter to Tyrrell, June 3, 1913, TNA, FO 371/1856/20021 and /21906.

States has been drawn.' The centenary, the manifesto concluded, was 'an example and an ideal that we earnestly hope may be followed and pursued in the years to come. What nations have done nations can do.'[73]

In keeping with contemporary custom, the visiting delegations were lavishly wined and dined. At a banquet at New York's City Hall, Joseph H. Choate, the former US ambassador to Britain, amplified the message of the centennial manifesto, with all its internationalist, racial and religious connotations:

> Let Britannia and Columbia join hands across the Atlantic, and their outstretched arms will form a sacred arch of peace. It will excite the admiration of the nations, and will proclaim to the world the hope that with God's help the earth shall never more be deluged with blood shed in fratricidal war.[74]

In Britain, *The Economist* echoed such sentiments. The forthcoming centenary, the magazine observed, provided an opportunity 'for cementing those close ties of blood and friendship which unite us in common love of freedom.'[75]

The conference was an undoubted success, reported C.W. (later Sir Courtney) Bennett, the British consul-general in New York: 'the general feeling of the United States is, with certain exceptions, strongly in favour of closer and more cordial relations with Great Britain.' But he warned against 'too much talking about peace', a view that was reciprocated in London.[76] Indeed, in light of the still unresolved Panama question, senior officials at the Foreign Office 'scarcely welcome[d] the expenditure of public money on effusive displays of friendliness for the US.'[77] Nor was the Isthmian canal the only potential trouble spot. Mexico's descent into civil war in the course of 1913 pitted private British interests, largely of a commercial kind, against American political interests. Indeed, both Wilson and Bryan were driven by a strong ideological commitment to eradicate the power of foreign financial interests in Central and Southern America. With that aim in view, in a public speech in Mobile, Alabama, in October 1913, Wilson declared European commercial penetration of the region an obstacle to its political development. This was, in effect, a new corollary to the Monroe Doctrine, but it

---

73   *British-American Peace Centenary: Official Report of the International Conference held at New York, USA, May V to IX, MCMXIII, including Proposals for International Programme and Manifesto to the Nations* (London, 1913), 5-6; for a Canadian account of the conference see Dandurand et al. to Perley, May 24, 1913, Borden Mss, vol. 26.

74   As quoted in [P. Stanhope] Lord Weardale, *British-American Peace Centenary: Visit of the British Delegation to the United States, May MCMXIII* (London, 1913), 2; Edward S. Martin, *The Life of Joseph Hodges Choate* (2 vols, London: Constable, 1920) vol. 2, 352-54.

75   'The Peace Centenary,' *The Economist* (May 24, 1913), 1286.

76   Bennett to Grey (no. 9, confidential), May 19, 1913, and min. Spicer, May 27, 1913, TNA, FO 371/1856/24129.

77   Min. Sperling, n.d. [July 4, 1913], on Perris [Secretary, British committee] to Grey, July 4, 1913, TNA, FO 371/1856/30661.

was couched in vague terms: 'The speech is full of vague moralisings', commented a Foreign Office clerk.[78] Wilson's pronouncements, nevertheless, left a somewhat bitter aftertaste; and this, combined with the Mexican complications, induced the British government to remain aloof from the centenary movement.

Official reservations in London notwithstanding, the activities of the three committees went into overdrive in the autumn of 1913. Indeed, all manner of other groups now sought to attach themselves to them. The emphasis laid on the principle on arbitration and the peaceful settlement of disputes attracted the sympathies of the American peace activists. For them, the forthcoming centenary was an opportunity to advance the cause of 'rational internationalism' and 'of arresting the rivalry of nations in building up armies and armaments.'[79] Various Christian denominations endorsed the planned peace celebrations, and the organizers of the 1914 Canadian National Exhibition sought to jump on the centenary bandwagon, too, 'to show the advancement of Arts, Manufacture and Agriculture ... during the past one hundred years.'[80]

The racial bonds, reinforced by a shared heritage, remained the organizing theme of the work of the centenary committees. American, Britons and Canadians were 'knit [together] by ties of blood and relationship, by language and institutions.'[81] During the build-up to the planned celebrations the burgeoning Anglo-Saxonism produced some strange flowers. There were proposals for an 'Anglo-Saxon' or an 'English Speaking Holiday', principally 'for the glory of the English races or Protestantism.'[82] A provincial Canadian journalist, the editor of *The Moose Jaw Evening Times*, even submitted a detailed proposal for changing the calendar by limiting each month to twenty-eight days and inserting a 'peace month' during the summer period, 'contain[ing] at least twenty notable anniversaries' and combining the Anglo-Saxon with the internationalist theme.[83]

---

78   Min. Knatchbull-Hugessen, Nov. 18, 1913, on 'Hands off says Wilson – warning European greed,' *The Picayene* (Oct. 29, 1913), TNA, FO 371/1859/51008, reprinted in James B. Scott, ed., *President Wilson's Foreign Policy: Messages, Addresses, Papers* (New York: Oxford University Press, 1918), 19-26. For some of the background see J.A. Spender, *Weetman Pearson, First Viscount Cowdray, 1856-1917* (London, 1930), 163-204; John Womack, *Zapata and the Mexican Revolution* (Harmondsworth: Penguin Books, 1972), 227-64.

79   Hudson and Smith [Secretaries, 4th American Peace Congress] to Borden, Mar. 10, 1913, Borden Mss, vol. 26.

80   Stillman [Secretary, Methodist Church, British Columbia] to Borden, June 11, 1913, Orr [General Manager, Canadian National Exhibition] to Buxton [President of the Board of Trade], Nov. 12, 1913, and McNaught [Member of the Ontario Legislative Assembly] to Borden (personal), Mar. 12, 1914, Borden Mss, vol. 26 (quote from latter).

81   Tel. Parker et al. [US Centenary Committee] to Borden, Dec. 22, 1913, ibid.

82   C.F. Hamilton [Secretary, Canadian Association] to Blount [Private Secretary to Borden] (private), Mar. 3, 1914, ibid.; J.W. Hamilton to Borden, Oct. 31, 1914, ibid., vol. 49.

83   Editor, *Moose Jaw Evening Times* to Borden, Apr. 25, 1914, ibid., vol. 26.

More fantastical than practical, these proposals nevertheless are suggestive of the extent to which the forthcoming centenary had captured the popular imagination. In London, efforts to acquire Sulgrave gathered pace. In June, the British committee had obtained an option on the purchase of Sulgrave, which was eventually extended until early 1914; and the committee decided to launch a general appeal for funds. J.F.H. Baily, the editor of the *Connoisseur* magazine, offered to open a separate fund for the maintenance of the property. Crucially, thoughts turned to the future function of the manor house. William B. Howland, editor of the influential New York journal *The Outlook*, considered that the 'fruitful and successful use of the property would depend largely upon its visitation by Americans.' For that reason he suggested the creation of an 'International Body of Trustees', consisting in equal proportions of US and British citizens, and chaired *ex officio* by the American ambassador. As for Sulgrave's eventual use, Howland advised 'that it should be furnished with a view to making it a repository of documentary, pictorial, and general records of the peaceful relations of the two peoples since 1814 and that later it might be advisable to build a hostel in the grounds at which guests might be lodged.'[84]

In the autumn of 1913, the British committee launched its public appeal for £8,000 for the purchase of the manor, raising £6,600 in a matter of days. The campaign enjoyed the support of a long list of politicians, religious and civic dignitaries and university principals. *The Times* endorsed the purchase of Sulgrave 'as a place of pilgrimage for Americans in England as a symbol of the kinship of the two peoples.'[85] During the winter of 1913/14, the campaign kept up its momentum through a series of events on the West End dinner circuit. At one such banquet an appreciative letter from James Bryce, who had now joined the British committee, was read out. The former ambassador expounded on the unique nature of the forthcoming centenary. It was rooted in 'the deep underlying feeling of a common ancestry and a common share in noble traditions' and 'the sense each [nation] entertained of its responsibility to set the world an example of preferring arbitration to war.'[86]

Sufficient funds had by now been raised to justify the appointment of a committee of management along the lines indicated by Howland earlier, to oversee the purchase, restoration and furnishing of Sulgrave Manor. At its first two meetings, in October 1913, the idea was raised once more of eventually transferring the day-to-day running of the manor house to the National Trust. Even so, the committee decided that, whatever arrangements might be come to with the National Trust,

---

84   Minutes of the Executive Committee of the British Peace Centenary Committee, June 10 and July 11, 1913, Sulgrave Manor Trust Archive [hereafter Sulgrave Mss], CP Box 15/7e; also Perris, 'Milestones of Anglo-American Friendship' (unpublished TS), Perris Mss, fo. 25.

85   'One Hundred Years of Peace,' *The Times* (Oct. 7, 1913). For the public appeal see ibid. (Oct. 7, 13 and 14, 1913).

86   'Hundred Years' Peace,' *The Times* (Dec. 6, 1913).

'the entire control of the property must be retained in the hands of the Committee of Management ... as the property was being purchased not simply as a relic, but for specific purposes in the future.'[87] Having inspected the property, Donald and Brittain reported favourably in November; and the committee invited the well-known architect Reginald Blomfield, FSA, to submit plans for the restoration of the property and for the design of a sixteenth- century style knot-garden.[88] At last, on 22 January 1914, the British peace centenary committee purchased Sulgrave Manor, 'a somewhat dilapidated shrine' and nine acres of adjoining land 'in a very rough and unkempt condition' for £8,398 from the proceeds of the previous appeal, with some additional assistance from their bankers.[89]

By the standards of the day, the committee was adept at modern media management. The Duke of Teck, Queen Mary's brother, had now taken over the presidency of the committee from Earl Grey, who had resigned on health grounds. On the day following the purchase, the Duke and Bryce, now very much the public face of the committee, visited Sulgrave, accompanied by a posse of press photographers.[90] And in early February, the acquisition of Sulgrave was announced by prime minister Asquith at a major event at the Mansion House. As in the previous year, there had been some to'ing and fro'ing behind the scenes in Whitehall about the precise nature of the government's endorsement. The precarious situation in Mexico suggested caution; and so did a frisson of tension between London and Washington following British pressure on the Wilson administration to repeal the Panama Canal tolls. Indeed, Spring-Rice at the Washington embassy argued that the 'sanctity of treaty engagements does not have much weight with the [US] public' and that Wilson would not act against public opinion.[91] Enthusiastic backing of the centenary movement, possibly followed by the United States reneging on perceived treaty commitments, had the potential to embarrass the British government. In consequence, the Foreign Secretary, Sir Edward Grey, settled for expressions of goodwill, but refused to associate the

---

87   Minutes of Committee of Management, Oct. 15 and 19, 1913, Sulgrave Mss, CP Box 15/7e; Smith, *Sulgrave Manor*, 207.

88   Minutes of Executive Committee of the British Peace Centenary Committee, Nov. 19 1913, Sulgrave Mss, CP Box 15/7e. Blomfield's 1892 treatise, *The Formal Garden in England*, introduced the term 'formal' into horticultural writing, see Miles Hadfield, *A History of British Gardening* (Harmondsworth: Penguin Books, 1985), 365.

89   Brittain, *Pilgrims*, 122; Minutes of the executive committee of the British Peace Centenary Committee, Jan. 22, 1914, Sulgrave Mss, CP Box 15/7e (Cowdray had signed an additional £1,000 overdraft facility); 'Schedule of Title Deeds,' ibid., CP Box 6a; 'The Home of the Washingtons,' *The Times* (Jan. 23, 1914).

90   Minutes of the Executive Committee, Jan. 22, 1914, Sulgrave Mss, CP Box 15/7e; 'The Peace Centenary Memorial,' *The Times* (Feb. 23, 1914). For Earl Grey's resignation see also Teck to Asquith, n.d. [Jan. 6, 1914], TNA, FO 371/2151/881.

91   Spring-Rice to Borden (private), Feb. 10, 1914, Borden Mss, vol. 30; Page to Wilson, July 5, 1914, in Burton J. Hendrick, *The Life and Letters of Walter Hines Page* (3 vols., London: W. Heinemann, 1926) vol. 3, 61-64.

government more closely with the centennial activities 'unless and until the United States Government takes some action in the matter.'[92]

The public meeting at the Mansion House was nevertheless a great success for the centenary campaign. Speeches by Asquith, Bryce and Randall Davidson, the Archbishop of Canterbury, stressed the internationalist, racial and religious aspects that had influenced the movement's activities. The prime minister praised them as 'the spontaneous expression on the part of two great peoples of a feeling which is equally firmly and equally deeply rooted on both sides of the Atlantic.' The 1814 peace treaty, he averred, was 'the earliest formal embodiment' of the ties between 'the two great kindred races.' Bryce followed that same line, but also established a link with disarmament by emphasizing the significance of the 1817 Anglo-American Rush-Bagot treaty, which had effectively demilitarized the Upper Lakes region. For his part, Davidson, the first Anglican primate to visit North America, stressed the common Anglo-American heritage. Praising the 'revived respect and affection for what was old in English life', he attributed them to a 'common caring for even the outward vestiges of a past which belonged to them both.' The events of 1776, he opined, had 'divided the great stream into two channels [but it] did not make either channel forgetful of its common source and its long identity.'[93]

A stream of high-profile events kept the activities of the committees in the public eye. Preparations were made for joint celebrations, including a commemorative dinner in Ghent on 5 January 1915; a second appeal was launched to raise further funds for the restoration, furnishing and permanent endowment of Sulgrave Manor; and a reception room and inquiry office were opened at the Anglo-American Exhibition then taking place in Shepherd's Bush.[94] Various voluntary associations came forward to assist with the furnishing of the house, most notably the National Society of the Colonial Dames of America, reflecting, both, the more prominent role played by women in preservation efforts in America as well as the role of American-born, British-wed women in similar efforts in Britain.[95] An original autograph letter from Washington, written during the 1782 campaign, was donated to the British committee for permanent deposition at Sulgrave.[96] In the spring of

---

92   Min. Grey, n.d. [Jan. 8, 1914], and Selby to Duke of Teck, Jan. 8, 1914, TNA, FO 371/2151/881 (quote from latter).

93   'Peace Centenary,' *The Times* (Feb. 5, 1914); for Davidson's 1904 visit to North America see George K. Bell, *Randall Davidson, Archbishop of Canterbury* (2 vols, Oxford: Oxford University Press, 1935) vol. 1, 442-53.

94   Teck et al. to Asquith, Mar. 25, 1914, TNA, FO 371/2151/17028; 'Anglo-American Peace,' *The Times* (Feb. 14 and June 5, 1914). The principal subscribers to the Centenary Fund are listed in Smith, *Sulgrave Manor*, 223-26.

95   Minutes of the 1st Meeting of Sulgrave Management Board, Feb. 20, 1914, Sulgrave Mss, CP Box 15/7e; for the role of women in preservation in America see Patricia West, *Domesticating History: The Political Origins of America's House Museums* (Washington, DC: Smithsonian Institution Press, 1999).

96   'Peace Centenary Gift,' *The Times* (Apr. 4, 1914); Smith, *Sulgrave*, 134.

1914, aided by a $500,000 subsidy by the American peace centenary committee, the management board agreed to establish the Sulgrave Manor Institute, 'in order to utilize to some permanent advantage Sulgrave Manor and the sentiment which surrounds it.'[97] As London's Society season drew to a close, the events organized by the British committee reached their peak in a historical pageant, part of an Anglo-American Peace Ball at the Albert Hall in mid-June. And on 27 July, the Duke of Teck, accompanied by the American ambassador, Walter Hines Page, and several members of the British committee, visited Sulgrave for the ceremonial opening of the manor house. That 'delightful gathering in the heart of England'[98] was the last of its kind for some time. Eight days later, Britain was at war.

Already during the first half of 1914, there were indications that official American interest in the planned centenary celebrations was waning. The appropriations bill had still not passed Congress. Indeed, the amount to be voted for had been drastically reduced to $25,000. Earlier centennial enthusiasm, reported the *chargé d'affaires* at the Washington embassy, 'appears now to have lapsed into indifference.' The 'original Anglo-Saxon taint of the movement still clings to it', but resentment at Britain's supposed dictation in the matter of the tolls repeal could easily turn 'openly hostile.' The current moment, appeared 'not favourable for anything but anti-British speeches.' In consequence, both governments decided to kick the question of public funding into the long grass.[99]

There it remained while the United States stayed neutral in the European conflict. Indeed, the activities of the Sulgrave Management Board ceased as well, a brief initiative by Howland, in March 1915, to recommence the work of the Sulgrave Manor Institute notwithstanding.[100] The work was only resumed in mid-May 1917, some six weeks after America's entry into the war.[101] Indeed, little progress was made during the remainder of the war. There was an abortive attempt to complete the centenary committee's original educational plans; and there were some organizational changes with regard the Board of Governors. In October 1918, at a special meeting of the centenary committee and leading subscribers to the Sulgrave Institute, 'recognising that the work for which the committee was originally founded must now be carried on amidst different conditions and through different agencies', the affairs of the committee were wound up.[102] But it was

---

97   Minutes of 2nd and 3rd Meetings of the Sulgrave Management Board, Mar. 9 and June 19, 1914, Sulgrave Mss, CP Box 15/7e.

98   Brittain, *Pilgrims*, 122; for the pageant see Smith, *Sulgrave Manor*, 208.

99   Quotes from Barclay to Grey (no. 223), July 2, 1914, and min. Langley, n.d. [July 15], TNA, FO 371/2151/31805. For the British government's position see Grey to Barclay (no. 426), July 28, 1914, ibid./34967.

100   Minutes of the Sulgrave Sub-Executive, Apr. 14, 1915, Sulgrave Mss, CP Box 15/7e.

101   Minutes of the 4th Meeting of the Sulgrave Management Board, May 15, 1917, and of Sulgrave Executive Committee, June 11, 1917, Sulgrave Mss, CP Box 15/7e.

102   Minutes of Special Meeting, Oct. 17, 1918, ibid.

not until 1925 that the management of the manor house was consolidated in a new organization, now substantially endowed through American donations.[103] The refurbished house was reopened to the public in June 1921, and soon became the destination for weekend excursions by the swelling ranks of the suburban middle classes in the post-war years.[104]

The aim of the organization remained the same: to advance the cause of Anglo-American unity. The tone of its public events, however, had changed. The pre-war optimism, espousing ideals of universal arbitration, had gone; the focus on Anglo-Saxon ties had hardened. The war-time alliance was seen as the natural coming together of what history had once parted. And Sulgrave itself was 'the old homestead of the American branch of the common stock, the best stock.'[105] When unveiling the Houdon statue of George Washington in London's Trafalgar Square, originally planned for Westminster Hall, Henry Louis Smith, president of Washington and Lee University, summed up such sentiments, before concluding with a rousing appeal to Britain to stand with America against the new threat of Bolshevism:

> The English-speaking nations ..., with a common racial kinship, a common religion, and similar ideals of character and conduct, obeying the same general code of laws, accustomed to the same modes of self-government, and utilizing the same methods of business organization, constituted today the most homogeneous group of nations ever known on earth.[106]

ii

The acquisition and subsequent restoration of Sulgrave Manor as an 'English Mount Vernon' offers an instructive case-study from the early history of preservation in Britain.[107] In many ways, it reflected a well-established pattern in the preservation efforts of the late Victorian and Edwardian periods. A perceived external threat first alerted preservationists to the plight of this ancient manor house. Society figures were able to use connections with the worlds of the City and of Westminster to mobilize financial and political support. These also helped them to circumnavigate

---

103   Minutes of the Sulgrave Management Board, July 25, 1917; Dec. 17, 1919; and June 25, 1925, ibid.; for the endowment after the war see Smith, *Sulgrave Manor*, 212-21.

104   'Reopening of Sulgrave Manor,' *The Times* (June 20, 1921); Mandler, *Fall and Rise*, 231.

105   Lord Burnham, ex-proprietor of the *Daily Telegraph*, as quoted in 'Anglo-American Entente,' *The Times* (Apr. 25, 1919).

106   'Washington Statue in London,' *The Times* (July 1, 1921); for further details see Jean Hood, *Trafalgar Square: A Visual History of London's Landmark Through Time* (London: Batsford, 2005), 85-87.

107   'An English Mount Vernon,' *The Times* (July 22, 1920).

the shallows and eddies in the stream of international diplomacy. Indeed, they insulate the Sulgrave project against the vicissitudes of high politics, and allowed it to develop a dynamic of its own. A forthcoming centenary provided a fitting public platform for efforts to preserve an historic property as a tangible monument of a privileged version of the past. Finally, frequent references to Sulgrave as a 'shrine' or 'a place of pilgrimage' reflected the elevated position of history, indeed the 'sacralization' of the past, in public life before the Great War.

At the same time, there were also unique aspects to the Sulgrave campaign. Its preservation was driven by 'racial' concepts, prevalent on both sides of the Atlantic since the 1890s and based on notions of racial and other ties between the two kindred Anglo-Saxon and largely Protestant branches. 'Sulgrave was political and historical "truth embodied in a tale"', as the secretary to the British centenary committee later reflected.[108] However, the significant role played by the Canadian centenary association underlined the importance of a Greater British frame of reference. The preservation of Sulgrave, as part of the Anglo-American peace centenary celebrations, had symbolic value for the past, present and future relations of the peoples of the British Empire and the United States. Initially, the racial undercurrent in the efforts to preserve Sulgrave was aligned with the nascent pre-1914 peace and arbitration movement, which had particular resonance among ecclesiastical and Radical political groups in both countries. Indeed, Anglo-Saxonism allowed public figures of sharply diverging political persuasions to come together under the umbrella of the centennial committee.

While both the Anglo-Saxon and the internationalist aspects were to some extent forward-looking, the preservation of Sulgrave also reflected, at least in part, a reaction to the crisis of Edwardian politics and society. The restoration and preservation of Sulgrave thus presented perceived Tudor prosperity and stability as a counterpoint to the convulsions and uncertainties of contemporary Britain. In this, however, it was also more specifically English than British. Sulgrave, its Tudor associations and its recreated formal, Elizabethan garden certainly resonated with the 'historical-pastoral' renaissance in English cultural life before 1914.[109] Indeed, nestling in the rural seclusion of Northamptonshire, with its parks and enclosures, it was a 'much more typically English scene' than could be found elsewhere.[110]

---

108   Perris, 'Milestones of Anglo-American Friendship' (unpublished TS), Perris Mss, fo.24.

109   For the pastoral impulses see also Jan Marsh, *Back to the Land: The Pastoral Impulse in Victorian England from 1880 to 1914* (London, 1982), 245-47; Paul Harringron, 'Holst and Vaughan Williams: Radical Pastoral,' in Christopher Norris, ed., *Music and the Politics of Culture* (London: Lawrence and Wishart, 1989), 106-27; Robert Stradling and Meirion Hughes, *The English Musical Renaissance, 1860-1940: Construction and Deconstruction* (London: Routledge, 1993), 60-68.

110   William G. Hoskins, *The Making of the English Landscape* (Harmondsworth: Penguin Books, 1973), 169.

The Sulgrave project was at one and the same time very English and transatlantic in nature. It reflected the complex and often disparate political, intellectual, social and cultural influences that shaped the 1914 peace centenary movement. And as such it is a useful reminder for historians of preservation to be less insular and inward-looking: no historic building is an island, entire of itself.

PART II
Framing the Practice

Chapter 6

The Law's Delay? Preservation Legislation in France, Germany and England, 1870-1914

Astrid Swenson

'All civilized countries have, at the present time, like our own, a legislation that protects the monuments of the past,' declared Louis Grandjean, French Inspector of Historic Monuments during a public conference cycle organized by the French Ministry of Public Instruction at the *Exposition Universelle* in Paris to accompany the Exhibition of Historic Monuments. Grandjean outlined how global a phenomenon historic preservation had become by 1900:

> This is the case not only in the great European states: England, Germany, Austria, Prussia, Italy, but also in the secondary states: Portugal, Belgium, Netherlands, Greece, Bulgaria, Finland etc. The same applies outside Europe: not to speak of Algeria and Tunisia that are truly speaking only an extension of France, [but] we find regulations for the protection of ancient monuments in Egypt, in the Asian parts of Turkey, in British India, Japan and [reaching] as far as China. America did not want to stay behind. Brazil and Mexico also took preservation measures. As for the United States that were only born yesterday settling on a land untouched by previous civilization, they undertook to safeguard the gifts of nature.

Like Grandjean, this chapter does not have the intention 'to take you to all of the countries whose name I mentioned. [...] If you agree we will not leave Europe and inside Europe we will make a choice.' Yet, while the Inspector of Historic Monuments was full of faith in his ability to 'only select nations whose example can interest or educate us',[1] the focus here on legislation in three of the 'great

---

1   'Tous les pays civilisés ont, à l'heure actuelle, comme notre propre pays, une législation qui protège les monuments du passé. [...] Il est ainsi non seulement dans les grands états de l'Europe: Angleterre, Allemagne, Autriche, Russie, Italie, mais aussi dans les Etats secondaires Portugal, Belgique, Pays-Bas, Danemark, Grèce, Bulgarie, Finlande etc. Il en est ainsi également hors d'Europe: sans parler de l'Algérie et de la Tunisie, qui ne sont à proprement parler que des prolongements de la France, nous trouvons une réglementation protectrice des anciens monuments: en Egypte, dans la Turquie d'Asie, dans l'Inde Anglaise, au Japon et jusqu'en Chine. L'Amérique elle-même n'a pas voulu rester en arrière. Le Brésil et le Mexique ont pris de même des préservations. Quant aux Etats-Unis,

European' States' France, Germany and Britain is in many ways a more arbitrary limitation. Rather, the paper examines why certain nations' examples were seen as interesting and educative. Drawing attention to networks, collaboration and rivalry among European preservationists, it will demonstrate the role played by cultural transfer in shaping turn of the century preservation.

Over the course of the nineteenth century, the growing enthusiasm for preservation had acquired a complex legislative dimension. There had been sporadic calls for Monument Laws since the 1830s, but in the period from 1870 to the outbreak of the Great War, the preoccupation with legislation became almost obsessive. Voluntary action was no longer thought sufficient. As a German preservationist put it: 'Without misconceiving what the enthusiasm of individual persons and the activity of associations did to awake and divulge the understanding of monuments' the growing opinion was that, 'support and care on the part of the State are, however, necessary to preserve and protect the great heritage of the past.'[2]

Despite a number of State institutions being in place in most European countries by 1870, the existing royal rescripts, ministerial decrees, circulars and local by-laws, were judged to be insufficiently robust.[3] Spread across different areas of the law, from civil to penal codes, administrative law and buildings regulations, they had official authority, but not the force of a formal Monument Act. As Grandjean indicated, legislation was debated in virtually every European and a number of non-European countries, and Monument Acts were passed *inter alia* in Greece (1834, 1899), Sweden (1867), Hungary (1881), Britain (1882, 1913), Finland (1883), Turkey (1884), Tunisia (1886), France (1887, 1914), Bulgaria (1889), Romania (1892), Portugal (1901), Italy (1902, 1909) and India (1904). In Switzerland, the Cantons Vaud, Bern, Neufchatel, Tessin passed Laws between 1898 and 1907, as did the German States of Hesse (1902) and Oldenburg (1911),

---

qui sont nés d'hier et installés dans une terre vierge de toute civilisation antérieure [...] ils ont entrepris de sauvegarder les dons de la nature. [...] Je n'ai pas l'intention – cela va sans dire – de vous conduire dans chacun des pays dont je viens de citer les noms. [...] Si vous le voulez bien, nous ne quitterons pas l'Europe et dans l'Europe nous ferons un choix. Nous ne prendrons que des nations dont l'exemple peut nous intéresser ou nous instruire.' Louis Grandjean, 'Les Monuments historiques à l'étranger' (manuscript for a paper presented at the *Exposition universelle* Paris, 1900), in the Médiathèque de l'architecture et du patrimoine, Archives, Paris (hereafter MAP) 80/1/32, dossier 'Législation étrangère concernant les Monuments historiques,' folder 'Les Monuments historiques à l'Etranger.'

2   'Die Pflege der kunstgeschichtlichen Denkmäler beschäftigt gegenwärtig alle Culturstaaten. Ohne zu verkennen, was die Begeisterung einzelner Personen und die Wirksamkeit von Vereinen gethan haben, um das Verständnis für die Denkmäler zu erwecken und zu verbreiten, so bedarf es doch der staatlichen Fürsorge und Unterstützung, um das große Erbe der Vergangenheit zu bewahren und zu schützen.' Julius Kohte, *Die Pflege der Kunstdenkmäler in Italien* (Berlin: Ernst, 1898), 1.

3   Felix Hammer, *Die geschichtliche Entwicklung des Denkmalrechts in Deutschland* (Tübingen: Mohr, 1995), 29-31.

to which can be added various so-called 'partial laws' on aesthetic disfiguration and prehistoric monuments in other States, for instance in Prussia in 1902, 1907 and 1914. In countries with existing provisions, the aim was to simplify and fortify State intervention, in those lacking previous regulations, the new laws enabled a country finally to catch up with more advanced ones.

On an international scale, France, Germany and Britain were thus neither the first nor the last countries to pass legislation, yet the three 'great European States' offer an insightful comparison as proponents of legislation in each country needed to tackle different levels of need and had to overcome different forms of opposition. France already possessed an elaborate administrative apparatus since the July Monarchy, albeit with little legal power over monuments. When Germany was unified, some States had a monument administration, while no provisions existed in others. In Britain, the State was not involved in preservation at all prior to the 1870s.

The debates about legislation help scrutinizing the prevalence of 'heritage-consciousness', but it would be reductive to draw a direct link between legal provisions, or even support for a Bill, and the intensity of love for 'heritage'. Success and extent of a given Monument Act were shaped by several external causes. In order to compare the debates on legislation and their results across several countries, it is necessary to take into consideration that Monument Acts had to be drafted in accordance with conditions which varied from country to country, a step generally omitted in primarily national studies. It is crucial to consider the wider legal and administrative framework, from provisions for the protection of private property and expropriation on the grounds of public utility to the status of corporate bodies, such as Churches and communes. The power of the State over public corporations differed considerably, being much stronger in France and Germany than in Britain. Furthermore, one must factor in the ownership of the bulk of monuments that were considered to be 'national monuments': British Stately Homes were a different case from State-owned cathedrals and former royal palaces in France. The successful passing of an Act also depended on political majorities and on the place preservation legislation had in major socio-political conflicts of the time, from culture-wars to land-reform.[4]

While the basic reason in all countries for desiring a Monument Act was straightforward – to improve the protection of monuments and increasingly of picturesque sites and landscapes, which were ever more threatened by modernization, building booms, and lootings for a steadily developing art-market; often a threat to particular monuments or site worked as a trigger for the preparation or passing of a law – proponents of legislation also had often more complex social, political and economic reasons for desiring preservation legislation and hence of extending State control into previously private areas.

---

4   Cf. Christopher Clark and Wolfram Kaiser, eds, *Culture Wars: Secular-Catholic Conflict in Nineteenth-Century Europe* (Cambridge: Cambridge University Press, 2003).

Accordingly, opposition to preservation legislation was not necessarily founded on opposition to preservation. As Peter Mandler argues for the British case, preservationism could be considered independently of politics in scholarly discussions and to a certain extent even in the general press:

> Once it entered wider debate in Parliament, however, the emerging consensus among scholarly and amenity groups hit stormy water. Liberals had to decide whether they (and their voters) cared enough about castles to extend legislative protection – or perhaps whether legislative control over historic buildings was worthwhile simply as a stroke against the landowners [...] Tories had to decide whether they cared enough about castles to accept the infringements on private property – and more strokes against the land – that such legislation would entail.[5]

In France, concerns about infringements on private property were also voiced, but preservation legislation was primarily caught in the 'War of the Two Frances'. The 1887 Act was not only an Act for the better protection of monuments, but also a measure to enhance a centralist approach and weaken the influence of local communities and the Catholic Church. It raised corresponding opposition. After the Separation of State and Church, matters became even more complicated. While before 1905, the scheduling (*classement*) of churches could be interpreted as an anti-clerical measure, now the failure to schedule churches also became one. To allocate tax money for church restoration could be seen as a financing of religion by the State, forbidden by the terms of the Separation. Consequently, anti-clerical preservationists had to decide whether they wanted to risk being seen as pro-clerical by financing the repair of churches. Clericals had to decide whether preservation was worth a further extension of the State's influence into the sphere of religion.[6]

In Germany, the conflicts surrounding preservation were more multifaceted. As in France, State preservation was to a certain degree motivated by the wish to obtain control over Churches and local communities. Consequently, in the aftermath of the *Kulturkampf*, the Catholic Church vehemently opposed any attempts by the State to control buildings through preservation and successfully impeded the earliest German attempts at legislation in the States of Baden and Prussia.[7] The mere fear of opposition sometimes prevented the introduction of a Bill. Several conservators in Germany, especially in Bavaria and Baden opposed formal legislation as they felt they had more possibilities to informally negotiate with owners and to obtain the collaboration of the Churches if no Act was passed that might raise fears over

---

5   Peter Mandler, *The Fall and Rise of the Stately Home* (New Haven, London: Yale University Press, 1997), 188.

6   Jean-Michel Leniaud, *Les Archipels du passé: Le patrimoine et son histoire* (Paris: Fayard, 2002), 191-241.

7   Günter Hans, *Denkmalschutz in Baden im 19. und 20. Jahrhundert* (Freiburg: Univ. Hochschulschrift, 1985), 46-69, 77-93.

the State's interference. Moreover, in Germany preservationism was used, but also got caught in the struggles about competence in a federal system. All attempts to create an imperial Monument Act failed because of the opposition of the South German States, which feared a loss of their sovereignty in cultural policies.[8]

Hard-line philistine opposition thus created probably the least of problems (although philistines grew more organized and desperate in some cases. The German federation of Industrialists founded in 1911 a commission with the ringing name *Kommission zur Beseitigung der Auswüchse der Heimtschutzbestrebungen* – 'Commission for the Abolition of the Excrescences of the Preservation Movement').[9]

More crucially, everywhere, some ardent preservationists turned against preservation legislation if it was felt to endanger other political aims. Nowhere was passing a Monument Act was an easy task. In England nine years elapsed between John Lubbock's introducing a first Ancient Monument Bill and the passing of the much watered-down Act in 1882. Yet, in our three countries, it was the first successful Act. In France passing the first Monument Act took almost fifteen years. In Germany an imperial Bill never saw light, the Bills drafted in Prussia and Baden in the 1880s encountered so much opposition that they had to be withdrawn before even being introduced to the diets, and repeated attempts after the turn of the century in these two German States shared the same fate. Thereafter, Prussia refrained from attempting to pass a comprehensive act, but focused instead on partial aspects such as the regulation of excavations and the aesthetic appeal of historic towns through building regulations.[10] Only two comprehensive Monument Acts were passed in German States, the first in the Grand Duchy of Hesse in 1902, the second in the Duchy of Oldenburg in 1911.

Before examining the impact of the above conflicts on the shaping of legal provision, it is necessary to ask how legislation could succeed at all in the face of such considerable opposition. This brings us to the contribution of international exchange. Despite national differences, campaigners for monument legislation, as well as the legislators themselves, faced a number of problems which overstepped national boundaries. The passing of a Monument Act meant that the concept of a 'national heritage' was approved by parliament. Thus, in the first place it was necessary to establish that monuments were a subject fit for legislation, i.e. a national, public, concern rather than a private one. Secondly, it had to be decided who should be responsible for establishing an inventory of monuments (i.e. for defining 'heritage') and which of these were worthy of legislative protection.

---

8   Winfried Speitkamp, *Die Verwaltung der Geschichte: Denkmalpflege und Staat in Deutschland 1871-1933* (Göttingen: Vandenhoeck & Ruprecht, 1996), 154-63.

9   Ibid., 372.

10   German preservationists and legal experts differentiated between a comprehensive Monument Act (*Allgemeines_Denkmalschutzgesetz*) applying to immobile and movable prehistoric, historic and natural monuments, protecting their substance and appearance – as opposed to partial laws, so-called *Teilgesetze*, such as Excavation Act (*Ausgrabungsgesetz*) and Anti-disfigurement Act (*Verunstaltungsgesetz*).

Thirdly, the impact of legislation on private ownership had to be addressed in connection with issues such as expropriation or compulsory purchase. Further, it was necessary to fix penalties for the destruction or exportation of cultural goods, and to regulate excavations. Finally, Acts needed either to regulate the competences of different existing agencies or found new institutions.

Confronted with similar problems, countries made no attempt to conceal that they sought solutions in cultural transfer. It was accepted and even expected in all European countries that inspiration should be sought abroad. Most works among the rich and substantial literature on the history of monument preservation with a national focus[11] cite the main comparative studies[12] and point out that laws were

---

11   On Britain: Nikolaus Boulting, 'The Law's Delays: Conservationist Legislation in the British Isles,' in Jane Fawcett, ed., *The Future of the Past: Attitudes to Conservation, 1174-1974* (London: Thames & Hudson, 1976), 9-34; Christopher Chippendale, 'The Making of the First Ancient Monuments Act, 1882, and Its Administration Under General Pitt-Rivers,' *Journal of the British Archaeological Association* 136 (1983), 1-55; Andrew Downing Saunders, 'A Century of Ancient Monuments Legislation 1882-1992,' *Antiquaries Journal* 63 (1983), 11-33; Tim Murray, 'The History, Philosophy and Sociology of Archaeology: The Case of the Ancient Monuments Protection Act 1882,' in Valerie Pinsky and Alison Wylie, eds., *Critical Traditions in Contemporary Archaeology* (Cambridge: Cambridge University Press, 1989), 55-67; Timothy Champion, 'Protecting the Monuments: Archaeological Legislation from the 1882 Act to PP16,' in Michael Hunter, ed., *Preserving the Past: The Rise of Heritage in Modern Britain* (Stroud: Alan Sutton, 1996), 38-56; Peter Mandler, *The Fall and Rise of the Stately Home* (New Haven, London: Yale University Press, 1997), 153-91; Melanie Hall, 'Affirming Community Life: Preservation, National Identity and the State, 1900,' in Chris Miele, ed., *From William Morris: Building Conservation and the Arts and Crafts Cult of Authenticity, 1877-1939* (New Haven: Yale University Press, 2005), 129-57. For a detailed account on France see Paul Léon, *La Vie des monuments français: Destruction, restauration* (Paris: Picard, 1951), 125-64. More recent works evaluate the laws primarily with regard to current heritage policies: Leniaud, *L'Utopie française: Essay sur le patrimoine* (Paris: Mengès, 1992), 19-21; André Chastel, 'La Notion de patrimoine,' in Pierre Nora, ed., *Les Lieux de mémoire* (4 vols., Paris: Gallimard, 1994), vol. 1, 1454; Françoise Choay, *L'Allégorie du patrimoine* (Paris: Seuil, 1992), 108-11; Françoise Bercé, *Dès Monuments historiques au patrimoine du XVIIIe siècle à nos jours ou 'Les égarements du coeur et de l'esprit'* (Paris: Flammarion, 2000), 52-53, 58-60; Dominique Poulot, *Patrimoine et musée : L'institution de la culture* (Paris: Hachette, 2001), 121-22, 132-33; Leniaud, *Les Archipels du passé*, 231-41. The most detailed account on Germany is Hammer. A selective chronology of legal provisions is provided in Michael Siegel, *Denkmalpflege als öffentliche Aufgabe: Eine ökonomische, institutionelle und historische Untersuchung* (Göttingen: Vandenhoeck & Ruprecht, 1985), 35-40, 291-94; and a strong contextual analysis in Speitkamp, *Die Verwaltung der Geschichte*, 287-394. On eighteenth- and early nineteenth-century decrees see Norbert Huse, ed., *Denkmalpflege: Deutsche Texte aus drei Jahrhunderten* (Munich: Beck, 1996), 17-34.

12   See generally the works by A. von Wussow, *Die Erhaltung der Denkmäler in den Kulturstaaten der Gegenwart: Im Auftrage des Herrn Ministers der geistlichen, Unterrichts- und Medizinalangelegenheiten nach amtlichen Quellen dargestellt* (2 vols,

passed in other European countries at the same time, yet the omnipresence of comparative argumentation has not been noted, and the impact of the exchange only marginally incorporated.[13] Conversely, most recent studies tend to repeat assertions found in the nineteenth- and early twentieth-century comparative literature on the 'backwardness' of the author's country, deploring the law's delay without taking into account that assertions of this type were mainly made according to a rhetorical strategy and cannot be assumed to reflect reality. A transnational perspective is able to redress these imbalances. To explain the interest in and relative success of legislation in all three countries and indeed all over Europe, it is essential to take into consideration the cultural transfer in legislative campaigning and planning and to understand that the establishing of preservation legislation was fuelled by a heightened sense of international progress and competition.

As can be seen when looking at the chronologies in comparison, legislative waves in France, Germany and Britain followed broadly similar rhythms. In a first phase, circa 1870 to 1890, several Bills were prepared, resulting in the promulgation of formal Acts in Britain and France in the 1880s. In a second phase, circa 1900 to 1914, the pace of legislation accelerated considerably: factors including the formation of new associations and the greater density of international meetings and exchange prompted ever more Bills to be introduced by increasingly diverse authors, resulting in ever more wide-ranging Laws. These processes were not primarily top-down, as has sometimes been said, but State intervention often responded to a public demand.[14] All existing transnational networks of preservationists were mobilized and further enhanced towards this aim. In the endeavor to remain up-to-date in a constantly evolving legal development, collaboration among State agencies, private associations and scholars inside and across national boundaries was intense. Preservation legislation was on the program of the First International Congress for the Protection of Monuments and Works of Art in Paris in 1889, and a central topic at the multiplying international congresses of architects, archaeologists, and orientalists; of public art, landscape protection and many others; it was even touched upon even at the Hague Peace Conferences.[15] These meeting not only provided regular updates on international

---

Berlin: Heymann, 1885); Josef Alexander Freiherr von Helfert, *Denkmalpflege: Öffentliche Obsorge für Gegenstände der Kunst und des Altertums nach dem neuesten Stande der Gesetzgebung in den verschiedenen Culturstaaten* (Vienna, Leipzig: Braumüller, 1897); Gerald Baldwin Brown, *The Care of Ancient Monuments: An Account of the Legislative and Other Measures Adopted in European Countries for Protecting Ancient Monuments and Objects and Scenes of Natural Beauty and for Preserving the Aspect of Historical Cities* (Cambridge: Cambridge University Press, 1905).

13   With the exception of Salima Hellal, 'Les Etudes de législations comparées du patrimoine en France, 1887-1913' (Mémoire de Maîtrise, Université Paris-IV, 2000).

14   Speitkamp, *Die Verwaltung der Geschichte*, 315.

15   Charles Normand, *Ministère du Commerce, de l'Industrie et des Colonies. Exposition universelle internationale de 1889. Direction générale de l'exploitation. Congrès international pour la protection des œuvres d'art et des monuments, tenu à Paris*

developments, but attempted to establish universal principles. Efforts to create an international committee overseeing the establishment of national legislation failed, but congresses regularly issued guidelines and drafted petitions to be sent to national governments.[16]

Furthermore, the late nineteenth century produced a wealth of comparative studies on monument protection, commissioned by both private preservation associations and state agencies.[17] They were often published as parliamentary papers,[18] but the topic also became fashionable for doctoral dissertations, memorial lectures, and scholarly articles. These studies generally discussed both attitudes to 'heritage' and institutions established in other countries. Despite the presence of some extra-European delegates at congresses, the approach was largely Eurocentric, the countries were frequently titled 'the civilized nations of the present time'[19] and encompassed namely other European States as well as their colonial legislation and to a far lesser degree the United States of America.

Within the domestic scene, proponents of competing ideas sought to bolster their desired policy for the preservation of cultural heritage by citing examples culled from other countries. Legislation was the area in which intercultural transfer was discussed most openly, transcending intellectual and administrative circles. For instance, in a number of regional German newspapers, reporting on the

---

du 24 au 29 juin 1889. Procès-verbaux sommaires. Rédigés par le secrétaire général Charles Normand, architecte diplômé par le gouvernement, directeur de l'Ami des monuments, secrétaire générale de la Société des amis des monuments parisien (Paris: Imprimerie Nationale, 1889); Jules Challamel, 'Exposition universelle de 1889. Premier congrès international officiel pour la protection des monuments et oeuvres d'art. Des législations françaises et étrangères établies pour assurer la conservation des oeuvres d'art et des monuments,' *L'Ami des Monuments* 4 (1890), 225-300.

16 *Bericht über den VIII. Internationalen Architekten-Kongress, Wien 1908* (Vienna: Schroll, 1909), 255-57, 648, 680; Astrid Swenson, 'Conceptualizing "Heritage" in Nineteenth- and Early Twentieth-Century France, Germany and England' (PhD diss., Cambridge University, 2007), ch. 3.

17 MAP 80/1/32-33, 80/1/61; Geheimes Staatsarchiv Preussischer Kulturbesitz, Berlin (GStA PK) I. HA Rep. 76 Ve Kultusministerium, Sekt. 1, Tit. VI, No 141; The National Archives, London (TNA) WORK 14/134, WORK 14/2278; Society for the Protection of Ancient Buildings Archives, London (SPAB) case files 'Ancient Monuments Acts 1867-1913,' 'Ancient Monuments Acts: correspondence;' 'List of Books in the Office.'

18 'Reports from her Majesty's Representatives Abroad as to the Statutory Provisions Existing in Foreign Countries for the Preservation of Historical Buildings. Presented to the House of Commons by Command of Her Majesty in Pursuance of their Address,' *Parliamentary Papers* 72 (July 30, 1897), 367; 'Reports Showing the Systems Adopted in Certain Foreign Countries for the Preservation of Ancient Monuments,' *Parliamentary Papers* 73 (1912-13), 1; 'Reports from his Majesty's Ambassadors at Vienna and Paris Showing the Systems Adopted in Hungary and France for the Preservation of Ancient Monuments. Presented to both Houses of Parliament by Command of His Majesty,' *Parliamentary Papers* 14 (February, 1914), 295.

19 Exemplified in the title of von Wussow.

demolition of local monuments, a comparison with foreign examples was made. When the *Allgemeine Zeitung* reported the planned destruction of a gothic town house in Nuremberg, it lamented that if only Germany had adopted a Law of the French variety this could be prevented, and urged Germany to emulate the other country's example.[20]

The argument was a potent one, as the protection of monuments was declared to be a measurement for civilization: 'The care of ancient monuments, in the way we understand it today, so that it can be regarded as a yardstick of a people's cultural attainment, is a modern concept and a modern activity.'[21] Coming from a leading German preservationist, the statement may sound self-important and pompous, but it was the argument in favor of heritage legislation which was most often used in all three countries and which proved most effective. Preambles, treatises and legal comments time and again referred to superior monument legislation abroad, advising the legislator to 'catch up' with the other 'civilized countries', namely the chief European powers.

To be surpassed by less civilized countries in such a civilizatory task was declared to be even less acceptable: the General Assembly of Historical Associations of Germany expounded in 1899 the necessity to improve monument legislation in Germany, as 'with regard to the legal protection of monuments, the North African States are at the present moment way superior to Prussia, Saxony, Wurttemberg: that is a dishonorable and untenable state of affairs.'[22] Civilizatory competition worked equally well on the intra-national level. Within the *Kaiserreich,* smaller States, such as Hesse, used the passing of a Monument Act to prove their modernity and fight off Prussian hegemony,[23] whereas Prussian proponents of legislation argued that the Iron Kingdom could not drop behind other European powers and

---

20 'Mittheilungen und Nachrichten: Denkmalpflege in Deutschland,' *Allgemeine Zeitung* (March 7, 1899). See newspaper clippings in GStA PK, VI. HA, Rep. 92 Nl. Schmidt-Ott (M), A X 1, fol. 2.

21 'Die Denkmalpflege in der Art, wie wir sie jetzt auffassen, so daß sie als Grandmesser für den Kulturzustand eines Volkes bezeichnet werden kann, ist ein moderner Begriff und eine moderne Kulturbetätigung,' in Adolf von Oechelhaeuser, *Wege, Ziele und Gefahren der Denkmalpflege: Festrede bei dem Feierlichen Akte des Rektoratswechsels an der Großherzoglich- Technischen Hochschule Fridericiana zu Karlsruhe am 20. November 1909* (Karslruhe: Braun, 1909), 5.

22 'Die nordafrikanischen Staaten stehen heute, was gesetzlichen Schutz der Denkmäler betrifft, weit über Preußen, Sachsen, Württemberg: das ist ein unwürdiger und unhaltbarer Zustand.' 'Die Verhandlungen über Denkmalschutz und Denkmalpflege auf der Hauptversammlung des Gesamtvereins der deutschen Geschichts- und Alterthumsvereine in Straßburg,' *Die Denkmalpflege* 1 (1899), 106-7.

23 *Verhandlungen der zweiten Kammer der Landstände des Großherzogtums Hessen in den Jahren 1900/1903. 31. Landtag, Protokolle* (Darmstadt: Leste: 1902), vol. 3, no. 71, 1808 (Ministerialrat Freiherr von Biegeleben).

smaller German states.[24] Countries advanced in legislation, such as Italy, argued that they had to further improve their system to maintain their status and provide the world with an imitable example.[25] Monument Laws in colonies were defended following the same logic. Lord Canning, Viceroy of India, declared already in 1862 that neglecting the care of ancient monuments 'will not be to our credit as an enlightened ruling power'. The protection of monuments soon became a yardstick of a nation's ability at colonial rule.[26]

The view that the preservation of the nation's monuments was an index of civilization, however, conflicted with an argument popular in other circles that the extent of a people's civilization could be measured by its provisions for the protection of private property: 'It has been remarked that the larger and more advanced the state the less easy is it to frame and to pass a satisfactory monument law'.[27] In order to be effective, a Law had to interfere with the rights of private property, 'and in advanced communities the individual has considerable self-assertiveness, and actively contests such proposed interference.' Consequently, 'for really drastic enactments, we have to go to countries [...] where the personal rights of the highly civilized man are almost unknown, such as in 'orientally governed'[28] States, or the British and French colonies.[29]

The French Senator Emile Combes[30] thus objected during the reading of a Bill in 1886: 'I must confess, gentlemen, that I am only mildly touched by the examples borrowed from foreign countries. That Sweden and Norway, Greece and Turkey, that Spain and Austria-Hungary... '. The former Minister of Public Instruction Bardoux, upon whose report the session was based, interrupted, shouting: 'and Italy! and England!', while Combes continued: 'From the fact, that other European States have overtaken us on this way, that they were the first to have such an exceptional law, I will not conclude that their behavior imposes that

---

24   *Stenographische Berichte über die Verhandlungen des Preußischen Hauses der Abgeordneten, 19. Legislaturperiode, Session 1901, March 12, 1901,* (Berlin: Moeser, 1901) vol. 3, cols. 3253-4 (Abgeordneter Seydel); "Entwurf eines Ausgrabungsgesetzes," *Stenographische Berichte über die Verhandlungen des Preußischen Herrenhauses in der Session 1914, Drucksachen* (Berlin: Sittenfeld, 1914), no. 9, 8-17.

25   'Camera dei Deputati, Legislatura XXIII, Sessione 1909, Documenti, Disegni di Legge e Relazioni, Nr 61 Disegno di Legge presentato dal Ministro dell'Istruzione Pubblica (Rava), di concerto col Ministro del Tresor (Carcano) et col Ministro di Grazia e Giustizia e dei Culti (Orlando V.E.) Per le antichità et le belle arti. Seduta del 30 Marzo 1909,' preamble, *Atti Parlamentari,* MAP 80/1/32.

26   As quoted in Brown, *The Care of Ancient Monuments,* 232-35.

27   Ibid., 44.

28   Ibid.

29   Cf. ibid, 235.

30   On Combes see Gabriel Merle, *Emile Combes* (Paris: Fayard, 1995).

we imitate them. I won't go to the Turks or even the Spaniards to look for models of liberty.'[31]

Combe's objection to the Bill stemmed less from a general disdain for ancient monuments than from his senatorial role as a defender of provincial interests, in this case protesting against encroachments on the authority of provincial mayors. (Somewhat unwillingly he equally protected the interest of the Church, despite being one of the most outspoken anti-clericals of the Third Republic.)[32] But the fact that he did so by attacking the foreign examples brought forward by Bardoux shows that, although at the moment of elaboration the broadest possible group of countries was compared, it was thus important to choose examples not only according to the efficiency of their legislation, but also according to the general stereotypes that existed about them. Patterns of citation of different countries varied from the private to the public, as it was inconvenient to refer to certain countries in public debate which were ranked low in terms of 'civilization' and hence could not be adduced as examples to emulate. As a general rule, British preservationists favored to refer to France and Germany, German ones to France and Italy, and French 'heritage-makers' to Italy and England.

For this, the debate over the listing of private property is a salient example. In the elaboration of the French Monument Act one of the main reasons for the delay between first draft and promulgation was the interference with the rights of property which made it a so-called 'exceptional law'. Whereas the first draft only included compulsory clauses for monuments owned by the State or by public corporations, the State-Councilor Courcelle-Seneuil, one of France's leading liberal economists, decided to include privately owned monuments in the scheme.[33] In his 1881 report to the State Council, Courcelle-Seneuil legitimized this choice by quoting 'an English Act from February 1880'.[34] He referred to John Lubbock's project, adopted in 1880 but then rejected, presenting it erroneously as an Act that has already been passed. Lubbock's project precisely foresaw a

---

31 'J'avoue messieurs, que je suis très médiocrement touché des exemples empruntés à l'étranger, que la Suède et la Norvège, que la Grèce, que la Turquie, que l'Espagne, que l'Austrie-Hongrie.' (Bardoux 'et l'Italie ! Et l'Angleterre') 'que d'autres Etats de l'Europe nous aient devancés dans cette voie, qu'ils aient eu la primeur de cette loi d'exception, je n'en conclurai pas [...], que leur conduits s'impose à notre imitation, et je ne m'aviserai pas d'aller chercher chez les Turcs ou même chez les Espagnols des modèles de liberté,' in *Journal Officiel* Annexe 83, Séance du Sénatdes 10 et 13 avril 1886, Première délibération, 599.

32 By a certain irony of fate, the Separation of Church and State (1905) of which Combes was one of the main instigators led to a major reshuffle in monument politics that led to even stricter provisions imposed on local authorities.

33 MAP 80/1/19, dossier 'Avant-Projet Rousse.'

34 'une loi anglaise de février 1880.' *Documents parlementaires* : 'Conseil d'Etat, sections réunies de l'intérieur et de législation. Annexe au no 364 du projet de loi pour la conservation des monuments et objets d'art ayant un intérêt historique et artistique. Rapport présenté par Courcelle-Seneuil, conseiller d'Etat le 28 février 1881.'

commission, composed of Trustees of the British Museum, to have authority over monuments in private hands. From the comparative material at his disposition, Courcelle-Seneuil could have invoked a Swedish Law of 1867, limiting private property on monuments and prohibiting owners from destroying or degrading the monuments as well as requesting the State's permission prior to any works. However, the reporter chose the English example, as England had a particular reputation concerning the precautions that were usually taken to protect private property.[35] The case was taken up by subsequent reporters, including Bardoux. Responding to the report of the latter at the Senate, the aforementioned Senator Combes objected again:

> The example of England suffices not to win my opinion. I will definitely not be persuaded that a nation so jealous of her rights of public and private liberties as England could voluntarily have sacrificed them in the interest of morality, of art and history, without making the act a most passionate one: for art and history have never, that I know, impassioned so sober a race as the English.[36]

Combes' objection did not convince the Senate to change the clause. It maintained the State Council's decision to include both public and private monuments in the *classement*. It proved to be a productive misunderstanding. The presumed 'English Law' had been a rejected Bill, a rejection stemming from precisely the motives Combes had guessed. When the Ancient Monument Act was finally voted in 1882, it only effectively protected prehistoric monuments and all compulsory clauses had been deleted.[37] However, the categories were subsequently enlarged and other types of monuments were included in the following decades. Those promoting the widening of categories constantly quoted the French law of 1887 as a major model.[38]

This particularly fruitful misunderstanding exemplifies a more widespread process. In the multilateral transfers and retransfers between approximately twenty countries, Bills rejected in one country often served as a model for drafts in others European countries or in the colonies, where legislation could be much stricter.

---

35   Hellal, 'Les Etudes de législations', 33.

36   'L'exemple de l'Angleterre [...] ne suffit pas à me séduire [...] assurément l'on ne me persuadera pas qu'une nation comme l'Angleterre, aussi jalouse de ses droits de ses libertés, tant publiques que privées, les ait bénévolement immolés sans apporter à cet acte des tempéraments extrêmes, à un intérêt moral, à l'intérêt de l'art et de l'histoire qui n'a jamais passé, que je sache, pour passionner un peuple aussi positif que le peuple anglais,' *Journal Officiel*, Annexe 83, Séance du Sénat des 10 et 13 avril 1886. Première délibération, 599.

37   See Chippendale, 'The Making of the First Ancient Monuments Act', 1-55; Saunders, 'A Century of Ancient Monuments Legislation', 11-33.

38   *Parliamentary Papers* 72, 367; National Trust, *Annual Report* (1897), Appendix 3, 21-25; Hall, 'Affirming Community Life',149; Society for the Protection of Ancient Buildings, *Annual Report* 31 (1908), 92-103.

With a certain time-displacement, ideas were then often retransferred – frequently via another intermediate stop – to the original countries.

Ideas were thus clearly developed in a shared process. Their implementation in different countries, on the other hand, was also shaped by pre-existing administrative structures and political climates: anti-aristocratic reform affected the debate in Britain, secularization and centralization in France, federalism and the *Kulturkampf* in Germany. To a certain degree, the personal preferences of activists also played a role.

Consequently, provisions varied with regard to monument types (prehistoric, historic, natural, surroundings); with regard to owners concerned by legislation (the State, public corporations, private persons); with regard to the level of compulsion introduced and to the administration installed, i.e. the budget put at the disposition of said administration.

Provisions	Great Britain	France	Germany		
			Prussia	**Hesse**	**Olden-burg**
Pre-existing Administration	No	1830s	1840s	No	No
Major Acts	1882 1913	1887 1906 1913	Failed 1887 Failed 1901 (1902 /07 *Verunstaltungs-Gesetz*) (1914 *Ausgrabungs-gesetz*)	1902	1911
Prehistoric mon. Historic mon. Natural mon. Surroundings	Yes (limited, 1900) (No) No	Yes Yes Yes, 1906 Limited, 1906	Yes, 1914 (Yes), 1902 Yes	Yes Yes Yes Yes	Yes Yes Yes Yes
Owners concerned: State Public corporations Private owners	Yes No (Yes)	Yes Yes (Yes)	No Yes (Yes)	(No) Yes (Yes)	No Yes Yes
Buildings in use: Dwelling Houses Churches	No No	Yes Yes	Yes Yes	Yes Yes	Yes Yes
Compulsion	Low, 1913	High	Medium	High	High
Commission	Yes, 1908	Yes	Yes, 1890	Yes	Yes
Participatory model	No	No	Yes	Yes	Yes
Budget	Low	High	Medium	High	Low

Figure 6.1   Legal provisions in comparison by 1914

The existence of earlier decrees, as well as the anchoring of a preservation administration within government, had an impact on how far reaching later Monument Acts were. When comparing France and Britain it is tempting to conclude that the existence of the *Commission des Monuments historiques* (Historic Monuments Commission) at the heart of the Ministry of Fine Arts and Public Instruction in France as opposed to the absence of a Fine Arts Ministry in Britain was at least in part responsible for a the greater extent of legislation in France. However, adding Germany into the equation muddies the water. Here, only those States that had neither any long established preservation institution nor a *Kultusministerium* managed to implement legislation.[39] Across the board it was, however, important, for both passage and implementation, to have a government relatively in favor of ancient monuments. That being said, it is noticeable that toward the end of the period, demands from certain sectors of the public exceeded government's willingness to legislate in all three countries.

Keeping in mind the limitations set by larger political concerns, the debates about preservation legislation allow analyzing the influence of associations and public opinion. Although in France, the role of a centralized State administration in fostering legislation was unique among the three countries and resulted in France having the most extensive legislation, it is noteworthy that after the turn of the century, demands from the public – by as ideological opposed figures as the nationalist writer Maurice Barrès and the anti-clerical founder of the Society for the Protection of Landscape, Charles Beauquier – for a stronger protection of the local 'heritage', went much farther than the aims of the Ministry of Fine Arts. Although ultimately unsuccessful in achieving the protection of the local 'heritage', they nevertheless left a mark by enlarging the patrimonial field.

In Germany, despite several attempts by distinguished individuals and the historical and architectural associations – no Monument Act was established at the imperial level because of the opposition of the individual States fearing a loss of competences. Preservation remained the domain of the different States. Yet collaboration among preservationists in the different States was intense. Comparatively, German public opinion was probably most strongly in favor of legislation. Moreover, the campaigns by preservation associations were viewed favorably by the different governments (unlike in Britain and France). Many a government attempting to pass legislation used the approval of the associations to back its schemes;[40] however, at the same time, opposition came from the most diverse quarters, including other government resorts, the Churches, local authorities, but also many State Conservators who felt that a law might rather reduce their leeway than enlarge it. In a sense, State officials were more reluctant about legislation than the preservation movement or the legal profession. By 1914 nearly all German States had established preservation institutions, but only

---

39 Speitkamp, *Die Verwaltung der Geschichte*, 338.
40 Kultusminister von Studt to Innenminister, December 21, 1900, GStA PK, I. HA Rep. 77 Ministerium des Inneren, Tit. 1215, Nr. 3, Beiheft, Bd. 1, fols. 1-2.

a few had passed formal Monument Acts, all of them relatively small territories. Many States instead used existing provisions spread across the law, and some States, particularly Bavaria, encouraged local by-laws. The strong emphasis on a participatory model, in which civil servants, learned experts and owners collaborated, was to a certain degree a pragmatic way of dealing with potential opposition, but also reflects trust in the preservation movement and the belief in decentralized solutions.

In England, the 1882 Ancient Monuments Protection Act did not contain any compulsory clauses and initially only included prehistoric monuments. Subsequently, more categories of monuments were incorporated, but dwelling houses and churches in use remained excluded. State institutions with some legal leeway were created since the Edwardian years, through the foundation of Royal Commissions and Advisory Boards. Unsurprisingly, England placed the strongest emphasis on the role of the private sector with the foundation of the National Trust.

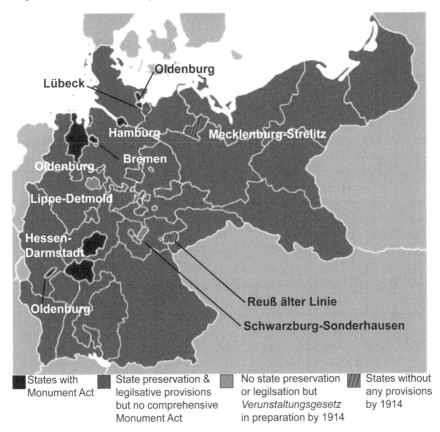

| States with Monument Act | State preservation & legilsative provisions but no comprehensive Monument Act | No state preservation or legilsation but *Verunstaltungsgesetz* in preparation by 1914 | States without any provisions by 1914 |

Figure 6.2   **Map showing forms of state preservation and legislation in Germany**

(At the time a unique model in Europe it was imitated by France and Germany during the 1990s in order to privatize monument preservation, remedying these countries' alleged 'backwardness'). Yet, again, it was the private sector – not the government – that was most strongly in favor of State intervention. The number of people actively campaigning for legislation was much smaller than in Germany, and they were more divided as to the solutions they favored. But, it is proof of the influence of this 'small knot of cultivated people'[41] on the political elite that legislation ultimately succeeded at least partially.

The comparison shows that judgments on 'delay' and 'backwardness' are highly problematic. For categories of 'pioneer' and 'latecomer' are generally dependent on perspective and subject to constant change.[42] Assessment varies whether we ask: delay with regard to whom? At which precise moment? With regard to what? Any government intervention as such? A proper Monument Act? A Law protecting all monuments? A Law with compulsory clauses? The application of the Law? The reaction of the public? A Law that seemed advanced in its own time or advanced today? For contemporaries, highlighting their own 'backwardness' was mainly a rhetorical strategy used to convince opponents. The choices of countries portrayed as superior were linked to ideas about civilization. At the same time, a reading of comparative studies also shows that authors sincerely admired different aspects of other countries' 'heritage' system at different moments.

The question of 'failure' versus 'success' is equally subjective. If comparing original drafts with final Acts, every Act could safely be called a failure. To quote Gerald Baldwin Brown: 'It has been pointed out already that in draft acts we find a completeness and a logic which measures actually passed after parliamentary struggles have seldom retained.'[43] All Acts passed in Germany, France and Britain had to make compromises with regard to the respect of property rights and budgetary constraints. Yet to interpret this as a failure of preservation would be misleading. Despite considerable resistance from diverse opponents in all three countries, the idea that the State had to protect the 'national heritage' was largely accepted on the eve of the First World War. At the beginning of the period it was by no means taken for granted that State intervention should be the favored course of action, not even by ardent preservationists. Yet, by 1914, it was a truth almost universally acknowledged, even by those whose interests were not particularly antiquarian. Through the international collaboration and competition, preservation, and especially formal Monument Acts, had become an attainment thought to define the modernity and advancement of a given State. In short, a status symbol.

---

41 William Morris, quoted in Chris Miele, "'A Small Knot of Cultivated People:" William Morris and the Ideologies of Protection,' *Art Journal* 54 (1995), 77.

42 Hartmut Berghoff and Dieter Ziegler, eds, *Pionier und Nachzügler? Vergleichende Studien zur Geschichte Großbritanniens und Deutschlands im Zeitalter der Industrialisierung. Festschrift für Sidney Pollard zum 70. Geburtstag* (Bochum: Brockmeyer, 1995), 28.

43 Brown, *The Care of Ancient Monuments*, 107.

Chapter 7

Heritage and its Communities: Reflections on the English Experience in the Nineteenth and Twentieth Centuries

Chris Miele

There are many reasons for conserving old buildings, and many different mechanisms. Ultimately, the systems which have arisen over the last the last two centuries have all relied more or less on political justifications that find their way back to notions of community. This chapter looks at the way one institution, the Church of England, used medieval parish churches to articulate ideas of community at the local level. I offer these observations in the context of a new community agenda which is shaping the environmental thinking here in England and across our political culture more generally.

Heritage has always had a special relationship with community, and this sets it apart from other parts of the environmental movement. Nature conservation has practical uses. It supports public health. It is also part of sustainable resource management. It is, though, harder to quantify the operative benefits of heritage. Reflecting on Benedict Anderson's pioneering analysis of cultural identities, one could say that heritage provides an opportunity for groups to imagine themselves beyond the bounds of daily, face-to-face social and economic relationships.[1] The power of heritage to shape shared identities extends beyond national boundaries, and this was particularly true of the English Gothic Revival which provided a sense of common purpose far outside the geographical boundaries of the United Kingdom, amongst English speaking peoples united by the Anglican (Episcopalian) Communion in the far flung corners of the Empire.

Until recently in this country official justifications for regulating the historic environment have rested on notions of its benefit to a generalized, imagined community expressed through the words 'our', 'us' or 'we' or through broad concepts such as 'inheritance' or 'patrimony'.[2] Such words assume a shared

---

1   Benedict Anderson, *Imagined Communities: Reflections on the Spread and Origin of Nationalism* (London and New York: Verso, 2006), esp. 10-13, 37, 44-45, 52-56, 67-69, 200-206.

2   On post-war legislation in the UK, see John Delafons, *Politics and Preservation: A Policy History of the Built Heritage, 1882-1996* (London: E & FN Spon, 1997), 62-70, 77-115, 136-88. National planning policy on the historic environment was formerly set out in

interest limited to white, British Christians. This generalized, abstracted audience sharing a common set of values probably never existed in quite such simple terms and certainly does not now. 'Our' can no longer stand on its own as a universal shorthand for shared interest sufficient to justify the conservation and regulation of the historic environment. In a multi-cultural, pluralist and more affluent and democratic society, where citizens are empowered consumers of services, there needs to be a more direct and accountable linkage between cultural patrimony and what matters to individuals at the local level. This transformation is part of a larger change in public policy which is encouraging political and social renewal at the local level. Politicians in the UK are considering how to empower local communities directly, without the intermediary of local authorities, and so entrenched is this idea that it is hard to imagine changes in government will dislodge it any time soon. The reasons for this change are numerous. A leading Marxist writer on town planning and urban development, Yvonne Rydin, put her finger on one when she observed that the community empowerment agenda is part of a more wide ranging re-orientation of political priorities that emerged later in the 'New Labour' project.[3] Before then, British Liberal Democrats pioneered the notion of community politics, and the new coalition government is seeking to push the community engagement agenda even further.

The community agenda is certainly now firmly embedded in the practice of town and country planning, the all-embracing civil administrative system which regulates the use and physical characteristics of the land in the UK and which also secures the regulation of the historic and natural environments. Indeed, the government ministry responsible for planning was re-branded as Department for Communities and Local Government or 'Communities'.[4] The Department for Culture, Media and Sport – a government ministry which shares responsibility for the historic environment with DCLG – and its non-departmental, public body, English Heritage, have successfully repositioned heritage as a community activity in line with this larger policy agenda.[5]

Planning Policy Guidance Note 15: Planning and the Historic Environment (Department of the Environment; Department of National Heritage, September 1994; www.communities. gov.uk/planningandbuilding), esp. section 1.0, paras 1.2-1.6. This was revoked in March 2010 and replaced with *PPS5: Planning for the Historic Environment (DCLG)* which combines above ground heritage and archaeology (treated previously in PPG16). The new PPS aligns with the community and localist agendas.

3   Yvonne Rydin, *Urban and Environmental Planning in the UK* (London: Macmillan, 2003), 67, 68-82.

4   See www.communities.gov.uk.

5   The emerging orthodoxy is set out in 'Conservation Principles,' *English Heritage* (May 2008; www.english-heritage.org.uk). Other important policy statements include 'Power of Place,' *English Heritage* (1998), and the *Heritage Matters* series (www.english-heritage.org.uk), published annually. More recently the Department for Culture, Media and Support (which has taken over the heritage competency formerly discharged by the Department for National Heritage) has published a draft 'Heritage Bill' (April 2008). This

Whatever the reasons, the official embrace of community empowerment in all policy sectors represents a major change especially when one considers that as recently as the early 1990s 'community' was still a grass roots concept not official policy. The conservation movement we have today in the UK was the creature of community, grass roots activism in the sixties, seventies and eighties.[6]

Heritage and its communities have, however, never been fixed. If we assume an historical perspective, heritage appears first as an elite pastime which served the interests of a small and very narrowly defined community. The monuments of English antiquity gained a wider appeal in response to the French Revolution, when antiquaries and artists in England came to think of their native historic buildings (parish churches, cathedrals, abbeys and castles) as symbols of national identity, standing endorsements of gradual, evolutionary change against revolutionary innovations surging across the Channel.[7] This native heritage constituency expanded rapidly during the nineteenth century as greater literacy, increased leisure time and higher wages made historic buildings culture accessible to an increasingly wider audience. This audience had access to a publishing network. All of these developments, in turn, led to a recognizably modern, leisure-tourist economy.[8] The Reform Act of 1867 made more British government democratic and on the back of that there came the first generation of environmental pressure groups.

Throughout this period the several heritage cultures that promoted monument preservation (municipalities, the landed elite and the clergy) and the related tourism industry had explicit ideological agendas. Finally, towards the end of the century, one particular community came to assert its right to act on behalf of the community in a more neutral and even-handed way, dissolving, perhaps even disguising its own ideological agenda in a generalized sense of 'public interest'.[9]

---

seeks to integrate heritage more fully into local communities through greater emphasis on consultation at every level. The planning ministry website, www.communities.gov.uk, provides more information on the new community planning agenda.

6   Lionel Esher, *A Broken Wave. The Rebuilding of England, 1940-1980* (London: Allen Lane, 1981), 139-71. For an overview see Barry Cullingworth and Vincent Nadin, *Town and Country Planning in the UK* (13th ed., London: Routledge, 2002), 18-33.

7   Simon Bradley, 'The Gothic Revival and the Church of England,' (PhD diss., University of London, 1996), 97-100, 157, 234-37.

8   Elite interests have not been considered as such in any one study to date, but the best overview remains Charles Dellheim, *The Face of the Past: The Preservation of the Medieval Inheritance in Victorian England* (Cambridge: Cambridge University Press, 1982) and Peter Mandler, *The Fall and Rise of the Stately Home* (New Haven: Yale University Press, 1997), esp. 21-69, 75-76. See also Chris Brooks, *Gothic Revival* (London: Phaidon Press, 1999), and the more general observations of David Lowenthal, *The Past is a Foreign Country* (Cambridge: Cambridge University Press, 1985), and Raphael Samuel, *Theatres of Memory* (London: Verso, 1994), 5-48.

9   Melanie Hall, 'Affirming Community Life: Preservation, National Identity and the State, 1900,' in Chris Miele, ed., *From William Morris: Building Conservation and the Arts*

It is this jump from the explicit ideological agendas of the Victorians, formulated around notions of particular communities, to notions of generalized public benefit in the early twentieth century that I consider in what follows.

What interest, then, did the Victorians think they were protecting when they conserved, modernized or refurbished historic buildings? A good place to start is with those images which A.W.N. Pugin added to the 1843 edition of *Contrasts*. Here is an archetypical English town in 1540 compared with that same town three hundred years later. Pugin portrayed the changes brought to medieval cities by industrialization as evidence for a morally bankrupt system. Based in materialism and the modern money economy, the modern world had broken traditional ties of community and replaced them with relationships based on personal and class interests, Pugin's solution was to fix the broken container: restore the monuments and design new ones according to ancient principles, and, with a wave of the wand, proper communal relations return.[10] The Ecclesiologists promoted this vision from an Anglican perspective, and it was of a piece too with the Tory Young England movement. The Gothic Revival in architecture drew strength from this critique which had parallel expressions in Victorian literature and painting.

Victorian ideas of community life directly affected the way that historic buildings were cared for, studied, used and interpreted in all media. In another publication Peter Mandler analysed this phenomenon in relation to 'stately homes', the ancient country estates of the aristocracy and landed elite. This group gradually opened up their historic houses to the public as part of a wider strategy that sought to maintain class privileges and power in a time of progressive and broadly democratic reform.[11] Landed elites expanded access to their private domains in order to persuade a wider constituency that their interests aligned with the national interests.

Ancient country houses were a perfect symbolic opportunity. Gradually over the nineteenth century their owners remade these buildings as images of national identity, and imbued them with an aura of common, cultural property without ever actually ceding ownership. This transformation came about when the particular history of an estate, house or family was related to larger nation-forming events in antiquarian literature. In this way once closed entities became part of a national story. Buildings were opened for the enjoyment of genteel tourists. Estates were made available from time to time for leisure visits. The message that this opening up delivered was, simply, that these private fiefdoms were not bastions of reactionary power but part of a common national identity, illustrating traditional English values. And so one narrow class promoted heritage as a shared community

---

and Crafts Cult of Authenticity (New Haven: Yale University Press, 2004), 129-57.

10   Brooks, 232-58; Rosemary Hill, *God's Architect: Pugin and the Building of Romantic Britain* (London: Allen Lane, 2007).

11   Mandler, *Fall and Rise*, 71-106, esp. 81-82, 90.

resource in order, effectively, to disguise their interests and so to preserve their power or prerogatives.

A similar process affected the care and appreciation of historic churches, abbeys and cathedrals. A widespread interest in religious monuments as sites of national commemoration emerged in the second half of eighteenth century. There is ample evidence for this early heritage movement in the pages of *The Gentleman's Magazine*, a literary and political miscellany. Cultural tourism began to emerge during this same period, notably at Westminster Abbey.[12] These interests influenced the internal literature of the Church of England in the 1820s and 1830s and spread rapidly across a wider segment of society through cheaper periodicals, books, journals and mass-produced images which made ancient parish churches, abbeys and cathedrals accessible to a wider and more diverse audience. Through this new information, the articulate, educated public at large came to understand the historical position of their local parish church and through it to appreciate the linkage between it and wider currents in regional and national histories. At first the styles of English architecture were tied to the reigns of English monarchs, but very quickly publications related the history of particular buildings to notable people and events of the parish or region. In this way the two narratives – national and local – were combined in single buildings used for communal worship.

By the 1850s early guidebooks make it clear that these buildings had become popular tourist destinations, many tied together by new railways. In this same period, from 1830 to the 1850s, the several agents of the Church of England were busy transforming these buildings into symbols of parish renewal. Improving accommodation for local people whilst restoring the ancient style of sometimes decrepit buildings was part of a deliberate policy intended to make the Church more relevant at the local level.

In place of private pews and monuments to individual wealth came clean, warm, well-lit open interiors decorated with biblical narratives and references to local and national sacred histories. National saints such as St. Augustine of Canterbury or the Venerable Bede mingled with images of the local patron saint. These changes, which I shall come back to, were intended to give all parishioners an equal opportunity of participating in services without reference to their class or station. Equality of all in the eyes of God, in a revived ancient setting, supported

---

12   See, for example, [David Henry], *An Historical Description of Westminster Abbey, its Monuments and Curiosities ... Designed Chiefly as a Guide to Strangers* (1753; London [1778]), which was much reprinted, including in inexpensive, abridged versions such as *A Description of Westminster Abbey, its Monuments and Curiosities* (Brentford, n.d. [1770]). See Ian Ousby, *The Englishman's England: Taste, Travel and the Rise of Tourism* (Cambridge: Cambridge University Press, 1990). I am grateful to Melanie Hall for drawing my attention to these early sources. On the Abbey as a site for national commemoration see Philip Connell, 'Death and the Author: Westminster Abbey and the Meanings of the Literary Monument,' *Eighteenth-Century Studies*, vol. 38 (Summer 2005), 557-85.

**Figure 7.1   Church of St. Mary, Stafford. Interior restored and reseated for
communal worship by the young George Gilbert Scott**

Source: Rev. John Masfen, *Views of the Church of St. Mary Stafford* (London, 1852)

the established order of things in society, a linkage which the promoters of church restoration made expressly.[13]

The task of Victorian Churchmen who sought to reassert the centrality of religious buildings in community life was far easier than the owners of country estates. As places of worship, legally open to all, ancient churches were, and in some places are still, obvious symbols of community life, and in England they also have a very direct association with national identity. At the Reformation the Crown appropriated church buildings and estates, creating a national church, the Church of England, headed by the monarch. Anglican bishops took a hand in shaping public life and communities by direct involvement in the legislative process through *ex officio* seats in the House of Lords.

At a more practical, day-to-day level many would have seen their local parish as represented by a church, and hence the Church by law established. The parish itself remained an important administrative entity for most of the century. Through the activities of the local vestry, the parish (a Saxon survival) had an important civic role which continued in towns until the Municipal Corporations Act of 1835. Later acts of 1888 and 1889 (which created County Councils and County Boroughs) introduced secular local government in urban and rural areas. The parish church continued to be the place where many came to be christened, married and buried (excepting the minority populations of Jews, Catholics and 'Dissenters' as non-Conforming Protestants and Quakers were known). The established Church provided the first rudimentary, state-subsidized education through fee paying National Schools run at the local level and in some cases the basis for (very) rudimentary health care through local dispensaries and workhouse infirmaries. In many districts vicars took a leading role in public improvements and often ran or served on various charitable committees, building and endowing the cottage hospital, say, or promoting public health measures. In some parts of the country this customary role in local governance persists; in rural districts of England and Wales, parish councils are still recognized public bodies with a defined role in local administration, undoubtedly one which is set to increase in line with the new local empowerment agenda.

Even amongst atheists or non-Conformists, parish churches were appreciated as community buildings at a symbolic level. These structures were often the most prominent buildings in any settlement. Crowded with memorials and monuments, or surrounded by markers and monuments, they represented history. Their styles of architecture, whose precise dating was widely understood by the 1850s, linked the church fabric itself to larger historical events: Norman Conquest as represented in the Romanesque style (or 'Norman' to use the common English term for 'Romanesque), the Age of Monasticism and reform in the Early English, the Hundred Years War in the Decorated style, the Wars of the Roses and

---

13   Chris Miele, '"Their Interest and Habit": Professionalism and the Restoration of Medieval Churches, 1837-1877,' in Chris Brooks and Andrew Saint, *The Victorian Church. Architecture and Society* (Manchester: Manchester University Press, 1995), 151-72.

Reformation in the Perpendicular. There is very good evidence to show that non-Anglicans valued this aspect of Anglican churches, particularly in cities, though in places with a strong tradition of Nonconformity, there were instances of local people objecting to rates levied by parish vestries to maintain the physical fabric of ancient churches.[14]

So just as the owners of 'stately homes' made their estates more accessible in an attempt to preserve traditional privilege by accommodation to modern democracy, so did vicars, churchmen and the Church hierarchy offer up their ancient churches for public appreciation and use in order to defend the institution from charges of class bias and to shore up the Church's position in an age defined by rising secularism, the loss of traditional privileges and increased religious toleration. Church restoration and preservation, and in burgeoning cities new church buildings in the ancient styles, were part of a deliberate policy to sustain the position of the established church at the local level in the face of these challenges.

This attempt to represent England's medieval religious monuments as symbols of community renewal began in the visual arts during the late eighteenth and early nineteenth centuries, a period unsettled by revolutionary developments across the Channel. The artists Turner, Cotman and Prout portrayed England's ancient churches as proud and enduring objects, signs of durability and evolutionary change at a time of uncertainty. For Burke, turning against the French Revolution, the image of an ancient parish church as an historical object – standing solid against over centuries – was a metaphor for the English people and constitution. This view stood in defiance of the modern, radical innovations raging across the Channel under the aegis of a strong, centralizing authority. It was a commonplace then and in subsequent generations that the success of the nation derived from the vitality of local institutions.[15] These views come in antiquarian literature of the early nineteenth century and in Church publications.[16]

By the 1830s several tendencies converged. Art historical knowledge gave individual monuments a definite relationship to historical periods, at the

---

14   See for example the restoration of St. Mary Redcliffe Church in Bristol funded by a specially constituted society of Anglicans and non-Conformists, the Canynge Society (in memory of a fifteenth-century Bristol merchant who had previously contributed to the church fabric). This issued regular appeals from 1847 to the 1860s, and the editor of *The Builder* magazine (the architect of the restoration, see vol. for 1848, 114) promoted this method as a model for fund raising nationally. A similar venture was established to fund Cathedral restorations at York after the serious fire of 1829 and for Hereford Cathedral, as described in Robert B. Phillips, *Letter to the Landowners of the Diocese of Hereford ...* (Hereford: J. Head, 1847). Great minsters, abbeys and cathedrals were generally more likely to enjoy support across religious denominations than ordinary parish churches. For a general discussion of the local urban oligarchies which promoted restoration as part of local development initiatives, see Chris Miele, 'Conservation and the Enemies of Progress?,' 15-28.

15   Ibid., 13-14.

16   Bradley, 'The Gothic Revival'.

very moment when the owners of these buildings began to use them as agents of institutional renewal at the local level. The broad stylistic periods that were decoded by the architectural taxonomists Rickman and Bloxam, through their popular handbooks of medieval styles, gave any literate person the ability to relate the ancient parish church on their doorstep to the grand narrative of English history.[17]

The process started with the image of a typical parish church at the end of the Georgian period, stripped of its rich decorations during the Reformation, altered during the Commonwealth and turned, finally, during the eighteenth century, into a space that was divided up into private pews, let to wealthy and middle-class parishioners. Free seats for labourers were at the rear of the church or tucked into damp corners. The lighting was poor, the heating inefficient or non-existent. Ventilation was often inadequate and the build-up of burials outside was unhealthy. Services were irregular and lacklustre. Parish choirs were a rarity and there was no tradition of communal singing as had developed in non-Conformist chapels and meeting houses. Large pulpits with sounding boards or testers blocked views of the building. Galleries filled ancient arcades. All this made it hard to appreciate those architectural elements and features which linked the building and institution to that larger, common narrative of English history. Where obvious features were exposed, they tended to be in poor repair or badly weathered.

This is the picture that the Victorians like to paint of the ancient parish churches, cathedrals and abbeys which they inherited from the Georgians, and whilst it involved some exaggeration – the Georgians were not as irresponsible as later reformers like to claim – the gist of the criticism was fair.[18] And there can be no doubt that the condition of medieval churches at the end of the Victorian period was far better than at the beginning, accepting that this did entail the comprehensive modernization of church fabrics (a point of some later, and continuing, contention). This startling transformation was achieved through an impressive combination of institutional incentives and generous voluntary giving.

Some of the best early evidence for this new way of seeing heritage comes in the from 'Episcopal Charges' prepared by Anglican bishops for their clergy, documents

---

17   Chris Miele, 'Real Antiquity and the Ancient Object: The Science of Gothic Architecture and the Restoration of Medieval Buildings,' in Vanessa Brand, ed., *The Study of the Past in the Victorian Age* (Oxford: Oxford Books, 1998), 103-24. For Rickman see Megan Aldrich, 'Gothic Architecture Illustrated: The Drawings of *Thomas Rickman in New York,' The Antiquaries Journal* 65 (1985), 427-33; Megan Aldrich, 'Thomas Rickman (1776-1841)' (MA thesis, University of Toronto, 1983). The best ready summary of Rickman is Dr Aldrich's entry on him, in H.C.G. Matthew and Brian Harrison, eds, *Oxford Dictionary of National Biography* (Oxford: Oxford University Press, 2004).

18   C. Webster, '"Absolutely Wretched": Camdenian Attitudes to the Late Georgian Church,' and Simon Bradley, 'The Roots of Ecclesiology: Late Hanoverian Attitudes to Medieval Churches,' in Christopher Webster and John Elliott, eds, *'A Church As It Should Be:' The Cambridge Camden Society and Its Influence* (Stamford: Shaun Tyas, 2000), 1-21, 22-44.

which both set policy at the diocesan level and reported on its effectiveness. During the 1830s and 1840s these reports focus increasingly on the importance of church restoration in the encouragement of seemly, communal worship.[19] Architectural improvements increased capacity, made buildings more attractive and so would draw more regular worshippers. In this way, the Church could fight the growth of Dissent and non-belief, or so the argument ran. Bishops were generally concerned to avoid any greater interest in architecture being mistaken for adherence to Tractarian doctrines which smacked of Catholicism to many Anglicans of a Low Church – or more Protestant – stamp. A most important figure in this movement was the Right Rev. Samuel Wilberforce, Bishop of Winchester and then of Oxford (a relation of the more famous anti-slavery campaigner). Like many Bishops Wilberforce appointed Diocesan Architects to advise him and enforced the ancient system of church faculties (a kind of ecclesiastical planning permission or permit) to secure his reforms, often commenting directly on proposed works.[20] His particular interest was ensuring that the total volume of an ancient church (the sweep of space from nave and aisles to chancel) could be clearly seen from every seat, and equally that the design of new uniform benches (replacing privately rented pews) was conducive to kneeling and so encouraged group piety.

The attention that Wilberforce (and Bishops Denison of Sarum and Jackson of Lincoln) paid to the architectural details of a church restoration was unusual but the general principle of inclusiveness they put was a common theme: whatever else a church might be, it should have free seats for every parishioner. It should be welcoming, warm, orderly, tidy and attractive. There were other aspects to this policy: new parish schools, improved parsonages to encourage a higher proportion of resident clergy, and above all a more engaged clergy actively discharging their pastoral duties.

The attempt to make medieval churches more democratic and accessible reflected wider institutional reforms, and here the job which the Church hierarchy faced was particularly daunting.[21] As an institution, the Church of England was not democratic. It was the preserve, largely, of Oxbridge-educated sons of the gentry, aristocracy and the more successful of the mercantile classes. The stipends paid to vicars were generally not enough on to support the kind of gentility which was, for cultural reasons, expected of its ministry. Mindful of charges of favouritism and self-interest, the Victorian diocesan hierarchy tried to professionalize its

---

19   Chris Miele, 'The Gothic Revival and Gothic Architecture: The Restoration of Medieval Churches in Victorian Britain' (PhD diss., New York University, 1992), 339-47.

20   Ibid., 344-47; R.K. Pugh, ed., *The Letterbooks of Samuel Wilberforce, 1843-68*, Bucks and Oxon County Record Societies vol. 47 (1970); E.P. Baker, ed., *Visitation Returns for the Archdeaconry of Oxfordshire in the Year of 1854*, Oxon Record Society vol. 35 (1954).

21   On these changes generally see Arthur Burns, *The Diocesan Revival in the Church of England, c. 1800-1870* (Oxford: Clarendon Press, 1999), which provides the background to the institutional reforms discussed in this chapter.

clergy. This meant encouraging more frequent and seemly services, the creation of diocesan training colleges, the promotion of a professional literature to develop this new cadre and the raising of clerical stipends to make livings more accessible to people without independent incomes. Church finances were modernized, and bishops attempted to gain the 'patronage' of as many livings as possible, buying these from private individuals whose rights in the church derived ultimately from the transfer of Church property at the time of the Reformation. The decoupling of the Church of England from private landed interests was actively pursued as part of a modernizing agenda meant to put the Church back at the centre of community life, a position it was said to have lost to the several forces that defined modernity in the late seventeenth and eighteenth centuries: secularism, urbanization, industrialization.

To ensure that vicars were implementing these policies at the local level, bishops reinvigorated older systems of church governance. The increased use of faculties, already mentioned, was one example; another was the re-introduction of the offices of Rural Deans and Archdeacons to carry out regular inspections and comment on the physical improvements needed to improve the Church's profile within the community.[22] There was no particular sanction enforcing these reforms. It was open to parish clergy to ignore them. But the reforms worked because of shared customary practices and assumptions, and the age-old principle of wanting to please those in power to advance one's career. Thus the local clergy and particularly younger local clergy were all too eager to fall into line with this Anglican 'Counter-Reformation'.[23]

Traditional funding structures – church rates and occasional bequests – could not pay for the required far-reaching and often very costly physical changes. Consequently most bishops had by the 1860s established diocesan societies for building and restoring churches. These made grants which depended on the local vicar – the person generally in charge of promoting the work and commissioning it – raising 'match-funding' from other sources, and the most common source was the local parish community. The funding for these societies came from substantial voluntary donations from the diocese's leading Churchmen, an Anglican oligarchy consisting of merchants, aristocrats and gentry, and MPs. Whilst the actual grants were fairly small, the institution itself focused voluntary giving at the local level and so encouraged independent, private donations.

The Church Commissioners, who were in charge of Ecclesiastical estates and central to the modernizing agenda, promoted these same policy objectives at the

---

22   Miele, 'The Gothic Revival', 350-56; Richard C. Coxe, *A Cursory Survey of the Churches and Church Buildings within the Archdeaconry of Lindisfarne. A Charge Founded on the Reports of the Rural Deans* (London: Francis & John Rivington, 1860); W.B. Stonehouse, *The Office of the Rural Dean ... A Sermon ... Being the Annual Visitation of the Venerable Archdeacon of Stow* (London, 1841); W. Dansey, *A Letter to the Archdeacon of Sarum on Ruridecanal Charges* (London, 1840), 8-19, 39-45.

23   Alan Haig, *The Victorian Clergy* (London: Croom Helm, 1984).

national level, and particularly the provision of 'seats free and open to all', more systematically through an arm's length body, the national the Incorporated Church Building Society (or the Incorporated Society for the Promotion of the Enlargement and Construction of Churches and Chapels, hereafter ICBS).[24] Founded by statute in 1818 to distribute one million pounds allocated by Parliament specifically to bring the established church to rapidly developing suburbs and industrial centres (there was a half million pound allocation made in the 1830s), the ICBS redirected its efforts to established parishes from the 1830s through 1914 (and into the interwar years, on a much reduced level). The progress of its efforts can be followed through the ICBS' in-house journal *The Church Builder* (1862-1939).[25]

The ICBS' absolute requirement was that the proposed work should provide a significant increase in free seating and only where it could be demonstrated – statistically – that the church was under capacity relative to the local population. The Society encouraged the elimination of rented seats, but accepted that pew rents were in some cases an important source of income because many new churches

Figure 7.2   A rare survival of the standard plaque installed at as a requirement of ICBS funding to extend 'free seats' in parish churches, Church of St. John the Baptist, Pilton, Wiltshire

24   For a contemporary summary prepared by the Commission itself see *The Ecclesiastical Commission, or the Bishops as Church Reformers* (London: Society for the Liberation of Religion from State-Patronage and Control, 1860).

25   On the so-called 'Commissioners' Churches, see Michael H. Port, *Six Hundred New Churches: The Church Building Commission 1818-1856* (Reading: Spire Books, Ltd, 2006).

were in fact not endowed. In such cases, its grants were made on the basis that these would be indistinguishable from free seats. There was also a condition on

Figure 7.3 **Church of St. John the Baptist, Pilton, reseated in the 1860s with the help of an ICBS grant**

each grant that the new seats provided remain free in perpetuity.[26] The Society required grant recipients to display a notice in their churches confirming that all the seats were free and open to all, and attentive tourists today will still occasionally find hand-painted, timber notices to this effect in sleepy parish churches.

In its early years the ICBS did not have an architectural agenda. Its surveyor, J.H. Good (a pupil of Soane), was mostly concerned to ensure that designs were structurally sound and that new seats conformed to standard dimensions. By the 1840s, however, the ICBS came to support schemes which were also well considered architecturally, in the Gothic style. Practically this meant that the Society only funded schemes which retained and revealed significant architectural features or restored lost ones in a correct medieval style. New extensions or features were likewise, increasingly, considered from an architectural point of view. Good was replaced by a Committee of Architects consisting of professionals who had demonstrated skill in the design of new buildings in an authentic Gothic style. This included George Gilbert Scott, R.C. Carpenter, Anthony Salvin, and Benjamin Ferrey. The Committee was expanded a year later to provide better coverage.[27] In this way 'church extension' – as the work of providing free seats was sometimes called – was joined to the Gothic Revival and associated ideas of monument care. It was not enough for the poor and working classes to be welcome in churches. The setting for this worship had to be in the authentic historic style of the national church.

Applicants for ICBS funding invariably described how the poor condition of their churches was destroying the parish community. The usual faults – the lack of free seats, combined with poor ventilation, dank, ill-lit and cold interiors, the absence of a modern organ and dedicated choir seating, the lack of spaces for children – were preventing local people feeling comfortable in church. Meanwhile, invariably, applicants described how the beneficiary was one or another thriving non-Conformist chapel nearby. The standard application forms sought information on population, church provision and the number of Dissenting meeting places in the locality, and a percentage formula was applied to calculating the number of seats a church ought to have given the local population. The applicants described their situations as dire, and in alarmist terms because, of course, grant funding was dependent on need and threat. The Society had no way of verifying the information

---

26   Tim Parry, 'The Incorporated Church Building Society, 1818-1851' (DPhil. thesis, Trinity College, Oxford, 1984). The ICBS is also discussed at length in Miele, 'The Gothic Revival', 359-68; and Miele, 'Their Interest and Habit,' 166-70, which reviews of the processes and mechanism discussed in this paper. The ICBS papers (plans, drawings, letters and application forms) are available in their totality on line at www.churchplansonline.org, maintained by the Lambeth Palace Library. This is a remarkable resource for the study of nineteenth-century Anglicanism, church building and restoration.

27   Christopher Webster, *R.D. Chantrell (1793-1872) and the Architecture of a Lost Generation* (Reading: Spire Books Ltd, 2010), 150.

submitted to them, and so modern historians must treat the descriptions which survive in the ICBS' archives with some care.

The published literature that accompanied restoration appeals reflected these same ideas. The authors of fundraising pamphlets used historical narratives to describe the local ancient church, abbey or cathedral as repositories of collective memory and symbols of local identity. The venerable church yard with its burials or the church interior filled with tombs and memorials were the solid reflections of generations. The object of these sentimental descriptions was clearly to encourage donations, but this literature should also be read as an attempt to assert traditional authority. As if to lend substance to their authors' claim for the Church's, and hence a church's, community identity, fundraising literature usually included a list of subscribers. Most list every contributor's name beside their contribution and the order adopted was very often alphabetical, by surname, not by size of donation or social status, to reinforce that this was a community undertaking. There are, it must be said, also examples of subscription lists that do differentiate socially, often by presenting the local clergy, bishops, local aristocrats and gentry at the head of the roll call, with the alphabetical list of 'commoners' to follow (though in my experience this approach was less common). As a matter of fact, however, the more substantial part of any restoration scheme was funded by a relatively small number of wealthy contributors.[28]

This same desire for the wealthy and powerful to submerge their personal identities and prerogative in a collective image influenced the nature of personal bequests. One very great art historical change illustrates this point. The Church of England is admired for promoting a magnificent culture of personal memorials in the eighteenth and early nineteenth centuries. This tradition changed rapidly and dramatically after 1840 such that by the 1870s the clear preference was for memorials which took the form of liturgical furnishings or decorations.[29] If one looks carefully at these later Victorian features, then you will almost always find a personal dedication. These, though, are subordinate to the design of the whole piece and so tend to have an anonymous quality. Indeed, many such bequests were made anonymously. This shift from private memorial to public benefaction reflects piety and commitment to service that characterized late Victorian religion

---

28   Miele, 'The Gothic Revival,' 393-407, 424-37. There are many examples published in contemporary accounts. See Phillips; also numerous pamphlets in the ICBS Archives, Lambeth Palace Library (www.churchplansonline.org), for example, *Public Subscription List for the Restoration of Newark-on-Trent Parish Church* (1852), *Restoration of Heckington Church* (1864), *The Restoration of the Parish Church of St. Nicholas, Great Yarmouth* (1848), or C.F. Watkins, *The Basilican Church of Brixworth* (1867).

29   James Markland, *Remarks on the Sepulchral Memorials of the Past ... with Some Suggestions for Improving the Condition of Our Churches, a Letter Addressed to the Oxford Architectural Society* (Oxford: John Henry Parker, 1840). Both the Cambridge Camden, later Ecclesiological, and Oxford Architectural Society were very influential in encouraging this trend.

and political culture. Instead of private monuments, people contributed liturgical features – pulpits, lecterns, reredoses, organs, pavements, choir stalls, and of course stained glass windows – benefiting the whole of the congregation by enhancing communal worship.

Perhaps the best visual image of how an ancient church, and hence the Church, might define and reinforce local communities are a series of prints made to mark the restoration of Kilpeck Church in Herefordshire, a building today known largely for its splendid interlace carvings, believed to reflect the survival of Anglo-Saxon craft traditions after the Conquest.[30]

INTERIOR of KILPECK CHURCH.

London, Published by G.R. Lewis. May. 1841

Figure 7.4a   An ideal image of the restored interior of the Church of St. Mary and St. David, Kilpeck, where social differences have been eliminated to enable humble parishioners direct access the site of the Sacrament

Source: G.R. Lewis, Illustrations of Kilpeck Church, Herefordshire (London, 1843)

One view shows local people making their way in a happy group to the newly restored building. Another view, of the interior looking towards the chancel, shows a person of modest means prostrate before the Chancel arch, the nave stripped of any liturgical furnishings. Another worshipper looks on from the door waiting to

---

30   George R. Lewis, *Illustrations of Kilpeck Church, Herefordshire* (London: G.R. Lewis and William Pickering, 1843).

Figure 7.4b   The restored church at the centre of community life, as the residents make their way to the ancient Church of St. Mary and St. David, Kilpeck

Source: G.R. Lewis, Illustrations of Kilpeck Church, Herefordshire (London, 1843)

take their turn. The pair of images taken together reflect the aspirations I have been describing in this chapter, the role a restored church could play in the promotion of community through individual piety as expressed in the act of worship, a shared experience overseen by a sign of historical continuity (and in this case since Saxon and Norman times, as reflected in the elaborate chancel arch).

The interest which the ICBS, diocesan societies, bishops and even vicars took in medieval architecture was, however, not simply operative. The promoters of church or cathedral restoration were interested in more than providing comfortable, dry, well-lit and attractive interiors. The finer points of medieval architecture mattered to them and for two reasons. First, their scholarly interests were to a certain extent cultural, part and parcel of the popularization of Romantic historiography, antiquarianism and the beginnings of modern archaeology. Second, and importantly, these promoters desired to give their modernizations authenticity of expression, and increasingly demanded that new works adhere to the canons of medieval, architectural form, language and style. The hoped-for restoration of the parish community would be best achieved by work which both increased the physical accessibility of an ancient church and provided an authentic architectural and decorative setting. It was this combination of greater functionality and art-historical authority which lent this new image of Anglicanism moral legitimacy in a modern, democratic age.

By contrast new, non-Conformist chapels rarely aimed to capture the appearance of an authentic medieval church; very often they simply did not have access to the sort of prime sites, in central or high street locations, which the Anglicans had historically or because of their greater wealth. In any event non-Conformist congregations were generally more receptive to stylistic experiment and variety precisely because they sought to express their identity as distinct from the established church. The most dramatic reflection of these divergent views is the contrast across Parliament Square between Westminster Abbey (Church of England, and resoundingly restored and authentic Gothic) and Methodist Central Hall (non-Conformist and in the modern French idiom). Roman Catholics were on the whole, at least during the first part of the Victorian period, more interested in achieving stylistic authenticity precisely because their Church could legitimately claim antiquity.[31]

The association between history, identity and visual appearance or 'style' are now so well established through education, publications and reproductive media, that it is hard to imagine a time when the visual language of a building, say, or indeed a painting, signified nothing in particular and certainly no sense of shared identity or community. In order for an old building to become historic, there needs to be a uniform and widely agreed set of reference points, key monuments illustrating general tendencies, by which the individual instance (a humble parish church in a remote part of England, say) can be tied back to an historical narrative and through this to some sort of shared identity at the national, regional or local level. Style has to be 'taxonomized', its sequence established and then, most important of all, publicized in an easily reproducible format, and this is precisely the process which shadowed the institutional reform of the Church of England.[32]

Working in parallel with the Church authorities were local, voluntary organizations which drew together historians, archaeologists and antiquarians, clergy, architects and Anglican oligarchs (including many potential architectural patrons) into a tight, cohesive unit. During the 1840s, 1850s and 1860s dozens of often county- or regionally-based architectural societies were founded to promote the study of medieval antiquity at the local level. The object of these societies often extended to the promotion of church restoration according to the correct rules of medieval style and church archaeology as it was then emerging as a distinct discipline. Through their periodical publications, church visits, activities and monographs, these groups organized at the regional or county level transformed the smallest and most out-of-the-way church into what Benedict Anderson has called a 'meaning-creating experience'. That meaning was established through a combination of local distinctiveness tied back wider historical continuities.[33]

---

31   On Roman Catholic architecture, see Christopher Martin, *A Glimpse of Heaven: Catholic Churches of England and Wales* (Swindon: English Heritage, 2006).

32   Miele, 'Real Antiquity and the Ancient Object' in Brand, 103-24.

33   Anderson, *Imagined Communities*, 53.

It was this last perspective, that of the tourist with a more generalized interest in art, local topography and history, which challenged the authority which the Church of England had asserted over so much of the nation's medieval cultural inheritance. As secularism spread and tourism increased in the later nineteenth century, a different community came to feel a proprietary right over the cultural patrimony which an Anglican oligarchy had claimed as its domain and the result was a move against the practices of Victorian church restorers. This opposition claimed it was acting in the interests of a wider community, the nation at large, and also future, as yet unborn generations. There is a clear trail of evidence, stretching back to the 1840s, of this more recognizably modern approach to heritage, one which put the case for careful preservation – as opposed to restoration – on the grounds that this approach recognized a wider community of interests.

During the mid-Victorian decades, even as Anglican groups began to come together to form so-called 'architectural societies' to promote the appropriation of medieval churches for their own benefit, there emerged other local societies which took a less ideological view of the antiquities, societies whose remit sometimes included natural history. At the time monument care that focused strictly on the preservation of authentic fabric or the elucidation of authentic style was referred to as 'antiquarian', reflecting that degree of relative detachment seen to characterize an older historical tradition. And there were of course writers who sought to combine these different perspectives.[34]

The two writers (both Oxford University based) who did most to define what we would now recognize as a 'modern' approach to monument care – that is, the position which sought ideological neutrality around generalized notions of community or public benefit – were Edward A. Freeman and, far better known, John Ruskin. So much has been written on the latter's contribution to this intellectual tradition, that I need hardly say very much more except to refer readers to that author's 'Lamp of Memory' in the 1849 critical work *The Seven Lamps of Architecture*, which contained a plea against the then prevailing practice of restoring ancient things. Ruskin's 'case' was ultimately one based on in aesthetics and his own rarefied literary intelligence, and critical response to this part of the *Seven Lamps* in journals that concerned themselves with church restoration was mixed.

Far more relevant for my purposes, and also somewhat earlier, was the work of Freeman, the great historian of the Norman Conquest, Oxford University Professor and later Liberal MP Freeman categorized approaches to ancient

---

34   For an overview of their activities, see Miele, 'The Gothic Revival,' 487-528; Philippa Levine, *Amateur and the Professional: Antiquarians, Historians and Archaeologists in Victorian Britain* (Cambridge: Cambridge University Press, 1986); J. Leatherbarrow, 'The Victorians Look at Their Churches,' *Trans of the Worcester Archaeological Society*, 3rd series, vol. 9 (1984), 88-90.

buildings based on their social utility and appeal to specific communities.[35] He recognized that buildings still in use had to be adapted for practical reasons, and argued that in such cases it would be wrong to take too strict an antiquarian approach to maintaining authentic fabric. He considered, however, that there were some buildings and structures, particularly those which were redundant or ruinous, that had value mostly to artists or poets, or to historians, and accordingly warranted a more preservationist approach. A similar argument was put by the Gothic Revival architect George Gilbert Scott in a book that promoted a moderate approach, and not surprisingly given that so much of Scott's professional practice relied on church restoration.[36]

These books – Scott's, Freeman's and Ruskin's – encouraged a wider ranging debate in the antiquarian community and in the internal literature of the Church too. By the 1850s and 1860s standards and expectations and architectural practice itself changed in line with our own ideas of preservation and our concept of the monument, an ancient thing retained for its sheer historical interest to the nation, even other nations and posterity, rather than to a sectional interest. This notion clearly gained greater currency in the 1860s, paving the way for a new and more radical approach.

The credit for promoting this more generalized, 'non-sectarian' concept of public benefit in relation to heritage belongs to William Morris (1834-1896), the noted polymath who is today revered for his seminal role in the Arts and Crafts Movement. Morris knew the practices he came to criticize intimately. He had studied, albeit briefly, with George Edmund Street, a Gothic Revival architect who made his reputation restoring ancient churches to the latest fashion and building new ones on improved historical models. The fledgling decorative arts firm Morris and his friends founded in 1861 also relied on stained ecclesiastical stained glass commissions, many forming part of heavily restored church interiors.[37] The attitude of Morris and some of his associates to the church restoration movement clearly changed during the 1860s, but the decisive period came a decade later when the publication of statistical information published by Parliament in 1874 on the amounts expended on church building and restoration prompted a wider debate on the merits of what was being done to the nation's medieval patrimony. The results of this survey, and the subsequent furore it caused, prompted John Ruskin to refuse the offer of the Royal Institute of British Architects Gold Medal specifically because architects routinely charged standard percentage commissions

---

35   Edward A. Freeman, *Principles of Church Restoration* (London: Joseph Masters, 1846). See also *The Ecclesiologist* 7 (1847), 161-68, whose review disseminated this work. For Freeman's ideas on monument care see Stephan Tschudi-Madsen, *Restoration and Anti-Restoration* (Oslo: Universitetsforl, 1976), 38-50.

36   George G. Scott, *A Plea for the Faithful Restoration of Our Ancient Churches* (London: John Henry Parker, 1850).

37   Chris Miele, 'Morris and Conservation,' in Miele, *From William Morris*, 31-65, esp. 38-41.

on this work, encouraging more costly makeovers when simpler repairs costing less would do.[38]

Morris entered the debate in March 1877 with a pithy letter to *The Athenaeum*, the literary and artistic journal whose editor had encouraged the budding anti-restoration movement. This lively polemic pitted the 'many thoughtful people who would be glad to sacrifice time, money and comfort' in defence of ancient buildings against architects and clergymen (and by implication their funders) who would turn what should be common cultural property to their own, narrow interests. Very shortly after Morris convened a meeting of close friends, associates and clients to form the new Society for the Protection of Ancient Buildings, the first public pressure group devoted specifically to architecture.[39]

In his biography of Morris, E.P. Thompson was certain that the anti-restoration crusade was born of his subject's frustration with contemporary politics, concluding that the SPAB provided the designer and poet with a forum to explore and develop a public persona that would shortly take him into an explicitly political arena, as a radical voice.[40] Fiona MacCarthy's more recent study makes much of the personal engagement Morris felt for ancient buildings, developing a theme articulated by Morris' first biographer Mackail in 1899.[41] Norman Kelvin's *Collected Letters* support both theses, revealing the passion of Morris' entanglement with this one issue and its close association with his growing radicalism. But what is more interesting about Morris' work in this area, for the purposes of this chapter, was the wider public justification he made to press the Society's case against that other community interest – promoted by the Church of England – which had up until this point used medieval architecture to its own end.

The new Society's 'Manifesto' – which Morris wrote jointly with the architect Philip Webb and his business manager George Wardle – develops that most important theme from the first *Athenaeum* letter: the absolute interest of a wide, generalized public in all ancient buildings cared for as monuments first and foremost, whether they were in active use or not.[42] The short text starts by pledging their new Society to defend this amorphous, unnamed constituency, and more than this even to protect the interests of as yet unborn generations. To do this the SPAB

---

38   Ibid., 43-44.

39   The Society's founding is described in detail in Chris Miele, 'The First Conservation Militants,' in Michael Hunter, ed., *Preserving the Past: The Rise of Heritage in Modern Britain* (Stroud: Alan Sutton, 1996), 17-37. See also Miele, 'Morris and Conservation,' 44-50. The SPAB Manifesto is reprinted in that publication, and is available online at www.spab.org.uk.

40   A classic account is in Edward P. Thompson, *William Morris: Romantic to Revolutionary* (New York: Pantheon Books, 1975), 226-42.

41   Fiona MacCarthy, *William Morris: A Life for Our Time* (London: Faber and Faber, 1994), esp. 375-79; J.W. Mackail, *The Life of William Morris* (2 vols, New York: Dover Publications, 1995), vol. 1, 338-46.

42   Reprinted as 'Appendix III,' in Miele, *From William Morris*; also available online at www.spab.org.uk.

would constitute itself as a committee of experts charged with a guardianship role. These terms of reference did not include any practical purpose. The Manifesto rejects the idea that the present should seek any productive use from an ancient building on the basis that to do so would inevitably require the disturbance of old fabric, in other words 'destruction'. If an ancient building, the text concludes, 'has become inconvenient for its present use ... [then] raise another building rather than alter or enlarge old one'. Morris was not a supporter of what the American preservation movement used to call 'adaptive reuse'. He took a more extreme line against all change. His object was to treat buildings as precious works of art and ancient monuments, not social commodities. The real needs of the local community mattered less, in other words, than the more abstract needs of Morris' imagined community of aesthetes, scholars, artists, poets and the nation at large, including as yet unborn future generations.

Much has been written about the theoretical justification behind this line of reasoning. The Manifesto explicitly rejects what it calls 'modern art', by which Morris and his collaborators meant the tradition of Revivalist architecture Morris' development of earlier, essentially Ruskinian ideas has also, rightly, been identified as the basis for his ideas about heritage.[43] But what has, I think, attracted far less comment are the assumptions about community, public interest and politics that underlie Morris' radical formulations. Morris and his colleagues put the proposition that the public at large had a moral right to historic buildings as common cultural property. And they did so expressly by attacking those narrow community interests which had previously shaped monument care. Ironically for one of the founders of British Socialism, Morris' gambit relied on a system of oversight which was not democratically accountable. His justification was, simply, that most people had too little time and energy to turn their minds to matters beyond daily existence. Professional architects and the clergy were no better because their self-interest did not encourage them to see ancient buildings as monuments to art and history first and foremost. The best guardians were high-minded, disinterested people with taste and education, and no particular interest in religion either.[44] From a modern perspective, the concept of a class without an interest seems naïve but Morris articulated it powerfully and that concept has shaped historic buildings culture since. Today official heritage agencies trade on the idea that their professional staff have no personal interest in the regulation of the historic environment other than the community's interest at large, and so too the staff and volunteers who run the national amenity societies.

And so what Morris rejected was half of a century of praxis whose object was to make ancient religious buildings into symbols for community renewal at the

---

43   For a discussion of his attitudes to contemporary design and its relationship with historic buildings, see 'Introduction,' in Chris Miele, ed., *Morris on Architecture* (Sheffield: Sheffield Academic, 1996).

44   Chris Miele, "'A Small Knot of Cultivated People:" William Morris and Ideologies of Protection,' *The Art Journal,* vol. 54, no. 2 (Summer, 1995), 73-79.

local level through the vehicle of the parish and its church. Morris asserted a kind of moral proprietary right over cultural patrimony, transferring the real property and professional rights of one group onto another, imagined community, or rather onto a group of experts or guardians who were acting in a completely disinterested way on behalf of a highly generalized beneficiary. Occasionally Morris argued a narrow nationalist line. Objecting to the proposed closure of Wren-school churches in the City of London, in response to rapidly declining congregations, Morris appealed directly patriotic sentiment. Writing to *The Times* in 1878, he called on 'Englishmen' to protest the consequent destruction of redundant buildings. What people failed to appreciate was that St. Paul's really was no different to many great Continental churches, and as such was 'scarcely ... an English design'. The smaller churches were far more interesting, 'examples of purely English renaissance architecture'. It was inconceivable that one of the richest capitals in Europe could not afford the small sites they occupied. For the Church authorities to promote these plans – which incidentally allowed for the redistribution of the old church endowments to the construction of modern suburban churches in outer London (a fact Morris failed to mention) – contradicted any claim the English had to artistic feeling, and would moreover be 'looked upon by foreign nations' as an amazingly short-sighted and selfish thing to do.[45]

More usually, however, he described the beneficiaries of the Society's work in more sweeping terms, as an international community. In that same letter on Wren's City Churches Morris lamented the harmful restoration of major minsters and cathedrals not just in England but also on the continent. It is remarkable that very soon after the Society's foundation, Morris pushed the SPAB into what it called 'foreign casework', the idea arising from a visit he and his family made to Italy to 1878 and in particular to Venice where he was concerned by the repairs he saw being made to the mosaics in the baptistery of San Marco. This turned out to be one of the most high profile, and widely supported of the SPAB's campaigns during Morris' lifetime and not surprisingly given the English fascination with Venice, from Canaletto through to Turner and Ruskin.

One letter Morris wrote to the *Daily News* in October 1879 is remarkable for introducing the concept of a monument which was not just important to a 'small minority' of English aesthetes and artists but was rather common cultural property of the all humanity, a 'wonder of the civilised world'. The particular object here was to challenge the decision to place the restoration of San Marco in the hands of a local architect accountable to a special committee of experts working under the aegis of the Italian Committee of Public Instruction, answerable to a politically appointed Minister.[46] An architect member of the SPAB, J.J. Stevenson, undertook to carry out his own survey, in association with a leading Italian art historian, and to make the findings available to the Italians nation at large as a courtesy. Here

---

45   Norman Kelvin, ed., *The Collected Letters of William Morris* (4 vols, Princeton: Princeton University Press, 1984), vol. 1, no. 523.

46   Ibid., no. 585.

was Morris transforming his general public into the whole of humanity, laying the groundwork for the post-war concept of World Heritage Sites, identified for their 'outstanding, universal significance'. Italy would remain, as Frank Sharp has written, the focus for the Society's overseas work and after that in Britain's overseas possessions which presented obvious channels of influence. [47]

Not surprisingly the owners and guardians who bothered to reply to badgering letters from Morris and his comrades, very often asked by what right a Society based in London could claim to be sticking its nose into local matters.[48] Other owners bridled at the suggestion that there were acting without that higher interest in mind, particularly where restoration works aimed at improving community access to a church or where the works were intended to emphasize the historical features of the building by reconstruction or lost features (a course of action which conflicted with Morris' theories).

After all, as the Bishop of Oxford wrote Morris about the conservation of the ancient church at Inglesham (a poor, tiny Cotswold hamlet), there was a normal way to go about these matters: first a local fund-drive within the parish, then an application to the ICBS or Diocesan Society, and after that a Church Faculty application approved by his Lordship, the Bishop himself, accompanied of course by appeals to the local aristocracy and gentry. These were, all of them, the necessary steps that bound the restoration of the building into a chain of community through the Church of England. What Morris and his Society were doing – which was to offer to fund and pay for the work themselves, with funds raised by Society-faithful (and generous anonymous contributions from Morris himself) – was unacceptable precisely, the Bishop observed, because they were seeking to work outside this established local network. After a while, because the incumbent at Inglesham, the Rev. Oswald Birchall, was a Socialist acquaintance of Morris' (and a local corresponding secretary for the SPAB), the Bishop gave way, though he had no sympathy for the ultra-preservationist approach applied by J.T. Micklethwaite, the architect-cum-archaeologist whom the Society instructed. In the end the building became a monument to SPAB-principles simply because no one in the local community had any interest in modernising the Church.[49]

Morris' concept of guardians representing the interests of a community of humanity was in its day novel and radical. There was not then, and would not be for some time, even the remotest chance that Parliament would step in to impose conservation regulations which would override local or private property interests, or the Church of England's mission as delivered through its historic buildings. The endless debate that dogged what became the first Ancient Monuments Act 1882 proved this beyond all doubt, and as a matter of fact really effective and

---

47 F. Sharp, 'Exporting the Revolution: The Work of the SPAB outside Britain 1878-1914,' in Miele, *From William Morris*, 187-98.

48 Miele, *Art Journal*, 73-79.

49 'Inglesham File,' letters 1883 to 1889, Archives of the Society for the Protection of Ancient Buildings, London.

comprehensive regulation of important historic buildings would only be recognized in the Town and Country Planning Act 1947, though the provisions here were minor.[50] Legislation since the late 1960s and mid-1970s has produced an effective, democratically accountable system for regulating the historic environment in the UK.[51]

The history of the Conservation Movement in twentieth-century Britain has yet to be written, although there are now many, very good studies of different areas of theory and practice. It is quite clear that the Movement would not have made the progress it has since 1945 and after 1967 (the date of the Civic Amenities Act) without the support of central government through legislative provision, grants and a centralized body of expertise (now concentrated in English Heritage, a non-department public body). More recently, in the last thirty years, local government has become increasingly important in building conservation; this tier of government is responsible for most planning and development decisions and it is at this level that planning policy for an area is set (through the development plan system, which encompasses conservation as one of its core competencies). Still, the experts in local government are part of a chain of command which has a stronger focus on the centre than their pure land-use planning colleagues who are oriented more towards local issues. These professionals, rightly, have a split allegiance. On the one hand is the local community they serve, who may or may not support conservation, and on the other is the general public in that wider sense Morris imagined. Conservation of the historic environment is achieved through private individuals working in concert with, and regulated by, professional heritage officers employed in local authorities, officers who work to statutory objectives which require them to think beyond the immediate interest of an individual or community and through local administrative structures to serve that abstract imagined community which Morris envisaged more than 150 years ago.

---

50   See T. Campion, 'Protecting the monuments: archaeological legislation from the 1882 Act top PPG16,' in Hunter, 38-56.

51   On the history of UK legislation, see Delafons, *Politics and Preservation.*

Chapter 8
America's Early Historic Preservation Movement (1850-1930) in a Transatlantic Context

Michael Holleran

The preservation movement in the United States emerged in the mid- to late nineteenth century. European preservationists got an earlier start, but by the end of the century, developments in many western countries moved in tandem, and the transatlantic flow of ideas and influence was not entirely one-way. This chapter briefly sketches a comparative framework for examining the emergence of the American preservation movement in the context of the similar movements in Europe. It identifies parallels, contrasts and unanswered questions. Some is not merely comparative, but a history of actual interactions across the Atlantic. Much is speculative, and I hope will suggest fruitful areas for further research. The chapter's organization is thematic rather than chronological or geographical. It uses Boston as its main example, and other American places where they add layers to the story. I do not attempt to summarize the whole story of the early preservation movement in the US, or in Boston. The European side of the comparison can even less pretend to comprehensiveness. Like Americans during this period, I look first to Great Britain and France.[1]

---

1   For general sources on the early period of American preservation history: Charles B. Hosmer, Jr, *Presence of the Past: A History of the Preservation Movement in the United States Before Williamsburg* (New York: Putnam, 1965); Charles B. Hosmer, Jr., *Preservation Comes of Age: From Williamsburg to the National Trust, 1926-1949* (2 vols, Charlottesville: University Press of Virginia, 1981); Lewis F. Fisher, *Saving San Antonio: The Precarious Preservation of a Heritage* (Lubbock: Texas Tech University Press, 1996); Michael Holleran, *Boston's 'Changeful Times': Origins of Preservation and Planning in America* (Baltimore: Johns Hopkins University Press, 1998); James M. Lindgren, *Preserving the Old Dominion: Historic Preservation and Virginia Traditionalism* (Charlottesville: University Press of Virginia, 1993); James M. Lindgren, *Preserving Historic New England: Preservation, Progressivism, and the Remaking of Memory* (New York: Oxford University Press, 1995); William J. Murtagh, *Keeping Time: The History and Theory of Preservation in America* (Hoboken: John Wiley, 2006); Max Page and Randall Mason, eds, *Giving Preservation a History: Histories of Historic Preservation in the United States* (New York: Routledge, 2004); Randall Mason, *The Once and Future New York: Historic Preservation and the Modern City* (Minneapolis: University of Minnesota Press, 2009).

National Identity

On both sides of the Atlantic, preservation played a role in constructing national identities. The US was relatively new to the family of nations, and in the mid-nineteenth century there was a new urgency to defining a national identity as its citizens sought first to avert and then to recover from Civil War. Similar endeavors were underway, for a variety of reasons, in other countries. In newly united Germany and Italy, preservation served as a tool to reinforce national identities among citizens who were more accustomed to identify with sub-national regions. For older nations such as France and Great Britain, consolidating the culture of the modern nation-state was a continuing project. All national identity relied upon historical identity, and therefore to a large degree on the invention of traditions, in Eric Hobsbawm's phrase.[2] Historic sites anchored tradition to place, and tangible historic fabric made the past present.

The first American preservation efforts were especially likely to focus on the common national origin story of the Revolution. The state of New York in 1850 purchased George Washington's headquarters in the Hudson River valley. In 1853, Ann Pamela Cunningham – a distant Washington relation – began an effort to preserve Washington's home, Mount Vernon, twelve miles down the Potomac from Washington, DC. Over the next five years, she organized the Mount Vernon Ladies Association of the Union, and led an unprecedented nationwide fundraising

---

European preservation has a comparatively vast literature. For surveys and overviews: Françoise Choay, *The Invention of the Historic Monument* (Cambridge: University of Cambridge Press, 1998); Wim Denslagen, *Architectural Restoration in Western Europe: Controversy and Continuity* (Amsterdam: Architectura & Natura Press, 1994); John Delafons, *Politics and Preservation: A Policy History of the Built Heritage, 1882-1996* (London: Spon, 1997); Cevat Erder, *Our Architectural Heritage: From Consciousness to Conservation* (Paris: UNESCO, 1986); Michael Hunter, ed., *Preserving the Past: The Rise of Heritage in Modern Britain* (Phoenix Mill, Gloucestershire: Alan Sutton, 1996); Jukka Jokilehto, *A History of Architectural Conservation* (Oxford: Butterworth-Heinemann, 1999); Rudy Koshar, *From Monuments to Traces: Aspects of German Memory, 1870-1990* (Berkeley: University of California Press, 2000); Peter Mandler, *The Fall and Rise of the Stately Home* (New Haven: Yale University Press, 1997); Chris Miele, ed., *From William Morris: Building Conservation and the Arts and Crafts Cult of Authenticity, 1877-1939* (New Haven: Yale University Press, 2005); Stephen Tschudi-Madsen, *Restoration and Anti-restoration* (Oslo: Universitretsforlaget, 1976). G. Baldwin Brown, *The Care of Ancient Monuments* (Cambridge: Cambridge University Press, 1905) is still useful as a systematic international compilation from this period.

For a comparative perspective: David Lowenthal, *The Past is a Foreign Country* (Cambridge: Cambridge University Press, 1985); David Lowenthal, *The Heritage Crusade and the Spoils of History* (Cambridge University Press, 1998).

2　Eric Hobsbawm, 'Introduction: Inventing Traditions,' in Eric Hobsbawm and Terence Ranger, eds, *The Invention of Tradition* (Cambridge: Cambridge University Press, 1983), 1-14; Melanie Hall, 'Affirming Community Life: Preservation, National Identity and the State, 1900,' in Miele, *From William Morris*, 129-57.

campaign that succeeded in purchasing the house in 1858. Washingtoniana was conveniently spread from Virginia to Massachusetts, and provided a way of distributing to many locations the forging of national identity through tangible traces. Later preservation efforts focused on the Civil War, on other presidents, and on other chapters of the national origin story. For example, adherents of a 'cult of the flag' adopted as one of its subjects Betsy Ross, who may have sewn the first stars-and-stripes, and found spatial expression in the 1890s through efforts to preserve a Philadelphia house where she may have lived.[3]

Other American preservationists aimed to save resources whose significance was more regional or local, such as churches and old family seats. Here these is room for nuanced comparative study: Europeans also pursued both national origin stories and regional heritage, in which the regional might be refracted through national identity, but often started with deep local rootedness that could be at odds with nationalism. Were American local efforts more often linked to a national identity (e.g. numerous local campaigns for homes of signers of the Declaration of Independence), perhaps as an outgrowth of greater mobility? There certainly were and are American regions with their own distinct identities reinforced by preservation efforts: the former Confederacy as a whole, the Hispanic Southwest, Texas, California. But even preservationists in New England and Virginia, with their strong regional orientations, presented their stories less as markers of regional separateness than as competing versions of a national origin story. This competition became particularly evident in the years leading up to the 1907 tricentennial of the settlement of Jamestown, Virginia.[4] New Englanders would have to wait a while before celebrating that anniversary of the Pilgrims' landing in 1620.

Material Culture

National identity could be fostered not only through grand narratives but also through the non-narrative vehicle of folk culture, and preservation could contribute through attention to domestic and vernacular structures. Such efforts embodied a romantic view of nationality – at its toxic extreme, a racial conception of nationhood; in its academic version, the ethnographic study of material culture. Steven Conn, in his history of American museums, points to an emphasis on collection and classification of material objects as a foundation of nineteenth-century culture, in science and art as well as history.[5] Preservation can be seen as a classification and collection approach extended to immovable objects. Similarly, old-world archaeology sometimes functioned as an acquisitions program for

---

3   Hosmer, *Presence of the Past*, 41-75, 88-91; Patricia West, *Domesticating History: The Political Origins of America's House Museums* (Washington: Smithsonian Institution Press, 1999), 5-37.

4   Lindgren, *Preserving the Old Dominion*, 91-136.

5   Steven Conn, *Museums and American Intellectual Life, 1876-1926* (Chicago: University of Chicago Press, 1998).

museums, but it also uncovered objects at a scale that could only be curated on site. Antiquarians in nineteenth-century America noted survivals of early Colonial buildings, and increasingly documented and studied them. Another route toward folk structures was the many filiopietistic ancestral societies that blossomed in the 1880s and 1890s, such as the Sons of the American Revolution. For descendants of progenitors not quite so prosperous as Washington, the search for the ancestral hearth would lead to more ordinary dwellings.

The Paul Revere House in Boston served as one pivot turning the preservation movement away from the dominance of narrative history. Revere ascended in the Revolutionary pantheon after Henry Wadsworth Longfellow in 1861 immortalized his 'midnight ride' to warn of British troops before the battles of Lexington and Concord. At the turn of the twentieth century a Revere descendant, John Phillip Reynolds, Jr, initiated an effort to preserve his house, in the immigrant North End. So far this story is a textbook example of a narrative elevated to new prominence in the service of patriotic tradition, and then reinforced through preservation of a tangible monument. But the house had been old even when Revere lived there, dating from the 1680s, and was perhaps the oldest surviving building in Boston. When restoration began in 1907, preservationists encouraged architect Joseph Everett Chandler to take it back not to the late eighteenth-century period of the Revere narrative, but to its very different original appearance in the seventeenth century.[6]

Figure 8.1   Paul Revere House, 1895 (center, with awning and plaque). Courtesy Bostonian Society, Old State House, Boston

---

6   Holleran, *Boston's 'Changeful Times'*, 216-18.

Figure 8.2 **Paul Revere House after the 1907 restoration.**
Courtesy Bostonian Society, Old State House, Boston

The Revere House campaign was led by William Sumner Appleton, Jr, who would go on from this first effort to become one of America's most important preservationists. Appleton entered preservation through involvement in ancestral societies – his was an old and elite Boston family. He had some exposure to architecture as a student at Harvard, where he studied with Charles Eliot Norton, the first American professor of Art History. After the Revere House effort, Appleton made preservation his life's work, beginning his new career with a self-directed tour of conservation practice in Europe. He kept in touch with developments there (and joined England's Society for the Preservation of Ancient Buildings). In 1910, Appleton founded the Society for the Preservation of New England Antiquities (today named Historic New England). SPNEA was a new sort of organization, dedicated to the built heritage of a region, focusing almost exclusively on domestic architecture, not as monuments to any historical narrative but as artifacts of intrinsic value. He was an important conduit to preservationists throughout the US, through the *SPNEA Bulletin* that he edited, and through his indefatigable correspondence.[7]

---

7 Lindgren, *Preserving Historic New England*; Holleran, *Boston's 'Changeful Times'*, 218-44.

One of Appleton's European models was in Sweden, where Artur Hazelius established at Skansen a new type of preservation institution, the outdoor museum. Skansen's collection consisted of actual buildings brought together from their original sites, exemplifying the regional variations in traditional Swedish vernacular construction. The outdoor museum spread through northern and central Europe, and then to the US and Canada – primarily areas where the vernacular was of wood rather than masonry, and therefore comparatively easy to relocate. Appleton unsuccessfully sought support for such a museum in New England. The outdoor museum would appear in the US first in 1928, as Henry Ford's Greenfield Village near Detroit, with an anti-scholarly posture but a deep conviction that material culture was the way to tell the story of American identity.

The Idea of Built Heritage

Before preservation could be enlisted in the service of national identity, there needed to be some agreement that parts of the built environment constituted a heritage to preserve. Here was the first and most important contrast between the US and Europe. From the earliest modern times, Europeans acknowledged a built heritage; indeed the advent of that consciousness serves as one of the markers of modernity. Americans did not initially see a heritage anywhere in their own built environment. Henry van Brunt, in the preface to his translation of Viollet-le-Duc, wrote in 1875 that 'we Americans occupy a new country, having no inheritance of ruins.'[8] Charles Eliot Norton in 1889 described an America 'barren ... of historic objects'; to his eyes, he said, 'Boston is in its aspect as new as Chicago.'[9] Americans looked to the same built heritage as Europeans: the classical past of the Mediterranean, and then the Gothic structures of northern and western Europe. This contrast reflects the respective developmental stages of the New World and the Old, former colonies and parent culture. It led to different kinds of preservation movements.

The first American attempts at commemorating heritage in the environment expressed themselves not through preservation but through erection of new commemorative monuments. The fiftieth anniversary of American independence was a fertile occasion for such schemes. Bostonians began to construct a monument to the battle at Bunker Hill (not completed until 1843), in the process destroying the last actual traces of the battle. Members of the Massachusetts Historical Society considered demolishing the 1712 Old State House to create

---

8   Henry Van Brunt, 'Introduction,' in Eugène Emmanuel Viollet-le-Duc, *Discourses on Architecture* (Boston: James R. Osgood, 1875), x.

9   Charles Eliot Norton, 'The Lack of Old Homes in America,' *Scribner's Magazine* 5 (1889), 638, 639.

Figure 8.3   Bunker Hill Monument, c. 1870. Courtesy Bostonian Society, Old State House, Boston

an appropriate site for a statue of George Washington.[10] Early attention to Mount Vernon concerned not the fate of Washington's house, but construction of a suitable monument at his grave.[11]

---

10   Holleran, *Boston's 'Changeful Times'*, 30-31; Jane Holtz Kay, *Lost Boston* (Boston: Houghton Mifflin, 1980), 129-33.

11   Blanche Linden-Ward, *Silent City on a Hill: Landscapes of Memory and Boston's Mount Auburn Cemetery* (Columbus: Ohio State University Press, 1989), 105-30.

Restoration

From differential consciousness of built heritage arose another important
Transatlantic contrast. In Europe, restoration efforts began long before the
preservation movement, or, viewed another way, they were the first form of
preservation, which only later developed clear philosophies and institutions.
Scattered efforts to restore ancient monuments began by the sixteenth century. By
the end of the eighteenth century, James Essex and James Wyatt had restored a
number of English cathedrals, and in the nineteenth century, industrial prosperity
unleashed a tide of restoration efforts on churches throughout the country. The
early nineteenth century saw cathedrals and monuments restored in Prussia,
France and Italy.

Much of European preservationism arose in response to restoration and its
perceived excesses, the destruction of original fabric and its replacement by
inauthentic inventions. John Ruskin wrote to debunk the very idea of restoration:
'The thing is a Lie from beginning to end ... Take proper care of your monuments
and you will not need to restore them.'[12] William Morris organized to put these
principles into practice through the Society for the Preservation of Ancient
Buildings, founded in 1877. In France, Eugène Viollet-le-Duc wrote to codify
good restoration practice. He is often cast as the anti-Ruskin, partly for his
interventionist enthusiasm at Pierrefonds, Carcassonne and other restorations,
partly for the spooky phrasing of his goal in restoring a building: 'to reestablish
it in a finished state, which may in fact never have existed at any given time.'[13]
Yet Viollet's writings called for thorough investigation and documentation, and
for balancing the values of accuracy, appearance and utility – all still familiar
exercises today.

Other parts of the European preservation movement arose not in reaction to
restoration but in anticipation of it, to catalog the heritage in order to rescue it from
neglect and deterioration. The work of antiquarians in recording and describing
historic structures became the work of official bodies such as France's Commission
des monuments historiques, created in 1837, listing structures deserving of
protection. The work of these cataloguers slowly but inexorably expanded the
scope of the movement, toward more recent and more vernacular resources.

American preservation came before restoration. It started out as advocacy. Its
practical aims were to protect structures from neglect or destruction, and in each
case preservationists first had to convince their countrymen that there were indeed
structures of historical value in the New World. Educated Americans read Ruskin.
What did they make of his exhortations on the value of built antiquity and the
problems with restoration? If 'a building cannot be considered as in its prime

---

12   John Ruskin, *The Seven Lamps of Architecture* (New York: Farrar Strauss and
Giroux, 1971), 185-86.
13   Eugène-Emmanuel Viollet-le-Duc, *The Foundations of Architecture: Selections
from the Dictionnaire raisonné* (New York: George Braziller, 1990), 195.

until four or five centuries have passed over it,' as Ruskin wrote, then Americans could follow this European conversation without thinking that it concerned their own buildings. Norton, a life-long friend and correspondent of Ruskin, arrived in England in 1868 to find that the patina of age gave scenes in that country 'a deeper familiarity than the very things that have lain before our eyes since we were born.'[14] Antiquity meant the Old World. Only later, after Americans began to acknowledge a built heritage on their own side of the Atlantic, could the question arise of how (or whether) to restore it.

Architects

Perhaps as a result of the paucity of restoration practice, architects in the US were not particularly involved in the early preservation movement. In Boston that movement began, not long after the success at Mount Vernon, with a campaign to save the house of John Hancock, the Massachusetts governor and ostentatious signer of the Declaration of Independence. The campaign failed and the house was demolished in 1863, a scrape-off to build two new mansions. When the Old South Meetinghouse faced a similar fate ten years later, Bostonians took the threat more seriously, eventually raising $400,000 from more than 50,000 donors, and saving the building even after demolition had already begun. Lawyers, merchants, clergymen – and their wives and daughters – were the stars and the supporting cast of these and other preservation dramas.[15] Architects seemed conspicuously absent even though Boston was one of the centers of the young profession in the United States. European preservation controversies, by contrast, could sometimes seem to be intramural disputes within the architectural profession.

Many American restorations were supervised by antiquarians rather than architects. After the success at the Old South, Bostonians tackled other preservation projects, including the surviving but tattered Old State House. A preservation effort beginning in 1876, and a restoration completed in 1881, were both led by William H. Whitmore, President of the Boston Common Council.[16] In the country's earliest recorded restoration – Philadelphia's Independence Hall in 1828 – architect William Strickland proposed an 'improved' brick tower; it was his lay

---

14   Ruskin, 183; Norton to James Russell Lowell, August 30, 1868, in Sara Norton and M.A. DeWolfe Howe, eds, *Letters of Charles Eliot Norton* (Boston: Houghton Mifflin, 1913), 1, 306.

15   Holleran, *Boston's 'Changeful Times'*, 91-104.

16   Ibid., 105-109. Boston city architect George A. Clough supervised the restoration, working under Whitmore's direction. Before beginning work, Clough in a letter to Whitmore referred to 'the restoring of the outside walls to conform with its original outline and appearance as shown by your sketches.' *Boston Evening Transcript* (June 24, 1881), 2.

clients who forced a more faithful reconstruction.[17] After the Mount Vernon Ladies Association took possession of Washington's home, and shepherded it through the Civil War years, Ann Pamela Cunningham and her successors directed its restoration, apparently without assistance from any architect before 1876.[18]

American architects did begin to practice preservation architecture, as the efforts of lay preservationists produced an inventory of buildings for them to work on. Architects began an important strain of pre-preservationism by studying American buildings as architectural history. Charles F. McKim, Stanford White and William B. Bigelow traveled together through New England in 1877 making measured drawings. Such research formed the basis for the Colonial Revival style, which in turn placed a value on the prototypes as objects worthy of preservation. [19] By the turn of the century, a small handful of architects had begun to make practices

**Figure 8.4   Massachusetts State House, Architect Charles Bulfinch.
Courtesy of Historic New England**

17   Charlene Mires, *Independence Hall in American Memory* (Philadelphia: University of Pennsylvania Press, 2002), 73-78.

18   Hosmer, *Presence of the Past*, 55.

19   Richard Guy Wilson, *The Colonial Revival House* (New York: Harry N. Abrams, 2004), 35-41.

around historic structures. Other mainstream architects included historic work in their portfolio, including some major figures such as Charles A. Cummings, president of the Boston Society of Architects, who consulted on restoration of the 'new' Bulfinch State House for its centennial in 1898.

Architects as a group first embarked upon preservation advocacy in the 1890s. At the national AIA convention in 1890, Richard M. Upjohn proposed and the convention established a Committee of Conservators of Public Architecture of the Country, which functioned for a few years with a focus mainly on New York.[20] In Boston, the Bulfinch State House had been threatened by demolition or radical disfigurement for an enlarged capitol, and became the first preservation effort in the United States in which architects took a leading role. The building's main significance was to architectural history – how many seats of government are known by the name of their architect? The Boston Society of Architects in 1894 established a State House Committee to promote its preservation. Architects could answer the arguments against the building's safety or practicality, and they were critical to the success of the campaign. The Philadelphia chapter of the AIA urged accuracy in restoration of the interior of Independence Hall, and later donated professional services toward that goal. The AIA in 1898 leased, and four years later acquired, the 1800 Octagon House in Washington DC as its national headquarters.[21] Architects' preservation efforts, even when effective, were still episodic, and leadership of the growing movement remained largely with non-architects.

Once American architects understood themselves to be operating on monuments or antiquities, the methods and the discourse of the best of them quickly converged with practice and debate on the other side of the Atlantic. Excesses of fanciful restoration, or excesses of degàgement that destroyed the context of historic structures, were perpetrated mainly by popular demand and not by architects (or at least not by the architects who were leading preservation debates). American architects had some familiarity with European practices and debates. The ones who were educated at the Ecole des Beaux-Arts in Paris, or who had studied in Rome, were themselves prolific educators and writers; many

---

20   *American Architect and Building News* editorialized: 'That so young a country as this should already stand in need of the services of a 'Commission for the Preservation of Ancient Monuments' is rather amusing, but there is certainly reason in the suggestion from a sentimental, if not from a practical point-of-view. How far such a commission can hope to receive the support of local architects in such case [sic] is open to doubt, as many might feel that in checking the American mania for changing and 'improving' they were possibly preventing themselves from earning fame and gain through creating the new buildings which the owners of the estate desired to substitute for the old ones.' ... 'They will in all probability find their efforts ... quite futile.' *AABN* (Nov. 1, 1890), np.

21   Holleran, *Boston's 'Changeful Times'*, 135-50; Hosmer, *Presence of the Past*, 86-88; George McCue, *The Octagon: Being an Account of a Famous Washington Residence: Its Great Years, Decline & Restoration* (Washington: American Institute of Architects Foundation, 1976).

other architects managed to travel on some version of a Grand Tour.[22] American architectural periodicals reported on European preservation accomplishments and controversies.

Institutions

A more complex comparison concerns the institutional vehicles that arose to accomplish preservation. In broad outlines, US preservationists mostly favored private organizations. European efforts were more likely to be through governments. To some extent this represents the voluntarism and associationalism that de Tocqueville found characteristic of American life, in contrast to European traditions of strong governmental action. This simple distinction masks a set of more complex stories. In the US, a tug-of-war was several times repeated, in which advocates would try without much success to make preservation a governmental function, and only then would organize privately.

The archetypal private preservation organization is the Mount Vernon Ladies Association of the Union, which has cared for George Washington's home for the past century and a half. When Ann Pamela Cunningham began her work in 1853, others had already spent years in fruitless attempts to bring the property into government ownership. Initially her goal was to transfer title to the Commonwealth of Virginia, and only later in the campaign did the association become the intended custodian. [23] The John Hancock House was demolished because the effort to save it relied on the state and city governments, and both proved ineffective. Preservationists defending the Old South Meetinghouse repeatedly turned to the Massachusetts state government for assistance, both legal and financial, almost always in vain. American preservationists did not start with a bias against government action, they learned it.

The European story varied from country to country, and NGOs arose both to advocate for and to supplement governmental action. In the UK, vocal and strong private organizations, such as the Society for the Preservation of Ancient Buildings, represented a common Anglo-American tradition of privatism. But the private role could be to advocate for public actions: the Archaeological Institute, founded in Britain in 1842, argued in its manifesto that 'to preserve from demolition or decay works from ancient times which still exist, is an object that should merit the attention of the Government.'[24] In France, similar voluntary associations encouraged and aided government action.

---

22   Isabelle Gournay and Elliott Pavlos, 'Americans in Paris,' *Journal of Architectural Education*, vol. 38, no. 4. (Summer, 1985), 22-26.

23   Hosmer, *Presence of the Past*, 45; West, *Domesticating History*, 5-37.

24   Timothy Champion, 'Protecting the Monuments: Archaeological Legislation from the 1882 Act to PPG 16,' in Hunter, 39.

Specific national circumstances helped shape institutional arrangements. The French Revolution turned royal properties into public assets, leading to the first serious attempt at a comprehensive listing of heritage properties for the purpose of managing them. In other countries, established churches put the government in the business of caring for some of the most important and most widespread category of monuments; even non-established churches were not necessarily walled off from state aid as in the United States. Each of these factors tended to put European governments in the business of managing historic monuments; each was absent in the United States. The US had no established church, and its secular monuments were not concentrated in the hands of the central government (which had not existed until the end of the eighteenth century). It would be illuminating (but beyond the scope of this chapter) to compare governmental and nongovernmental institutions not only across different countries in preservation, but also across different cognate spheres of action, for example housing reform and social housing.

Institutional arrangements varied in their degree of centralization. Preservationists organized at a national level in France and other European countries, but only rarely in the United States. This may reflect not only tradition but also differences in scale. The US, like Germany, is organized on a federal basis, with much of domestic governance left to the individual states. The US did organize at the national level in one sphere not found in any European state: in the exploration and settlement of the American West. Archaeological antiquities like other Western resources were the province of the federal government, and of NGOs of national scope. The Archaeological Institute of America was founded in 1879, with Charles Eliot Norton as its first president. Preservation efforts for American historical resources of European-American origin, however, were organized locally or at most regionally until the establishment of the (American) National Trust for Historic Preservation in 1949. The Ladies Association of Mount Vernon remained a unique exception, focused on a single building but with an organization of truly national scope.

Gendered History

An apparent contrast between American and European preservation was the prominent role of women in the United States.[25] Ann Pamela Cunningham specifically directed her first appeal 'To the Ladies of the South,' and many thousands of ladies from North as well as South eventually joined together in the Mount Vernon Ladies Association. The Old South campaign was not explicitly

---

25   West, *Domesticating History*; Lindgren, *Preserving the Old Dominion*; Barbara J. Howe, 'Women in the Nineteenth-Century Preservation Movement,' in Gail Lee Dubrow and Jennifer B. Goodman, *Restoring Women's History through Historic Preservation* (Baltimore: Johns Hopkins University Press, 2003), 17-36.

gendered, but women did most of the hard work of raising funds, and the single largest donor and most important leader of the effort was Mary Hemenway. The Daughters of the American Revolution, founded in 1890, included among its purposes 'the acquisition and protection of historical spots,' and its individual chapters quickly took up many such causes, for example taking over custodianship of Independence Hall from the city of Philadelphia.[26] In San Antonio, the Daughters of the Republic of Texas saved the Alamo. Randall Mason finds of the Women's Auxiliary of the American Scenic and Historic Preservation Society, in New York City, that 'these women were very successful in selecting a few preservation projects and implementing them – which contrasted with the [male] Society's strategy of advocating many things and implementing very little.'[27]

Patricia West examines the prominence of women in establishing house museums, starting with Mount Vernon, and its congruence with women's evolving roles in American society. Women were keepers of the home and raisers of children, responsible for inculcating in the next generation the values of morality and patriotism. It was a relatively simple extension to become the keepers of historic houses that could instill patriotic and filiopietistic virtue. A 'cult of domesticity' segregated middle-class women from economic production, making them instead consumers responsible for constructing their families' status through the appointments of the home. The house museum, and its careful furnishing and display, could become a communally-constructed exemplary model home. But these gender roles were not uniquely American; why did they evidently express themselves more in the US through preservation action?

In part this might be explained by the role of architects and of government in European preservation. Both were professionalized and almost exclusively male spheres, reducing the space available for action by women. But even when Europeans engaged in grass-roots organizing, advocacy and fundraising appear to have been carried out overwhelmingly by men, while in the US, women were frequently participants, and often the leaders, in these activities. Why the transatlantic difference? Or is this contrast merely apparent, with the gendered history of the European movement yet to be written? After all, Octavia Hill was among the founders of the British National Trust, and women were active in the organization and its branches.[28] Some scholarship in the American context has also suggested that gay men were prominent in the movement on this side of the Atlantic.[29] The question awaits comparative study.

---

26   Lewis Barrington, *Historic Restorations of the Daughters of the American Revolution* (New York: R.R. Smith, 1941).

27   Randall Mason, 'Historic Preservation, Public Memory, and the Making of New York City,' in Page and Mason, 138.

28   Alan Crawford, 'Supper at Gatti's: The SPAB and the Arts and Crafts Movement,' in Miele, 102; Hall, 'Affirming Community Life,' 142.

29   Will Fellows, *A Passion to Preserve: Gay Men as Keepers of Culture* (Madison: University of Wisconsin Press, 2005). See also Douglass Shand-Tucci, *Ralph Adams Cram:*

Preservation from What?

Against what threats did preservationists mobilize? In Old World countries, the threat had long been cumulative neglect and then its flip side, efforts to remedy neglect through overzealous restorations. This was punctuated by occasional destruction through war or civil unrest, and spasms of religious iconoclasm (mostly finished by the nineteenth century). In the US, the main threat was the unrestrained capitalism of the (mainly urban) land market. This was the actual threat at Boston's Hancock House and Old South Meetinghouse, and the implied threat at Mount Vernon, where the anticipated fate for the neglected site twelve miles from the city of Washington was its eventual purchase as 'the seat of manufacturers and manufactories.'[30] In 1931, Charleston, South Carolina, adopted the first historic district ordinance in the US in response to demolition of eighteenth-century houses for a Standard Oil station. The European preservation experience eventually converged (especially as more recent, secular, and ordinary buildings came within its purview) so that urban redevelopment came to be seen as a leading threat there as well. Capitalism as a threat was sometimes an invocation of outsider groups, in Europe, Jews, in the American South, Yankees. But the land market and redevelopment in Boston and New York could do plenty of damage at the hands of these cities' own elites.[31]

The early city planning movement in the US began to address some of these dislocations. Modern city planning developed hand in hand with preservation, and paid attention to integrating the old into the new. Massachusetts passed restrictions on the heights of new buildings, intended to maintain iconic views of the dome of the Bulfinch State House after its restoration in 1898. These and other similar Boston efforts eventually produced case law that became the basis for New York's adoption in 1916 of comprehensive zoning, the main form of public development regulation in American cities throughout the twentieth century. Zoning regulations drew on planning practice in Germany, particularly through the work of German emigré Ernst Freund. Planning in New York was pursued by the same business and civic elite who led the city's preservation movement.[32] The Municipal Art Society

---

Life and Architecture. Volume 1: Boston Bohemia, 1881-1900 (Amherst: University of Massachusetts Press, 1995).

30   Hosmer, *Presence of the Past*, 44.

31   For example, at the same time that John Phillip Reynolds, Jr., was saving the Paul Revere House, he led a real-estate syndicate that was trying to buy and destroy Boston's Park Street Church.

32   Mason, 'Historic Preservation', in Page and Mason; Max Page, *The Creative Destruction on Manhattan, 1900-1940* (Chicago: University of Chicago Press, 1999); Holleran, 182-93, 245-68; John R. Mullin, 'American Perceptions of German City Planning at the Turn of the Century,' *Urbanism Past and Present* 3 (1976), 5-15; Thomas Logan, 'The Americanization of German Zoning,' *Journal of the American Institute of Planners* 42 (1976), 377-85; Brian Ladd, *Urban Planning and Civic Order in Germany, 1860-1914* (Cambridge: Harvard University Press, 1990), 7-13.

in New York, and similar City Beautiful groups elsewhere, brought together the goals of preservation, parks and planning. In the 1920s and 1930s, zoning was used implicitly to protect historic neighborhoods in Boston and some other cities, before being adapted explicitly to the purpose by Charleston and then other Southern cities.

Landscapes

One of the most complex comparisons among countries concerns the relationship between preservation efforts directed at buildings, and contemporary efforts to save natural and scenic landscapes. These were kindred campaigns and often attracted the same individuals, but whether they behaved as a single movement varied from place to place and time to time. The US in the nineteenth century pioneered two forms of landscape conservation: the rural parks movement in big cities, and national parks in the West. In rural parks, such as Fredrick Law Olmsted's Central Park in New York and Prospect Park in Brooklyn, American cities built for themselves what many European cities had inherited in former royal grounds. Some cities did not create rural parks from scratch but adapted them as a way to save fragments of cherished historical landscape that were threatened by urban development: Boston's Jamaica Pond and Franklin Park, Philadelphia's Fairmount Park.

National parks were America's first great contribution to world preservationism. Beginning with Yellowstone in 1872, the federal government set aside remote landscapes that were conceived as untouched wilderness. What had been treated as a resource for exploitation became a subject for curatorial care. National Parks reinforced a particularly American national identity, rooted not in history but in the idea of a new civilization in communion with nature. After the establishment of Yellowstone, the national park idea was quickly adopted by other new Western countries (Canada, Australia, New Zealand), and in time by most old countries too. In many European countries, royal parks and other communal open land (such as German forests) played roles similar to national parks, but with an important difference: most of their landscapes were the product of human actions or the scene of long human inhabitation, and most were managed as historic landscapes. American national parks started out with a wilderness ethos that actively opposed preserving any human artifacts. That changed with the establishment of Mesa Verde National Park in 1906, though it was years before the National Park system focused on resources from the historical period.

American advocates of historic preservation – in the narrower architectural sense – did venture into landscape preservation, especially where the landscape was perceived as historic. Immediately after saving the Old South, many of the same Bostonians worked together to turn back threats to Boston Common and the city's historic burial grounds. Numerous campaigns sought to protect battlefields, of the Revolutionary War and the much more recent Civil War. Preservationists also worked to preserve landscapes whose value was primarily scenic, such as the New Jersey Palisades, and much of Boston's Metropolitan Park System. Even if there were overlapping constituencies for architectural preservation and the conservation of natural landscapes, for the most part they remained distinct enterprises, organized separately. Natural conservationists more quickly and successfully organized themselves into national NGOs: the Appalachian Mountain Club in 1876, the Sierra Club in 1892, and the Audubon Society in 1905. In European countries the two efforts were more likely to proceed in tandem through the same organizations – perhaps because even the wildest landscapes were understood as part of a long cultural heritage, rather than as wilderness unconnected with humans.

Historic and natural preservation efforts are entwined in the transatlantic story of the Massachusetts Trustees of Reservations, founded in 1891 by Charles Eliot, the young partner of Frederick Law Olmsted (and son of Charles Eliot Norton's cousin, who was Harvard's president). Eliot's family connections gave him the opportunity to participate in discussions about the early conceptualization of what was to become the English National Trust (as described elsewhere in this volume in Melanie Hall's chapter). Eliot acted first, and the Massachusetts Trustees thus became a trial run for the English group. Each of these organizations, according to their charters, was to collect and hold both landscapes and historic structures. Their histories, however, quickly diverged. The Massachusetts Trustees found that historic structures required maintenance beyond their means, and concentrated instead on landscapes, becoming the original prototype of the land trust in America. The English National Trust acquired both architectural and landscape properties, eventually including vast tracts of unique lands that in the US would more likely have become national parks.

The American Scenic and Historic Preservation Society, founded in New York in 1896, likewise aimed to treat historic structures and landscapes together. It acted with explicit attentiveness to European models. One of these was the British National Trust (making American Scenic a second-generation descendant of the five-year-old Massachusetts Trustees); another was the centralized governmental model of France.[33] The society's role was mainly advocacy, particularly through its politically influential (male) leadership. While its ambitions were national, its actions were mostly confined to New York state.

By the early twentieth century, the historic preservation and landscape conservation movements in America had largely come delaminated, with preservationists seldom focusing on landscapes larger than the gardens attached to historic houses. The two began to be reunited as the National Park Service in

---

33   A.D.F. Hamlin, *American Scenic and Historic Preservation Society Annual Report* (Seventh annual report, 1902). American Scenic was founded as the Trustees of Scenic and Historic Places and Objects; the organization changed its name in 1901.

the 1930s turned its attention to historic sites and to parks created from historic landscapes, such as Colonial National Monument and the Blue Ridge Parkway.[34]

Transatlantic Exchanges

A final subject for review is direct transatlantic exchanges, of which there were many kinds. We have already touched upon American architectural students in Paris, the general cultural influence of Ruskin and other authors, German emigré planners and the Trustees of Reservations. What others? Most importantly, the European Grand Tour was not just for architects, but a rite of passage for much of the American elite. European treatment of monuments was influential in American preservation from the beginning: some of the women involved in the Mount Vernon effort were emulating efforts they had seen in Europe, such as the preservation of William Shakespeare's House in Stratford.[35] Later, the First World War gave more than a million young Americans an involuntary sort of Grant Tour, including first-hand experience with European cultural heritage, and sometimes efforts to protect it.

The Grand Tour invoked the very old idea that would later come to be called World Heritage. Americans could not understand their own culture without experiencing, firsthand, its Old-World roots. By the mid-nineteenth century, Ruskin had articulated the preservation corollary, a shared responsibility for preserving this shared heritage, without regard to national boundaries. This idea found its way into practice in various ways. Sometimes its implementation, as through the advocacy of SPAB's International Committee in Italy, was received as unwelcome meddling. Sometimes, as in British architect C.R. Ashbee's speaking tours and correspondence in support of American causes, it probably bolstered preservation by adding an international endorsement to local efforts.[36]

The most direct expression of shared responsibility was financial. European preservationists attempted to make use of American identification with European roots to raise money in the New World for preservation efforts in the Old, especially England (a direct approach was the union of English titles with American heiresses, to the advantage of some country houses). Was Anglo-American fund-raising a special case? Did German, Scandinavian, or Czech preservationists attempt to raise funds among their American emigré communities? The flow of money was

---

34   Ian Firth, 'The Blue Ridge Parkway: Road to the Modern Preservation Movement,' in Charles Birnbaum and Mary V. Hughes, eds, *Design with Culture: Claiming America's Landscape Heritage* (Charlottesville: University of Virginia Press, 2005), 179-202.

35   West, *Domesticating History*, 12, 15.

36   Frank C. Sharp, 'Exporting the Revolution: The Work of the SPAB outside Britain, 1878-1914,' in Miele, 187-212; C.R. Ashbee letter in *American Architect and Building News* (May 2, 1896).

not entirely eastward, as Europeans provided funds for New World archaeological efforts, primarily in Central America.

Americans participated directly in documentation and conservation efforts in Rome and in the Near East, including some efforts that originated in the American missionary tradition. The educational exchange of American students in European institutions was supplemented by the establishment of the American Academy in Rome (by Charles F. McKim, who directed it from 1894) and the American Universities in Beirut and Cairo. The Archaeological Institute of America set up the American School of Classical Studies at Athens in 1882, and similar schools in Rome and Jerusalem by the end of the century.

The concept of shared heritage could express itself not only through the movement of people but also of artifacts, most often artistic or archaeological, but also architectural. Numerous rooms or interior components, and sometimes exterior components or whole standing structures, crossed the Atlantic to give wealthy Americans an ersatz sort of built heritage. P.T. Barnum reported that he almost succeeded in buying Shakespeare's house for removal to America, though from his own account his intentions may have been merely to profit from the threat.[37] Tattershall Castle was partly dismantled in 1911, reportedly purchased by William Randolph Hearst to be re-erected in America.[38] As with Barnum at Stratford, the story may have been somewhere between a hoax and an exaggeration. But here was a rare American contribution to European preservation practice, as American acquisitiveness, or the threat of it, spurred the passage of cultural patrimony laws.

American preservation is a distinctive tradition and it has distinctive origins. The preservation movement in the US developed under different circumstances than, and often in isolation from, European practice (which, though several different national experiences, grew up with much more interaction among them). Historical scholarship on American preservation is still at an early stage. Just a few years ago one could comfortably carry all the books on the subject. Great progress has been made in the past fifteen years. Several studies have found the productivity of a comparative approach among American cities or regions.[39] We are beginning to be ready for the deeper comparative perspective of American national experience with others. The differences often stem from different national experiences. The similarities and interactions are particularly complex and instructive. The ambiguities and the blanks are opportunities for exploration.

---

37   P.T. Barnum, *Strategies and Triumphs, or Forty Years' Recollections* (Buffalo: Warren, Johnson & Co., 1872), 365; Francis Teague, *Shakespeare and the American Popular Stage* (Cambridge: Cambridge University Press, 2006), 41-51.

38   John Harris, *Moving Rooms: The Trade in Architectural Salvages* (New Haven: Yale University Press, 2007); Mandler, 184-88.

39   Lindgren, *Preserving Historic New England*; Lindgren, *Preserving the Old Dominion*; Judy Mattivi Morley, *Historic Preservation and the Imagined West: Albuquerque, Denver and Seattle* (Lawrence, Kansas: University of Kansas Press, 2006).

Chapter 9
Conservation and the Professions: The Swedish Context 1880-1920

Ola Wetterberg

Introduction

In the years around 1900 interest in the conservation of historic buildings and their place in the urban environment was developing in Sweden as in many other countries. Historical factors were considered to be of significance not only in restoration and preservation activities, but also in building management, in new architecture, and in town planning.[1] This development has been interpreted from an ideological point of view in terms of humanistic responses to contemporary development on the one side, and as a tool for, and a reflection of, a nationalist hegemonic agenda on the other.[2] Both of these interpretations have validity in certain respects. In this context it will be enough to state that conservation, and the identification of certain buildings, or certain aspects of buildings, as historic and different from the present, is an aspect of modernity. Nevertheless, the fact remains that the tasks in conservation and in planning activities with specific reference to history increased, and this raised demand for specialized education, experience, and competency in the field of historic buildings in their social and environmental contexts.

In this chapter I draw attention to the formation of new professions related to building preservation, and to their relationships to new fields of specialized practice in restoration and conservation. I suggest that these tangible aspects of conservation, and not only its ideas, provide an important context for our understanding of how and why conservation practice came to be what it is. While

---

1 Ola Wetterberg, *Monument & miljö,* Institutionen för arkitekturhistoria, PhD diss. Göteborg Chalmers tekn. högsk, 1992.

2 See e.g. Mia Geijer, *Makten över monumenten: restaurering av vasaslott 1850-2000,* Nordiska museets förlag, PhD diss. Stockholm Kungliga tekniska högskolan, 2007; Richard Pettersson, *Fädernesland och framtidsland: Sigurd Curman och kulturminnesvårdens etablering,* PhD diss. Umeå universitet, 2001; Peter Aronsson, *Historiebruk: att använda det förflutna* (Lund: Studentlitteratur, 2004); Lilian Andersson, *Mellan byråkrati och laissez faire: en studie av Camillo Sittes och Patrick Geddes stadsplaneringsstrategier [Beyond bureaucratic and economic rationality: a study of Camillo Sitte's and Patrick Geddes' town planning strategies]* (Acta Universitatis Gothoburgensis, Göteborg, 1989).

my focus is on Sweden, it is clear that there were international influences, notably from France, Germany, and Britain as well as from other Scandinavian countries. Swedish practice, and particularly the example of Skansen, is well known to have provided influence abroad. While Skansen internationally has come to stand for Swedish preservation as this chapter demonstrates, it was only a small part of long-standing and far more extensive campaign to preserve the face of the past. State-supported preservation developed early in Sweden. From the early nineteenth-century efforts focused on ancient castles and early churches which can be seen as defining the historic nation.

By the turn of the twentieth century, extensive criticism of preservation's practice led to a broadening of interest in historic domestic buildings and, gradually, to urban environments. Skansen's collection of buildings represents part of an early broadening phase which, at the same time, encapsulates a rise in folkloric and museum interests. At issue was not only which buildings should be preserved but, also, how far authentic and scholarly historical evidence should determine the aesthetic appearance of a building; whether the buildings was considered 'living' or 'dead', and how 'the community' should be represented. Importantly, debates were framed between the two long-standing state offices: the Office of the Superintendent of Public Buildings and the Royal Academy of Letters, History and Antiquities, and between the creative flexibility of architects and the more scholarly interests of antiquarians. With a new generation of museum professionals, the type of buildings that were saved greatly broadened. But, this 'protection' had to take place outside the existing preservation structures. These new interests amplified debates; at issue was not only whether more folkish buildings should represent 'the nation' but, also, where the physical place of the past within modern 'community life' would be located; that is, focus changed from the open-air museum to preservation *in situ* through the planning process. Entrenched interests in early structures proved beneficial to Sweden but, also, problematical.

The Swedish Scene

Michael Holleran points to the differences in professional development between the United States and Europe in Chapter 8. He argues that American antiquarians seem to have had a stronger position in the conservation field than their European counterparts, noting particularly those in France and England where the restoration architects were at the forefront. A similar situation developed in Sweden, where antiquarians (i.e. art historians, archaeologists, ethnologist, etc.) played a more prominent role than architects in building conservation during the early twentieth century though the reasons for this were somewhat different. It is tempting to try to explain this situation by the long-standing existence of monument and buildings protection in Sweden, which dated back as far as 1630 when the office of Custodian of National Antiquities [Riksantikvarien] was first established.

National aspirations and the personal antiquarian interests of the monarch during the seventeenth century had led to a range of activities both for the creation of inventories and for the protection by law or statute of ancient remains. Sweden's first law for the protection and conservation of old monuments and antiquities, dating from 1666 is well known.[3] However, during the eighteenth century interest in ancient monuments and buildings declined and it did not gain real strength again until the second half of the nineteenth century.[4] Although there is an early history of conservation and protection in Sweden this interest was not continuous during the ensuing centuries. The relative strength of antiquarians in relation to architects within the preservation institutions in Sweden instead can be explained by events during the formation of the heritage sector in the late-nineteenth and early-twentieth centuries.

In principle there were two public bodies dealing with historic buildings in the second half of the nineteenth century. The Office of the Superintendent of Public Buildings [Överintendentsämbetet] had the responsibility for the maintenance of state-owned buildings, including buildings of architectural and historical interest. Until this time, the care of old buildings was mostly seen as a natural part of the work of architects and craftsmen, but by the end of the century this was beginning to change and the area was considered to require specialized historical knowledge. The second public body involved was the Royal Academy of Letters, History and Antiquities [Kungl. Vitterhets, Historie och Antikvitetsakademien] and its secretary and official, the National Antiquary [Riksantikvarien]. Until the late 1800s the Academy only showed a faint interest in building conservation, and in post-medieval monuments, but at that time the Academy began to challenge and intervene in the responsibility of the Office of the Superintendent of Public Buildings. The renewed energy with which the debate and activities around monument conservation was addressed after the turn of the twentieth century came partly from events and changing approaches to the material evidence of the past within these institutions themselves. As in England, this interest was first seen in the churches.

The Superintendent's Office had responsibility for the 'living' state-owned buildings that is, buildings still in use, in 1828 the Academy of Letters was given formal responsibility for the 'dead' buildings, the ruins. The work that came with these new managerial responsibilities also led to a renewed interest in historic buildings and monuments. During the ensuing years, the destruction of medieval churches was also extensive. These churches tended to be small and as the

---

3   Henrik Schück, *Kgl. Vitterhets historie och antikvitets akademien: dess förhistoria och historia,* Wahlström & Widstrand, Stockholm, 1932-1944, Bd 1-2, s. 264 ff, p. 343 ff., 360 ff.; *Kongl: Mayst:tz Placat och Påbudh, Om Gamble Monumenter och Antiquiteter* (printed by Georg Hantsch, 1666) (http://www.kb.se/f1700/Monumenter/Default.htm).

4   Schück, *Kgl. Vitterhets historie och antikvitets akademien* (Stockholm: Wahlström and Widstrand: 1932-1944).

congregations grew, larger buildings were needed.[5] Within the Academy of Letters, a concern for protecting interior medieval wall paintings grew to include interest in preserving the churches themselves. This had two significant outcomes. In 1840 the Academy was permitted to document churches of historical value that were to be demolished, and in 1867 they were granted limited powers to prevent the destruction of particularly important churches.[6] The number of restorations led by the Superintendent's Office increased. Restorations were mostly carried out in the manner advocated by the influential French architect, Eugène-Emmanuel Viollet-le-Duc. This aroused strong interest from the architects at the Superintendent's Office influenced by international developments. As the Office workload escalated so did the level of the public debate that shortly will be outlined below.[7]

A more specialized organization able to supply suitable expertise was needed to ensure that restorations were carried out in appropriate styles. The administrative procedures for church matters were revised in such a way that, at the turn of the century, all matters concerning churches of historic interest were referred not only to the Office of the Superintendent, but also to the Academy of Letters and to the National Antiquary. However, there had still been no more fundamental shifts in the organization for the preservation of monuments and buildings.

Demands for Change and Reorientation

Although gradual change was taking place within these public bodies, the main stimulus for a profound change of heritage practice was to be found elsewhere. The incentive for debate and change originated partly in reactions against the restoration of national and church monuments, partly through an interest in rural and pre-industrial culture and its artefacts, and partly in reaction to the destruction of urban environments. The initiatives for change came from cultural elites, architects, journalists, writers and academics, acting both as individuals and within historical associations. Initially, harsh criticism was directed both towards the architects at the Superintendent's Office, and to the antiquarians within the Academy of Letters. As the architect Sigurd Curman noted in 1908, The Academy's interest had so far 'directed its main activities towards the most ancient time and still often lacked particular interest in what should be at least as close to our hearts, namely those still living representatives of Swedish architecture of the past centuries, in which our own architecture has or at least should have its deepest roots.'

---

5   Siegrun Fernlund, *'Ett Herranom värdigt tempel': kyrkorivningar och kyrkobyggen i Skåne 1812-1912 [Kirchenabrisse und Kirchenbau in Skåne 1812-1912]*, PhD diss., Lunds universitet, 1982.

6   Wetterberg, *Monument & miljö*, 29.

7   Ola Wetterberg, *Monument & miljö*; *Kulturfastighetsutredningen. Bilagor,* Statens fastighetsverk (Stockholm: 2009), 61. (http://156.51.176.32/sfv/Kulturfastighetsutredningen_Bilagor.pdf).

['inriktat sin hufvudsakliga verksamhet på den allra gråaste forntiden och sakna oftast allt intresse för det, som dock borde ligga oss minst lika nära om hjärtat nämligen de ännu lefvande representanterna för de senaste århundradenas svenska byggnadskonst, i hvilken vår egen har eller åtminstone borde hafva sina djupaste rötter.'[8]]

Although this was a characterization made in a polemical situation, and brought forward in the characteristic nationalistic sentiment of the period, there were plenty of good arguments to support the critics' points of view discussed below. It was also clear that the critique was heavily influenced by the European discussion on restoration and preservation.

Restoration

The intense critique of restoration practice came to one of its peaks in 1894 with the pamphlet written by the noted novelist and poet, Verner von Heidenstam, *Modern Barbarism: some words against the restoration of historic buildings* [*Modern barbarism: några ord mot restaurerandet af historiska byggnader*] and the following 'battle' in the vicinity of the contested restoration of Gripsholm Castle, widely regarded as one of Sweden's most important historic monuments.[9] This fortified former royal residence, mainly dating from the sixteenth century, was well known, partly because of its location in Mariefred, only sixty kilometres south-west of Stockholm. The Castle had already become a tourist attraction. In the early 1890s it underwent extensive restoration and alteration by the architect Fredrik Lilljekvist, a graduate of the Royal Academy of Arts who was heavily influenced by Viollet-le-Duc. While a student at the Academy (1884-87), Lilljekvist had documented the Castle under the guidance of art historian, Gustaf Upmark (Sr.) director of the Nationalmuseum [The National Museum of Fine Arts] who was the driving force behind the restoration. Upmark's aim was to achieve adequate exhibition space for the Swedish State's collection of portraits at the castle which was dilapidated and had been sharply transformed during the 1700s.

Lilljekvist had drawn up plans for the Castle's restoration, which were shown at the Royal Academy of Arts in 1888. On the basis of this, he was appointed architect of restoration work at Gripsholm, where he carried out his radical restoration in a somewhat modified form. Although endorsed by the exhibition at the Academy, as well as by authorities and architectural expertise, Lilljekvist's

---

8   Sigurd Curman, 'Byggnadsminnen i privatägo', *Arkitektur* 12 (1908), 133-35.

9   Verner von Heidenstam, *Modern barbarism: några ord mot restaurerandet af historiska byggnader* (Stockholm: Bonnier, 1894); Lars Ljungström, *Aendnu gamblare: Fredrik Lilljekvists restaurering av Gripsholms slott och 1890-talets restaureringsdebatt* (Södermanlands museum, Nyköping, 1987); Henrik O. Andersson and Fredric Bedoire, 'Lilljekvist, Johan Fredrik', *Svenskt biografiskt lexikon*, 23 (1980), 158-61.

approach and work was criticized and ridiculed in an article in the newspaper *Dagens Nyheter* by Verner von Heidenstam who suggested that a sign should be erected at the Castle in mock-archaic Swedish, explaining that in 'Anno Domini 1893 gjordes detta gambla hus ændnu gamblare' ['in 1893 this old house was made even older']. Lilljekvist was regarded as having destroyed the architectural history and integrity of the building by taking it back to a particular date, deemed its original style, and destroying any later additions and alterations.[10]

The restoration was certainly dramatic but not in comparison with other contemporary restorations, which invariably resulted in immense alterations to the building's historic fabric. In fact, the approach to restoration at Gripsholm Castle represented a slight move towards a more thorough historical study and respect for the original materials of the castle, and a more didactic approach to restoration. The restoration of the exterior was broadly organized to recreate the character of the former castle. The restored castle was to be used as an exhibition and storage space for the national collection of portraits. For that reason the interior restoration aimed to create through a mix of preservation, reconstruction and new additions a series of historic styles that reflected the periods and styles of the paintings. There was a vast suite of rooms in renaissance style from the Vasa time organized around the authentic Hertig Karl's kammare [Duke Charles' Chamber]; some baroque interiors were created in Danska rummen [The Danish Rooms] which still contained some remains from this time; other parts were dedicated to the rococo style from the reign of Gustav III.[11]

Nevertheless, the approach was not entirely purist; when important architectural or artistic features were found that did not fit the dominant interpretive model they were retained *in situ*. For instance original sixteenth century mural wall decorations were found in the courtyard and were preserved during the course of restoration and allowed to modify the work. Upmark himself initiated a discussion of restoration principles in a published lecture about 'Stilenlighetens faror' ['the dangers of unity-of-style'], and later his historical studies of architecture were expanded and published in the book *Swedish Architecture 1530-1760* [*Svensk byggnadskonst 1530-1760*].[12]

This large-scale and high-profile restoration project, together with other restoration works as the cathedral in Uppsala, had a significant impact in several areas. It not only led to an ideological debate, but also to a far greater interest

---

10   An ironic twist was that the expression 'this house was made even older' was written in archaic spelling taken from a joke by Lilljekvist himself. In a house in Stockholm he had practiced old mural painting techniques and put up a sign with the wording 'An. Do. 1889 gjordes detta gambla huus ännu äldre. Architecten var F. Lilljekvist ...' ['in 1889 this old house was made even older. The architect was F. Lilljekvist ...'], Ljungström, *Aendnu gamblare*, 217-18.

11   A detailed account of the restoration is found in Ljungström, *Aendnu gamblare*.

12   Gustaf Upmark, *Svensk byggnadskonst 1530-1760* (Norstedt: Stockholm, 1904). The book came in a German edition 1897-1900 and was translated into Swedish 1904.

in architectural monuments in their own right; significantly, this resulted in new research and publications in architectural history as part of an attempt to achieve more 'correct' restoration interventions.

Up to this point, in practice there had been a consensus between the architects within the Office of the Superintendent and the antiquarians of the Academy of Letters about the way in which so-called 'unity of style' restorations were executed.[13] However, in the aftermath of the Gripsholm Castle restoration debates a second academy, the Royal Academy of Arts, began to take a professional interest in such matters. In 1902, that Academy made a proposition for a more comprehensive change of the preservation of buildings and monuments. We will soon return to this.

Folk Culture

The debates about heritage and restoration in the closing decades of the nineteenth century was also supported by professionals concerned with language and dialect, folk customs and material culture, as their interests turned towards 'vernacular' buildings. Artur Hazelius, a professor of Scandinavian languages and dialects, is the most internationally famous member of this group; his international reputation is largely derived from his exhibits at world's fairs, and to Skansen, his open-air museum founded in 1891.

Thus, a further cause for criticism of the public heritage bodies was their lack of interest in Nordic folk cultures. In a spirit of romantic nationalism, the latter part of the nineteenth century had seen a growing enthusiasm for and interest in folk cultures and those 'vernacular' buildings that provided material evidence of those cultures. This enthusiasm stimulated a wide range of activities that were scholarly, political and romantic of which Skansen is the most famous. This open-air museum occupies one of Stockholm's smaller islands that was already being used as a pleasure garden for the city and was a sequel to Hazelius's Nordic Museum, founded in 1873, which displays Scandinavian ethnographic collections. In bringing together buildings and traditional artefacts from indicative areas of what was then until 1905 the Kingdom of Sweden and Norway to Stockholm, Hazelius promoted an interest in Nordic folk cultures by pedagogical projects.

Hazelius's collection of buildings is interesting in that in addition to increasingly redundant farm buildings, he also included buildings associated with peripheral ethnic groups, the Lapps and Finns; a Norwegian storehouse; and the summerhouse of Emmanuel Swedenborg, Sweden's famous scientist, philosopher, Christian mystic and theologian. Other patriotic and civic traditions were also represented, such as the house of a soldier-come-teacher (a familiar feature of

---

13   Wetterberg, *Monument & miljö*, 31-32; Ljungström, *Aendnu gamblare*, 64-66; Per-Olof Westlund, 'Byggnadsminnen', in: *Ad patriam illustrandam: hyllningsskrift till Sigurd Curman 30 April 1946* (Uppsala, 1946).

village life), and a maypole. Together, these buildings emphasized similarities in Swedish folk cultures, civic order and religion, at a modern leisure site within the capital city. Hazelius's projects were ideologically intertwined with the Pan-Scandinavistic and Nordistic cultural and political movements, and inspired by its Gothicistic forerunner.[14]

Hazelius's collections had become well-known outside Sweden as he exhibited his 'tableaux', with 'costumed interpreters' at the 1878 Paris World Exhibition, inspired by earlier Swedish contributions to the World Exhibitions in Paris 1867, Vienna 1873 and Philadelphia 1876.[15] However, Skansen was not the only

Figure 9.1 The Open Air Museum Skansen in Stockholm, Morastugan

Source: Nordiska museet archives, Stockholm

14	For Hazelius and Skansen see, Magdalena Hillström, *Ansvaret för kulturarvet: studier i det kulturhistoriska museiväsendets formering med särskild inriktning på Nordiska museets etablering 1872-1919*, PhD diss., Linköpings universitet, 2006; Bo Grandien, 'Grogrunden', in Arne Biörnstad, ed., *Skansen under hundra år* (Wiken; Höganäs. 1991); Sten Rentzhog, *Friluftsmuseerna. En skandinavisk idé erövrar världen*, (Stockholm, 2007); Cecilia Hammarlund-Larsson, 'I denna tid af slapp nationalkänsla: om Artur Hazelius, vetenskapen och nationen', in: *Samhällsideal och framtidsbilder : perspektiv på Nordiska museets dokumentation och forskning*, 2004, [11]-66.

15	Hillström, *Ansvaret för kulturarvet*, 157-59; Bjarne Stoklund: 'Between Scenography and Science. Early Folk Museums and Their Pioneers', *Ethnologia Euroaea: Journal of Ethnography*, 3 (2003), 21-36.

or indeed the first Scandinavian open-air museum. An inspiration for Hazelius was the collection of buildings at Bygdøy in Oslo, Norway, founded ten years earlier by King Oscar II; and among the more large-scale parallels in Sweden was Kulturen in Lund that was begun about the same time in 1892, and with the same intentions. But even more important was the fact that the founding of regional and local associations dedicated to an interest in folk cultures, and with both open-air museums and other related activities on the agenda, grew to the dimensions of a popular and national movement.[16] A parallel and intertwined development occurred of professional interest in, and a popular concern with heritage practice.[17]

Interest in collecting this national vernacular past at museums and sites raised a lot of distinct questions about the selection and management of artefacts, and the kinds of information to be provided. It is not surprising that Axel Nilsson, the first head of the Historical and Ethnographic Department at Skansen, wrote some of the earliest scholarly articles about vernacular buildings in Sweden. He had initially been responsible for collecting buildings for Kulturen in Lund but eventually moved to work with Hazelius at Skansen. Nilsson published only three articles; but at Skansen he was also the teacher of the most productive and well-known researcher of Swedish vernacular buildings, Sigurd Erixon. Erixon's work is still to a certain extent the foundation for contemporary practice in heritage management of the vernacular.

Professional Groups: Antiquarians and Architects

One result of this widening of the historical perspective to include more and more historic buildings of different types and scale was the emergence of a new kind of architect and professional building conservator: that is, people who specialized in the techniques and characteristics of historic buildings.

We will first consider the non-architects involved in the preservation field. In Sweden the employment title for this broad category of professionals was and still is antiquarian (antikvarie). The first professional association in Sweden for people in museums and other heritage practices was the Swedish Association of Museum Employees [Svenska museimannaföreningen] founded in 1906. The members of this association formed a rather diverse body. The association gathered older and younger members, scholars and autodidacts, employees and museum volunteers.

Some of the tensions that occurred during the formation of the new museum profession can be identified. On a personal level these tensions did not follow unambiguous lines, they were more mixed, even if a shift in generations seems to have been important. The *younger generation* was obviously more impatient, more offensive and expansive, than the already established employees of the

---

16 See e.g., Maria Björkroth, *Hembygd i samtid och framtid 1890-1930: en museologisk studie av att bevara och förnya*, PhD diss., Umeå universityet, 2000.

17 Wetterberg, *Monument & miljö*, chapters 3-4.

Royal Academy of Letters, History and Antiquities. One of the more significant reasons for impatience was the focus of interest within the existing research and heritage institutions. The new and widened direction of historical interests of the younger generation had no real equivalent within the Academy of Antiquities. Its critics thought that the Academy seemed to have only a rather faint interest even in medieval monuments, with the exception of church inventories, church paintings, and a few ruins.[18]

Strong *regional opposition* towards the national and central heritage institutions also came from museum employees and historians. This opposition grew apace with the development of regional museums, and their own needs for economic support. This aspect of the tension within the professional formation was profound and already had a long history.[19]

This regional discontent gradually mixed with *opposition from the universities*. For many years the Royal Academy of Letters, History and Antiquities had been the centre for research in both archaeology and art history; as new university disciplines in the humanities were established different patterns of rivalry evolved.[20] This growth and differentiation of the humanistic disciplines, in combination with the process of establishing a new profession and legitimacy in the growing heritage field, underlined the complicated relationship of the professionals to the *amateurs in the field*. Formalized criteria for heads of museums, for instance, led inevitably to conflicts of a personal character, but also to conflicts of a more structural nature because of the dependency of the national and professional heritage institutions on the many voluntary associations and their members that added to the growing popular heritage movement.

A closely related question was the relationship between a scientific and *scholarly approach* to curating, and a more populist, *community-building* agenda. For example, younger academics found Hazelius's displays at Skansen and the Nordic Museum too romantic, and thereby discrediting to the advancement of an academic museum profession.[21] But there was certainly ambivalence about the relationship between normative claims and more detached academic scholarship within the profession. It would be misleading to suggest that all curators relinquished a normative bias; some indeed showed ambitions towards fostering

---

18   See e.g., Curman, 'Byggnadsminnen i privatägo'; Ewert Wrangel, 'Våra fornminnen och deras vård', *Föreningen Heimdals populärvetenskapliga tidningsartiklar* 9, Uppsala, 1908; Ture J. Arne, *Våra fornminnens vård*, Studentföreningen Verdandis tidningsartiklar 193 (Uppsala, 1908).

19   Wetterberg, *Monument & miljö*, 41-49; Hillström, *Ansvaret för kulturarvet*, 103-29; *Svenska fornminnesföreningens tidskrift*, 1871-72, 121 f.

20   Sverker Janson, *Arkeologisk vardag*, interview with author, 1990; ms in author's possession, 21 ff.

21   Bo Lagercrantz, 'Nordiska museet, Skansen och hembygdsrörelsen', in Biörnstad, ed., *Skansen under hundra år*. Lagercrantz writes about the schisms after Hazelius death, and how the science-based position on exhibitions came to be dominant under new management.

taste in modern architecture that is, in new architectural development. These ambitions to take an active part in contemporary development increased during the first decades after 1900 in parallel to the growth of formal academic training and research. One of the officials and folklorists from the Nordic Museum, N.E. Hammarstedt, suggested that a proper academic study of vernacular architecture was badly needed, because without it no 'viable rebirth of our wooden architecture' would be possible which meant that the contemporary 'corrupting, distorted and cluttering' urban development could continue.[22]

A short note on Hammarstedt's view on Skansen illustrates the inherent ambivalence towards the role of education and research within the museum community. He was eager to point out the demarcation line between the more 'scientific' museums and the small local museums. To characterize these local museums Hammarstedt uses the Swedish term 'fornhem', which most readily translates as 'ancestral home' or homestead as a characterization of the purpose of local preserved heritage buildings. In his view, the importance of preserving historic, vernacular buildings as local museums should be to bring about a reverence for and interest in local history by showing a 'home, organically grown out of the ground'. But in his view it would be harmful to mix this patriotic agenda with ambitions to be a 'real' museum. Thus, according to Hammarstedt's

Figure 9.2   (Members of) The Swedish Association of Museum Employees under formation in 1906

Source: *Idun*, 8 (1906), 85

22   *Fataburen: kulturhistorisk tidskrift, Nordiska museet* (Stockholm, 1914), 182, 184.

principles, from the outset every local open-air museum would be destined to be unfulfilled as a museum in the more scholarly and scientific sense of the profession. By this reasoning, Skansen should be regarded as the 'fornhem' [ancestral home] of the whole Swedish people.[23] So even if Hammarstedt distanced himself from Hazelius and the local heritage movement, with their romantic nationalism, he found no difficulty in reconciling his position with the museum's professional ambitions.

In 1906 thirty-two museum representatives met to form the new professional body, the Swedish Association of Museum Employees, in order to stimulate a more systematic approach to collecting and curating and to encourage better cooperation within the field.[24] On the agenda during the early years were museum practicalities, a professional handbook for members, and the development of principles in exhibitions and collection management. Moreover, they were concerned with historic buildings and restoration principles, with the relationship between professional heritage institutions and local preservation movements, as well as with contemporary preservation cases. In addition, they were concerned to underline the need to institute a new Swedish body for the preservation of historic monuments.[25]

A new organization for historic monuments seemed to be necessary in order that the museum profession could gain influence over the preservation of historic buildings. In a speech at the annual meeting of the Museums Association in 1911, the director of the Nordic Museum, Bernard Salin, asserted that this influence was essential as museum professionals were the people with the most informed understanding and the best sense of the historic values of buildings as well as an awareness of academic development in the study of historic buildings.

Salin also underlined the importance of building the new organization upon existing structures. His other concern was that professionals did not fall under the influence of other countries. This was said in clear awareness of several propositions for a *separate* office for the preservation of building – bypassing the National Antiquarian – propositions built upon careful studies of European legislation and organization.[26] In fact, Salin was acting directly within the framework of the debate initiated by the proposals from the Royal Academy of Arts in 1902 in the aftermath of the Gripsholm restoration. This debate continued for decades and can be traced through the various professional journals as well as in a series of public inquiries on the matter of preservation and restoration of historic buildings.

---

23 *Fataburen* 1914, 182-84.

24 'Cirkulär', 1905, Svenska museimannaföreningens arkiv, koncept 1905-1933, BI:1, Riksarkivet [National Archives of Sweden, hereafter NAS].

25 See *Fataburen* several issues:1908, 169 ff; 1909, 186-87; 1911, 173-74; 1913, 122; 1914, 182, 184; 1916, 262 f.

26 Fornminnesvårdskommitténs arkiv, vol. 11, Riksarkivet [NAS].

A Separate Office for the Preservation of Historic Buildings?

In 1902 a letter from The Royal Academy of Arts to the King in Council addressed the issue of restoration; seen retrospectively, this was an important milestone. Architects and artists dominated this Academy. Even though their letter did not lead to any immediate changes in legislation or organization, it has become iconic in the Swedish understanding of when the modern thought in restoration was introduced. The letter tackled the registration, protection and conservation of historic buildings, and how these issues should be organized. The Academy did not comment on which authority would handle this that is, whether it should be one of the existing offices – i.e. the Superintendent's Office or the Royal Academy of Letters – or a newly created independent central authority. But they asserted that whichever authority received the responsibility, in addition to registration, documentation and supervision of laws, should also provide advice and guidance and try to gain public interest.[27]

Figure 9.3   **The National Antiquarian Hans Hildebrand, with Ms Andersson and Lillot Hildebrand. Courtesy of Antikvarisk-Topografiska Arkivet, the National Heritage Board, Stockholm**

27 *Underdånig skrifvelse i restaureringsfrågan* (Stockholm, 1902).

The National Antiquarian, Hans Hildebrand reacted very strongly against the proposal in the following year, a reaction published in the Royal Academy of Letters' own journal in 1903. Although the Academy of Arts said it would not take a position, it was clear that this Academy thought that a newly created authority for monument care was needed. Hildebrand thought that their unstated position was incomprehensible, impractical and unnecessary; he was of the opinion that the Royal Academy of Letters had long collected and indexed records of buildings in the Antiquarian-Topographical Archive, and moreover that the National Antiquarian already supervised the law effectively enough.[28] Hildebrand also discussed the role of architects in monument preservation. He thought that the architects were not qualified for a leading role even if they had an important task and they had expressed interest in preserving the old buildings: they too often showed a desire to 'improve' historic buildings, which meant changes and additions. For Hildebrand, in order to come to terms with the architects' deficiencies there was only one way, education: 'They ought to be educated not only in the knowledge of the laws of art and art history, but also to understand the delicacy required to undertake a restoration.' ['De böra uppfostras icke blott till kunskap om konstens lagar och konstens historia, utan ock till insikt om den grannlagenhet, som erfordras för att företaga en restaurering.'[29]]

Education – which had not been treated in the Art Academy's letter – was the core of the whole issue, said Hildebrand. It was not enough to have a *general* knowledge of art history; experience had shown that art exhibited very different developments in different parts of the country, and even within diocesan borders. In one example, he showed that the Superintendent's Office did not always take this into account.[30] His thus clearly demonstrated that neither the architects, nor the Superintendent's Office, nor the Academy of Fine Arts were qualified to assess the cultural aspects of restoration; it was an opinion with which the Royal Academy of Letters concurred.[31]

Hildebrand's essays were not only an attack on the architects, but also a defence of his own activities that had come under severe criticism. There was a widespread mistrust of the 'existing structures', e.g. the Royal Academy of Letters, History and Antiquities, and whether it could fulfil the demands in the practical management of buildings. This mistrust constituted a threat to the new profession of historic buildings curators. This mistrust was exemplified by Ewert Wrangel, the first professor in art history in Lund, who demanded a separate commission for the preservation of monuments.

In 1901, Wrangel had argued that cultural historians should work alongside the architects on church restorations, as the lack of such cooperation caused both

28 *Kungl. Vitterhets-, historie- och antikvitetsakademiens månadsblad.,* Kongl. Vitterhets, historie och antiqvitetsakademien, Stockholm, 1903, 310 f.

29 *Kungl. Vitterhets-, historie- och antikvitetsakademiens månadsblad,* 307.

30 *Kungl. Vitterhets-, historie- och antikvitetsakademiens månadsblad,* 307 f.

31 *Kungl. Vitterhets-, historie- och antikvitetsakademiens månadsblad,* 296 f.

long delays and poor restorations. He thought that architects were needed for their trained eyes and technical insights, as well as for their ability to give practical advice to the congregations. Wrangel's argument was based on international comparisons: in Germany, the organization of restorations was extremely rational with so-called provincial conservators; in Denmark a special commission decided all restoration cases. Common to both countries was that art historians and archaeologists worked closely with architects.[32]

Wrangel proposed that cooperation was also needed at government level in Sweden. On the contrary, he thought the Royal Academy of Letters and the Superintendent's Office acted as rival authorities. In 1908 Wrangel refined his critique still further. In an article published for an influential student association, he argued that the heritage from the Middle Ages had been neglected and diminished even in recent times, which he considered was due to a lack of interest in medieval culture. Wrangel thought that the interest was rising, but was often overshadowed at the Royal Academy of Letters and by the National Antiquarian. The situation was no more satisfactory when it came to preserving remains from more recent times. Wrangel further asserted that the organization for monument protection was totally inadequate and primitive. In reality, the National Antiquarian was one of the busiest officials in the country. He had two very demanding roles, as museum director, and as guardian of the national monuments. It was simply not possible for him to cope with these tasks when he also was secretary of the Academy. His three assistants were deployed on museum services: archives, libraries, and monument protection across the country. Theirs were not 'summer resort jobs' but involved travel and study throughout the country and throughout the year. Wrangel was surprised that the new positions at this time suggested by the Royal Academy of Letters were only intended for the museum. No new organization for the management of historic monuments had yet been proposed.

It was Wrangel's opinion that a *division* between the two tasks, museum work and monument preservation was necessary and further, that monument protection needed an independent and well-dimensioned Central Commission with permanent representatives around the country. He outlined a proposal for a Central Commission that would consist of historical expertise together with practical and technically trained people. He suggested that at least one architect should be part of the Commission to reduce the slow process of restoration and rebuilding, and that a scholar and an architect should also be engaged to make inspections around the country. In addition to the inspectors from a Central Commission, he proposed that there should be people stationed in the regions with a fixed salary and a degree of autonomy.

Wrangel summarized his criticism:

> ... under the present organization both the Superintendents' Office and the Royal
> Academy of Letters shall be heard and, at least until a few years ago, those

---

32   *Svenska fornminnesföreningens tidskrift* (Stockholm, 1902), 270, 277.

two authorities have often been of different opinions – and you have witnessed the edifying spectacle that two authorities giving different opinions of the same thing fighting each other with The King in Council as a buffer!

[... enligt den nuvarande organisationen skall Öfverintendentsämbetet jämte Vitterhetsakademien höras, och dessa båda myndigheter ha ofta, åtminstone ännu för några år sedan, titt och tätt varit af olika meningar – så har man fått bevittna det uppbyggliga skådespelet, att två om samma sak sig yttrande myndigheter bekämpa hvarandra med k.m:t som buffert!⁣[33]]

With this tarnished image of the responsible authorities, it is understandable that Wrangel proposed a separate Commission for the protection of historic monuments!

Wrangel's discussion was heavily influenced by international developments, mostly in Germany, and he wrote a report on a meeting on 'Denkmalpflege' to the State Department the same year.[34] The attention to international developments in preservation intensified during the early years after 1900. The protection of historic buildings had been discussed at the Congress of Architects in Madrid 1904, in London 1906 and in Vienna 1908, all with Swedish participation in the meetings and the congress committees, and with articles in the Swedish architectural journal.[35] During the same period the young restoration architect Sigurd Curman had been travelling in Germany, France and Italy and in 1908 he authored reports on historic preservation in different countries, on behalf of The Office of the Superintendent of Public Buildings.[36]

I.G. Clason, the influential architect of the Nordic Museum as well as a leading official in the Superintendent's Office, had visited the Congress of Architects in Vienna in 1908. On the basis of a detailed statement from this meeting he advocated, like Wrangel, that protection of historic buildings should be placed under one authority. Clason considered that it could not be a benefit to the various components of a restoration proposal to be handled by different authorities; he summarized his views on the organization for historic buildings:

It would be desirable to have *one* powerful and versatile, well-equipped central authority for the referral of issues relating to the preservation of our historic buildings which would also have the opportunity and obligation to practice a truly effective control over the performance *during the work*, and not only the ability to make inspections *after* completion of the work. In addition, it should

---

33   Wrangel, 'Våra fornminnen och deras vård'.
34   Fornminnesvårdskommittéens arkiv, vol. 11. Riksarkivet [NAS].
35   Wetterberg, *Monument & miljö*, 69-72.
36   Fornminnesvårdskommittéens arkiv, vol. 11. Riksarkivet [NAS]; Curman, 'Byggnadsminnen i privatägo'.

make sure that such works should only be planned and carried out by competent persons.

[Önskemålet vore därför *en* kraftigt och mångsidigt utrustad centralmyndighet för handläggandet af frågor rörande vården af våra byggnadsminnen samt möjlighet och skyldighet för denna att utöfva en verkligt effektiv kontroll öfver de *under utförande varande* arbetena, ej endast inspektion *efter* arbetets slutförande. Vidare att dylika arbeten endast må planläggas och utföras af kompetenta personer.[37]]

Remarkably, when the National Antiquary drafted a proposal for a law for the protection of historic monuments on behalf of The Royal Academy of Letters in 1912, he suggested no radical organizational change.[38] The Swedish Association of Museum Employees therefore demanded that the bill should include a comprehensive review of the organization for monument protection. The demand for an independent National Antiquarian with a focus on historic monuments was an important question:

The National Antiquarian should be released from several of his existing tasks, including the directorship of the Swedish Museum of National Antiquities, and instead only become a central authority which meets all the country's museums' interests and deals with the general welfare of the country's immovable and movable cultural monuments.

[Riksantikvarien frigöres från flere af sina nuvarande uppgifter, däribland också från chefskapet för Statens historiska museum, för att i stället uteslutande blifva en central myndighet, som tillgodoser landets samtliga museers intressen och handhar den allmänna vården af landets såväl fasta som lösa kulturminnesmärken.[39]]

Who was to be in Charge?

The fact that the Royal Academy of Letters and its senior officials demonstrated their unwillingness to address these long-standing organizational demands strengthened those who wanted to place the protection of historic monument under a new agency. In the coming decades the protection of historic monuments was

---

37   Isak Gustaf Clason, 'Våra byggnadsminnens vård', *Arkitektur* 2 (1910), 26-28.

38   *Betänkande med förslag till lag angående kulturminnesvård samt organisation av kulturminnesvården. 1, Historik, memorial angående minnesvårdens nuvarande ståndpunkt, utländsk lagstiftning samt bilagor* (Stockholm, 1922), 91.

39   'Ärade kollega …', call to the members of the Swedish Association of Museum Employees January 28 1913, Fornminnesvårdskommitténs arkiv, vol. 7, Riksarkivet [NAS].

investigated over and over again, and in several different public commissions. The question of how to organize monument preservation, and how to design a proper protection for historic buildings, was highly complex. It involved delicate questions about public interference with private property rights, as well as the relationship between the state and local autonomy.

Questions of how to regard and determine the level of historical value in different objects were at the centre of discussions, and a change in perspective from national to public motives for preservation was underlined. The reasons for preserving buildings in the minds of the involved actors had changed from the glorification of the state, to architectural quality and historical testimony.

Discussions about organizational matters revolved around the question of where to draw the dividing line between the responsibilities of the Academy of Letters and the Office of the Superintendent, and conflicts of interest between architects and conservationists appeared. In part these tensions can be explained by personal aspects of power and control. In a polemic discussion in one of the committees, Bernhard Salin asserted that separate bodies for building preservation in other countries merely resulted from the incompetency of foreign museum employees. This was hardly his true meaning, but it showed how bitter the situation had become.

On a more fundamental level, the relationship between architectonic (altering) and historical (preservative) will was at the focal point of the disagreements. For the rather diverse group of people who formed the new museum profession the tasks relating to the management of historic buildings represented an *expansion* of their traditional territory. This expansion included practical tasks for which they had no training and no efficient organization. On the other hand they had a general and wide-ranging training in history and its different academic disciplines (or sub-disciplines), and in their own opinion they had certain sensitivities to historic values. For the architects the situation was quite the reverse. They regarded treatment of the historic qualities of buildings as a *specialization* within the architectural profession.

For restoration architect Sigurd Curman, the ideal way to understand the historic building was to integrate the practical, technical and historical aspects as closely as possible. Such a fusion of different aspects should preferably be integrated into one professional: the specialized restoration architect. Curman came to stand as an ideal; he was trained as an architect and restorer, and later also in architectural history as a doctor and professor. The antiquarians, the architects argued, where not able to understand or handle the *practical and artistic* sides of restoration.

The antiquarian ideal of integration differed, and they saw problems with the model advocated by Curman and other architects: Architecture could not, antiquarians argued, be separated from the broader historical and art-historical understanding. Instead they emphasized the need to integrate knowledge of architectural history with a more general historical understanding and scholarly training. Architects, they argued, were not able to understand historic values in

In the early decades of the twentieth century the new professionals of historic preservation worked to expand and modernize heritage institutions in many different ways.

Figure 9.4a   Axel Nilsson, the antiquarian at Skansen. Courtesy of Nordiska museets arkiv, Stockholm

Figure 9.4b   Restoration architect Sigurd Curman. Courtesy of Antikvarisk-Topografiska Arkivet, the National Heritage Board, Stockholm

depth, and moreover, their architectural training made them disposed to change and improve monuments. This disposition could in a fundamental way be regarded to be hostile towards the idea of preservation.

As noted, this occasionally heated discussion about professional attributes was inscribed into extensive committee work on the reorganization of the national heritage institutions: Who should have the main responsibility for the preservation of historic buildings. Architects at the Superintendent's Office? Antiquarians at the Royal Academy of Letters? Or a mixture in a new Central Commission? Meetings, discussions and excursions into different solutions seemed to be endless; decisions were pending, the arguments sharpened.

Committee work that continued during for a decade did not lead to a new central and unifying commission. When proposals where put forward in 1921, there were no economic possibilities for greater change. The organization was built on existing structures, and on cooperation between the public bodies.

In 1918, the responsibility for the state-owned historic buildings was organized in a special department of the successor of the Superintendent's Office: The

National Board of Public Building. In parallel The Royal Academy of Letters, History and Antiquities created its own organization for historic preservation that led to the constitution of a new public bureau, the National Board of Antiquities in 1938.[40]

Both these processes were led by Sigurd Curman, the architect who had been at the centre of most committee work since the return from his European tour in 1905. Curman became head of the department for historic monuments at the Board of Public Building in 1918, and in 1923 he was appointed National Antiquary, as the only architect ever. This was an interesting situation. Curman, who had been advocating a separate office for historic monuments, was now set to lead the development of historic preservation within the Royal Academy of Letters, History and Antiquities, an organization mainly consisting of humanistic scholars.

Conclusion

The development of monument preservation in Sweden was so tightly connected to the centuries old "antiquarian" institutions that the public bodies of historic preservation mainly has been led by antiquarians, archaeologists and art historians. The same situation is true for the regional organization, as well as for local municipalities. The typical head of a town or regional conservation department (if one exists) would still today be an archaeologist or an art historian. This is contrary to the situation in the Swedish neighbour countries; in Norway or Denmark the same official would nearly always be an architect.

The antiquarian stronghold in historic conservation institutions does not mean that conservation has only been a museological subject outside the realm of urban, landscape and architectural planning. However the integrated approach in the early decades of the twentieth century changed rather abruptly in the modernist era around 1930. And it was not until the sixties and seventies a new position for historic values in planning practice, organization and legislation emerged, and conservation was integrated at full scale in all facets of the public planning domain.

I began this chapter by referring to Michael Holleran who argues that the American antiquarians seem to have had a stronger position in the conservation field than their European counterparts. I have argued that even the European structure of relations between antiquarians and architects was more varied, and have continued to be so. Details of this history are the subject for another article, but its roots can be traced back to the early formation of the public historic preservation in Sweden.

---

40   Today this board has changed its name to the National Heritage Board. In conjunction with a reorganization of the National Board of Public Building in the 1960s, their department for historic buildings was transferred to the National Heritage Board.

Chapter 10

Rethinking the 'Powers of Darkness': An Anti-History of the Preservation Movement in Britain*

Peter Mandler

Any history that seeks to trace the origins of the preservation movement – to which historians naturally feel a close sense of connection – is bound to start out at least implicitly with some pretty whiggish assumptions.[1] The story we are telling is about the origins of a movement, which began marginal and unheralded (yet righteous), and ended up contemporary, triumphant, achieved. 'Preservation' has become in our own time so consensual, even dominant, so universally approved, that it is hard not to believe either that it always ought to have been so or even that (if we only look at its history properly) it really always was. Today we may be achieving some distance from those origins, and from the implicit whiggish assumptions of the earlier histories of the movement, but our field is still shaped by those historiographical origins. In this chapter I want both to describe that shape and to suggest some alternative shapes, starting from different origins, not from the origins of preservationism but the wider context (often hostile) within and against which the early preservationists had to work.

The specific context I am employing is that of England from the mid-nineteenth to the mid-twentieth century, though I think some of my generalizations should be more widely applicable. I start by discussing the terms in which early preservationists themselves talked about their relationship to their opponents. In general they viewed this relationship as one of a small, embattled, but enlightened elite (a moral but also sometimes a social elite), struggling against monolithic, possibly impersonal, certainly rather vaguely characterized 'powers of darkness',

* I am grateful to the audience at the 'International Origins of the Preservation Movement' conference for their discussion, to the Leverhulme Trust for funding my research in this area as part of the work of the Cambridge Victorian Studies Group, and to the members of the group (and especially Astrid Swenson) for further discussion. Tim Larsen and Sujit Sivasundaram helped particularly with the theology. Melanie Hall, of course, has been the presiding spirit.

1   This bias was explicit in early titles such as Martin S. Briggs, *Goths and Vandals: A Study of the Destruction, Neglect and Preservation of Historical Buildings in England* (London: Constable, 1952).

as C.R. Ashbee mockingly called them in correspondence with the BBC in 1929.[2] I then discuss the way the historiography has picked up this self-characterization of the preservation movement, but instead of seeking to shed light on the 'powers of darkness' has magnified instead the presence and influence of the preservationists themselves, so that a small and embattled elite is turned into the cultural mainstream; thus the 'powers of darkness' disappear altogether. Finally, and more substantively, I suggest some ways in which we might finally understand (or at least explore) the 'powers of darkness' in terms a little more historical and precise.

First, then, what were the early preservationists' own views of the forces with which they had to contend? In the early, late Victorian phase of the preservation movement, much of the public rhetoric came from the Left. This was not only true of obvious candidates such as Morris's Society for the Protection of Ancient Buildings but also of radical-liberal defenders of the people's right to the land (in the commons and footpaths preservation movement, for example) and of public-spirited New Liberals arguing for civic beautification. Even Ruskin's idiom – which represented preservation as the trusteeship of an educated elite for an unprepared and unenfranchised people – was more likely in this phase to be employed publicly by radicals and socialists than by paternalist Tories. In all of these early movements, the enemies of preservation – the enemies of the people's rights – were property-owners, especially at this point aristocratic landowners. It was not only that aristocratic landowners were, as the principal owners of historic buildings and open spaces of natural beauty, necessarily the targets of preservationist efforts (it was useful to preservationists, for example, to present the commons preservation movement as a struggle of 'Commoners' vs. 'Lords'); they were also seen to typify broader defects in the national culture – selfishness, covetousness, materialism and philistinism. These defects they shared with – perhaps even passed on to – the rising capitalists, those middle-class 'Philistines' whom Matthew Arnold portrayed as the natural allies in these respects of the aristocratic 'Barbarians'.

By the turn of the century, aristocratic and capitalist property-owners were seen from this left-wing preservationist standpoint to be merging into a plutocratic class, at the same time taking on still less savoury elements which only deepened their cupidity and rapacity – American millionaires, more exotic foreigners still (the Chinese, for example), and of course Jews. In his 1894 propagandistic novel, *The Splendid Paupers*, the radical journalist W.T. Stead personified preservationism's enemies in the figures of the property speculators Glogoul and Faulmann, respectively a malevolent American doctor (more or less the prototype of the mad scientist of early Hollywood film) and a Jewish manufacturer; they are said to be 'in the confidence of the whole Hebrew tribe in Europe. They seem to have an unlimited supply of capital, and wherever there is an ancient castle, or modern treasure house, or stately palace, or charming estate in difficulties, anywhere in

---

2   C.R. Ashbee to Hilda Matheson, Jan. 16, 1929: BBC Written Archives Centre, Caversham, 910 (RCONT1), Ashbee, Charles R./1a.

the three kingdoms, they are the first to hear of it, and it is very seldom it escapes their clutches.' Together with Ping Ying Yaloo, a soulless materialistic Mandarin plutocrat, they represent the 'money power' which is coming between the people, their land and their heritage. At the climax of the book, Ping Ying Yaloo has fenced off the whole of Derbyshire, taken up residence at Chatsworth, sold off its contents, converted the great conservatory into an elephant stable and the main wing into a harem, released tigers, water buffalo and other wild animals into the countryside, and more or less globalized old England out of existence.[3]

This demonization of, especially, the foreign elements of the 'money power' continues apace through the 1920s. It culminates in a series of mostly phony scares about Americans dismantling the national heritage and exporting it overseas – from 1910, when mythical speculators were said to be prepared to export Tattershall Castle in Lincolnshire to America, to a series of equally phony rumours about the export of buildings after the First World War, when England was indeed 'changing hands', but mostly from one set of English hands to another, usually from aristocratic to plutocratic or tenant farmer hands. In fact, at the time of the Tattershall Castle scare, the National Trust was worried that this story about phantom American vandals was being circulated by real English vandals – the property speculators who had actually bought Tattershall Castle and were hoping to whip up the Trust's interest in buying it by public subscription at an inflated price.[4]

This anti-capitalist or anti-plutocrat rhetoric never dies away, and we can still see it in the language of preservationism towards the end of the twentieth century – for example, in Wayland Kennet's history of preservationism published in 1972, which identifies Tory property interests as the principal enemy, and Labour's 1947 planning act as the turning of the tide, 'the first introduction into the law of any democratic country of the principle that society might forbid a man to do what he would with his own land, without compensation ... the foundation of our whole system of land use planning'.[5] Nor was this anti-capitalist rhetoric ever unchallenged in preservationist circles; more politically moderate, mildly progressive preservationists, better able to work behind the scenes and less interested in muckraking in public, were able to steer a course between socialism and capitalism in advocating a reformist preservationism, as Melanie Hall has nicely shown.[6]

---

3   Quotation, in William T. Stead, *The Splendid Paupers: A Tale of the Coming Plutocracy* (London: Review of Reviews, 1894), 57.

4   For the Tattershall Castle affair, see Peter Mandler, *The Fall and Rise of the Stately Home* (New Haven: Yale University Press, 1997), 184-89. The Trust's anxieties are amply illustrated in the correspondence over Tattershall between August 1911 and September 1912, National Trust Archives, 140.

5   Wayland Kennet, *Preservation* (London: Temple Smith, 1972), 44-45.

6   Melanie Hall, 'Affirming Community Life: Preservation, National Identity and the State, 1900,' in Chris Miele, ed., *From William Morris: Building Conservation and the*

But in the early twentieth century, as the political tenor of the preservation movement diversified, a new rhetoric emerged to reinforce and ultimately to widen the more febrile anti-capitalist arguments, which could now appeal to a paternalist or Ruskinian or aesthetic kind of Tory as well. In private, Ruskinian Tories were likely simply to reverse the polarity of the left-wing movement, to credit landlords with a far-seeing sense of trusteeship over their lands, and to blame the democracy for spoliation, either carelessly through taxation or deliberately through vandalism. On a trip to the beautiful medieval ruins of Roche Abbey in 1892, for example, the Ruskinian Tory George Wyndham deplored the heedlessness of the democratic picnickers and worried 'that many things now standing as the monasteries stood in 1530, may fare as they did at the hands of Democracy ... when I hear and read the kind of attacks made on aristocracies, churches, etc., I begin to think that Walter Map and the others who attacked the monasteries merely translated the envy of the envious into convenient platform charges.'[7]

As this kind of thinking percolated into the preservationist movement in the early twentieth century, the movement's most chiliastic rhetoric broadcast with increasing bitterness, even hysteria, to indict not just a capitalist elite but the whole of English culture – it was the people now who were at fault; in a democratic age, they no longer had the excuse of disfranchisement, and they didn't appear any longer to be willing to listen to their educated tribunes. The hardness, materialism and philistinism once associated only with an economic elite was now seen to have pervaded the culture and indeed the very national identity. This was the point at which a catastrophic diagnosis of the Industrial Revolution was introduced – industrialization had changed everything, had destroyed the natural balance between town and country, between money and art, between continuity and change. Urbanization had captured the soul of Britain in the nineteenth century and it was eating up the country in the early twentieth century by the endless extension of its suburban tentacles. Suburbanism, complained R.C.K. Ensor in 1903, had displaced the native culture of the land – the folk song – with 'the rubbish of the London music-halls'. '"Suburbanism"', he argued, 'does not mean taking to a country life; it means sleeping in the country and living your active life in town; hence the villa's incurable way of looking spiritually irrelevant to a landscape where it is optically permanent.'[8]

The difficulty with this indictment, of course, was that it pitted preservationism against nearly the whole of the electorate upon whom the movement would rely for government action. In part this more ambitious rhetoric reflected in a healthy

---

Arts and Crafts Cult of Authenticity, 1877-1939 (New Haven: Yale University Press, 2005), 129-57.

7   George Wyndham to his father, Oct. 3, 1892, in J.W. Mackail and Guy Wyndham, eds, *Life and Letters of George Wyndham*, (2 vols, London: Hutchinson & Co., [1925]), vol. 1, 266.

8   R.C.K. Ensor, 'The English Countryside,' in Lucian Oldershaw, ed., *England: A Nation* (London: R. Brimley Johnson, 1904), 120-21.

way the wider political ambitions of the preservation movement. No longer willing or able to rely on individual acts of private philanthropy, or small groups of likeminded folk such as had been gathered together in the early preservation societies, they sought to mobilize a national movement and to grapple with national political structures. No longer focused on isolated monuments and buildings, they sought to project a vision of a reordered and newly civilized society in which preservation and modernization could be balanced and reconciled. Yet it was difficult to express these ambitions in a way that satisfied both Left and Right without also tendering a sweeping condemnation of the very democratic electorate who had to be persuaded.

As the movement sought to get more politically serious, increasingly it had to grapple with this paradox, and by around 1930 it had developed new language to deal with it. One can see this new approach at work in two manifestoes of the left-wing architect Clough Williams-Ellis – revealingly titled *England and the Octopus* (those suburban tentacles again, 1928) and *Britain and the Beast* (1937) – and two by the increasingly conservative historian G.M. Trevelyan – *Must England's Beauty Perish?* (1929) and *The Call and Claims of Natural Beauty* (1931). In Williams-Ellis's version, Britain was held in the grip of a kind of madness. 'The fact is that English people need mass psycho-analysis. We know the morbid symptoms – false standards and values, blindnesses and callousnesses and such-like ... Money itself has somehow usurped in our desires the place of the good things that it can purchase, or things good in themselves have crowded out of our limited imaginations the things which are better and best.'[9] There *were* people who cared for history and beauty but the status quo didn't recognize them.

> A.'s quasi-religious zeal about the sacredness of natural beauty and our duty towards it in right building may seem as queer and wrong-headed to his neighbour B. as B.'s perturbation over blasphemy or sexual unconventionality may seem to A. Yet B.'s sort have all the might, majesty and power of the law and the Churches on their side. You may ravish and defile the most divine landscape in the world, and your children (being your children) will rise up and call you progressive. You are a 'lucky prospector' or a 'successful real-estate operator', a 'live-wire' and what local newspapers call 'a prominent and respected citizen'. By your exploitation of the land you have enriched yourself and your heirs. You have done very well ... for a deliberate act, brutally disregardful of natural beauty, essentially anti-social, sacrilegious and blasphemous, we receive the protection of the State, the accommodation of the banks, the approbation of our fellows, and the toleration of the Churches.[10]

---

9   Clough Williams-Ellis, *England and the Octopus* (London: Geoffrey Bles, 1928), 22.

10   Ibid., 19-20.

The only solution was propaganda and education, of the kind at which Williams-Ellis excelled, to try to snap the British out of their madness.

Trevelyan had a more emollient explanation. He agreed that 'those of us who care for preservation of natural beauty are still outnumbered and overborne by those who, though not all of them wholly indifferent to our cause, place other considerations whether of business or politics in a higher place than any such considerations'.[11] But he was more optimistic about the immediate impact of propaganda and education, and he thought that appreciation of historic and natural beauty could be seen to be a natural and normal part of modern life. The problem lay more in impersonal forces of development – the impact of technology, especially the motor-car, and 'the law of the machine-age': 'Our generation is placed under economic pressure to use the machines and methods which most rapidly destroy the lines of nature.'[12] These impersonal forces were so general and so powerful that they could only be slowed or halted by an equally ubiquitous force, the State; yet the State was still in the grip of Victorian individualism, reluctant to trespass on the sacred rights of private property. It was the State, therefore, rather than the people that was truly to blame, and to the State, too, preservationists must look for answers: 'Unless the State reverses the engines and instead of speeding up destruction, plans the development of the country so that the minimum of harm can be done to beauty, the future of our race, whatever its social, economic and political structure may be, will be brutish and shorn of spiritual value.'[13]

Before I move on to my own analysis of the so-called 'powers of darkness', I want to point up one curiosity. When preservationism moved into the driver's seat in the 1970s and '80s, historians in their whiggish way began to seek its origins and, as we do, began to find them everywhere. All of these outcries and broadsides and manifestoes from the early preservationists were dredged up and used to show how late Victorian and early twentieth century English culture was absolutely *saturated* with preservationism. The whole country had, it now seemed, been preservation-mad, practically from the moment the Great Exhibition shut up shop in 1851. The English abandoned the manufacturing and trading of industrial goods and started running tea-shops in converted barns. They fled the towns and started living in the countryside, or at least in the virtual countryside, the suburb. They launched mighty movements like the Society for the Protection of Ancient Buildings and the National Trust. They strangled their towns in green belts and swooned over country houses instead. Of course this required a highly selective reading of the preservationist literature of the day. The preservationists' own view of themselves as an embattled minority had to be ignored. The preservationists' own view of the English people – as either greedy materialists, or at the very least deranged captives of a modernizing psychosis or of a property-mad State – had

11   George M. Trevelyan, *The Call and Claims of Natural Beauty* (London: University College London, 1931), 27-28.

12   Ibid., 15.

13   Ibid., 31.

to be ignored, too. Only the preservationists' claim that they really had the best interests of the people at heart – indeed, that they were the virtual representatives of the people's true soul, true desires – was taken seriously. Thus an embattled minority becomes the soul of the nation.[14]

It is, of course, easy to find contemporary evidence for both characterizations of the preservationists: compare, for example, two cartoons by 'Pont', both of which appeared in *Punch* in the 1930s, one of which suggested that the English

(a)

THE BRITISH CHARACTER.
PASSION FOR THE ANTIQUE.

(b)

THE BRITISH CHARACTER.
DETERMINATION NOT TO PRESERVE THE RURAL AMENITIES.

Figure 10.1   'Pont' illustrates two sides of the English character in *Punch* in the 1930s: (a) 'Passion for the Antique' and (b) 'Determination Not to Preserve the Rural Amenities'

14   The *locus classicus* is Martin J. Wiener, *English Culture and the Decline of the Industrial Spirit, 1850-1980* (Cambridge: Cambridge University Press, 1981). For a survey and critique of this literature, see Peter Mandler, 'Against "Englishness": English Culture and the Limits to Rural Nostalgia, 1850-1940,' *Transactions of the Royal Historical Society*, 6th series, vii (1997), 155-75.

really do love their antiquities, more than any other people in Europe, the other that the English really do love to destroy their countryside, and they are better at it than any other people in Europe.

'Pont' enjoyed setting up such paradoxes, and this one does suggest that contemporaries were aware that the relationship between the English and their historic environment was more complicated than the preservationists' rhetoric suggested. In turning now to an anatomy of the 'powers of darkness' I would like to push beyond the stark choice offered by that rhetoric – the English are either destroyers or preservationists. Obviously they were both at different times and in different ways. But we will not be able to capture that balance unless we consider the ideologies and rhetoric of development and modernization as well as the ideologies and rhetoric of preservation. Here I can only sketch out some ideas, more a programme for future research than anything else.

First I consider the notion of trusteeship, or stewardship, so potent as we have seen in the rhetoric of preservation. It was also – probably equally – potent in the language of development. Long before Ruskin developed the idea of stewardship over the remains of the past – which derived in part from his revulsion from the productivism of the present – the idea of stewardship had been developed by evangelicals especially to *justify* productivism. It had been an ancient Christian (especially Protestant) trope that human powers to mould the Earth were God-given and imposed a duty to apply and develop them. Genesis 1:28 provided the key text: 'Be fruitful, and multiply, and replenish the earth, and subdue it: and have dominion over the fish of the sea, and over the fowl of the air, and over every living thing that moveth upon the earth', in the Authorized Version. Before the nineteenth century, this trope had been applied most systematically to agriculture. Following obvious lessons in the Scriptures – which showed early human communities dutifully causing the desert to flower – the transformation of barren or infertile or simply uncultivated landscapes into civilized, cultivated ones was considered by Christian thought to be the proper stewardship not only of the land but of those human powers endowed by God.

While agriculture is a human activity, it is also traditionally seen as a human activity in harmony with nature, and one could reasonably imagine that traditional Christian doctrine would only support a reasonably non-invasive agriculture that kept this balance.[15] That was indeed the doctrine of preservationists who argued that industrialization and urbanization had destroyed the natural balance. It is also central to the argument of James Winter in *Secure from Rash Assault*, that excellent account of Victorian environmentalism, which emphasizes the relatively low-tech and non-invasive nature of Victorian development before the age of electricity and

---

15   On the development of a Christian paternalist idea of stewardship over the land that counterposed itself to a supposedly selfish, anti-communal Whig idea of improvement, see Nigel Everett, *The Tory View of Landscape* (New Haven: Yale University Press, 1994).

the internal combustion engine.[16] But ideas of 'balance' are not immanent; they are culturally constructed. It is equally reasonable to imagine that Christian doctrine might be interpreted to impose a duty for *further* development – which in certain hands it did.[17]

Evangelicalism, at its early-nineteenth century peak, could be very pressing about duty. In its then powerful pre-millenial versions, it predicted the imminence of the Second Coming and exhorted believers to prepare.[18] As Sujit Sivasundaram has recently shown, one of the most powerful manifestations of this pre-millenial impulse was intense missionary activity to gather in all the world's peoples, and this was normally accompanied by the extension of economic development to previously 'uncultivated' parts. In this vision nature was not benign and harmonious but savage and impious; it required 'conversion'. 'The heroic coloniser', he writes, 'was the pious individual who rose against the land and put it in its place, firmly in the bounds of human control.' The Rev. John Williams, pioneering missioner in the South Seas, was described by his own biographer as not 'a sentimentalist or a poet', but a 'practical' man, who sought 'for something more than the mountain and the dell, the forest and the stream, the crested wave and the coral strand'. Here stewardship was very much *not* defined as seeking a harmony with nature. 'To live in unison with nature', as South Sea Islanders were doing, was in Williams' mind 'to oppose [the] divine plan'. 'Nature's beauty arose for him when the divine injunction to master and order the environment was obeyed,' concludes Sivasundaram. 'A tall chapel, an ordered settlement and an industrious people; these were the "bright lights" that brought a landscape to completion.'[19]

Some of the same rhetoric was applied throughout the nineteenth century to the 'conversion' of the home country, both its people and its land. Most obviously, it was applied to the development of the church – the need to bring the church to the mass of the people, who lived in greater concentrations and therefore needed more commodious and more useful churches. It was this sense of duty, of stewardship, that led to the notorious Union of Benefices Act of 1860, which authorized the closure and merger of small parishes into larger ones – and, of course, what went with that benign sounding process, the demolition of older and smaller churches and the building of new ones. It was this understanding of stewardship that led to

---

16   James Winter, *Secure from Rash Assault: Sustaining the Victorian Environment* (Berkeley: University of California Press, 1999).

17   Even Winter's discussion of 'stewardship', for example, assumes that it tended to act as 'a restraint on radical specialization' (i.e. on intensive development), ibid., 255.

18   Boyd Hilton, *The Age of Atonement: The Influence of Evangelicalism on Social and Economic Thought, 1785-1865* (Oxford: Clarendon Press, 1988), discusses this among competing versions of evangelicalism and its implications for economic thought, though without particular attention to the questions of stewardship developed here.

19   Sujit Sivasundaram, *Nature and the Godly Empire: Science and Evangelical Mission in the Pacific, 1795-1850* (Cambridge: Cambridge University Press, 2005), 121, 148.

the demolition of twenty of the City churches of Christopher Wren in the 1870s and '80s – just as a different understanding of stewardship led to the founding of the Society for the Protection of Ancient Buildings and the preservation of another twelve of those same City churches. To take one example, St. Dionis Backchurch, demolished in 1878, was seen as a small price to pay for the construction of a new church in a more populated district, St. Dionis, Parson's Green, built in 1886.[20]

The language of stewardship was not only deployed to justify the demolition of old churches and the building of new ones; it was also used to justify re-development of the land more broadly, not just converting the people but converting the landscape and developing the powers of those people who used and shaped it. The language of 'convenience' and 'comfort' that was often used to warrant modernization – and often decried as selfish and philistine by preservationists – was in this idiom warranted as pious and dutiful. Making the world 'convenient' and 'comfortable' was about bringing it to fruition, matching the environment to the advanced state of civilization. Development could also enhance the beauty of the natural world, for pious Christians did not necessarily share Ruskin's views on the absolute perfection of natural beauty – God had endowed nature with beauty but He had also endowed humans with the ability to create beauty of their own.[21] As James Winter and Harriet Ritvo have shown, in the debate over the conversion of Thirlmere, in the Lake District, into a reservoir for Manchester's use in the 1870s and '80s, *both* sides, preservationists and developers, invoked concepts of stewardship for beauty – and, allowing for differences of taste in 'beauty', it is not easy to say who had right on their side.[22]

It is tempting to speculate at this stage about differences between Anglicans and Nonconformists. Nonconformist evangelicals were particularly exercised about their Christian duty to develop their own powers, to advance the state of civilization, and to spread conveniences and comforts to their lesser brethren. (They were of course also the most vigorous missionaries.) What is more difficult to assess is whether in general their Christian mission to develop outweighed their Christian mission to preserve *more* than was the case with Anglicans. Certainly

---

20   Apart from documentary material, there is no scholarly account of the demolition of the City churches, nor indeed of the struggles to preserve them, except as part of the early history of the Society for the Protection of Ancient Buildings. Dr Ben Weinstein is currently engaged in research on this subject, however, and I am grateful to him for sharing his preliminary thoughts on it with me.

21   See, for example, the arguments of the immensely popular Free Church of Scotland preacher, Rev. William Arnot, that God left nature defective precisely to allow Man to exercise his talents in improving upon it: idem, 'The Earth, as it has been occupied, cultivated, and improved by the industry of man,' in *Lectures Delivered before the Young Men's Christian Association, in Exeter Hall, from November, 1862, to February, 1863* (London: James Nisbet & Co., 1863), 219-46, esp. 237-38.

22   Winter, ch. 9; Harriet Ritvo, *The Dawn of Green: Manchester, Thirlmere, and Modern Environmentalism* (Chicago: University of Chicago Press, 2009).

Anglicans felt that Nonconformists had little sense of tradition. As Bishop Stubbs wrote in 1859,

> I do not believe that a Dissenter could write a History of England ... the determination of the Dissenters to see nothing good before the Reformation is so obvious in all that they do, that I have begun to wonder that they allow that our Saviour lived before it – as certainly they believe the Bible was written about that time. One must not expect that one's forefathers were anything but rascals in such people's opinions. I suppose they had none of their own.[23]

The Nonconformists' own version emphasized the laxity and self-indulgence of 'one's forefathers' and the need literally to improve upon them. One 1852 history of Blackburn, for example, depicted in past times 'fields lying waste, or forests preserved for the barbarous sport of the wild hunter', only lately put to good use by 'the sons and daughters of honourable industry' as 'the axe of times has levelled every vestige of the ancient forests for the good and noble purpose of man'.[24] The 'civic gospel' pursued by many Nonconformists on provincial town councils very much put urban development – roads, gas works, tramways and the like – above preservation of the relics of the 'barbarous' (also not incidentally Catholic) past. Right up to the First World War and beyond, the rhetoric deployed by Nonconformist Liberals like Lloyd George against the great landowners continued to emphasize their unproductivity, their parasitism, and the imperative to develop the land that could only be delivered by a dismantling of the traditional structure of landed society and a parcelling out to a productive peasantry – this was an argument for the stewardship of productivity against the stewardship of conservation. And Lloyd George used the scriptural language – 'Land ownership is not merely an ownership; it is a stewardship', he tiraded in a famous speech at Bedford in October 1913, yet landlords have 'annexed Naboth's vineyard. There is now occasionally a little garden. Sometimes, as a matter of grace, [the workman] has a little row of potatoes, but he has no longer a right in the soil as his fathers had.'[25] Against this, however, there are the arguments of Patrick Joyce and Charles Dellheim that provincial nonconformists often had an especially powerful

---

23   William Stubbs to E.A. Freeman, Nov. 3, 1859, in William Holden Hutton, ed., *Letters of William Stubbs* (London: Archibald Constable, 1904), 75.

24   Tristram Hunt, *Building Jerusalem: The Rise and Fall of the Victorian City* (London: Weidenfeld & Nicolson, 2004), 154-56. P.A. Whittle, the author of this particular history of Blackburn, was, however, according to his Dictionary of National Biography entry, a Catholic! See, C.W. Sutton, 'Whittle, Peter Armstrong (1789-1866),' rev. Nilanjana Banerji, in H.C.G. Matthew and Brian Harrison, eds, *Oxford Dictionary of National Biography* (Oxford: Oxford University Press, 2004; http://www.oxforddnb.com/view/article/29332).

25   David Lloyd George, *The Rural Land Problem: What It Is; A Speech* (London: Liberal Publication Department, 1913), 11-12.

sense of place and a need to vindicate a distinctive local tradition against those propagated by metropolitan elites.[26] In the Thirlmere debate, too, Anglicans and Nonconformists appear on both sides – it was the Anglican Bishop of Manchester who was most keen to promote Thirlmere, not only for its convenience to the people of Manchester, but also as an improving act of stewardship, enhancing natural beauty by human agency. Clearly, the language of the stewardship of productivity was no more a monopoly of any one religious current than was the language of the stewardship of conservation.

Development did not, of course, depend upon an explicitly religious language for its cultural currency; it could also draw on a more secularized, but still spiritual, language of progress. This is (or should be) so obvious for late Victorian and indeed for early twentieth- century Britain that it tends to get neglected. Preservationists often tried to vindicate relics of the past by celebrating their ruinousness – the virtues of the fragmentary and the crumbling, the delights of loss and nostalgia, were all being rediscovered in the nineteenth century, as we are constantly reminded today by endless books about historical consciousness and places of memory. We ought not to forget, however, that they were often consciously seeking to counter a resonant language in which the past was portrayed as old and rotting and unpleasant, which needed to give way to the new, the beautiful and the useful. This language could, of course, be used as a superficial justification for money-making; but it had a power of its own. Old buildings could be depicted as 'mouldering', dark, dank, dangerous and unhealthful; they were, as one correspondent arguing for demolition in Burnley in 1896 wrote, 'fusty, musty, hobgoblin boxes'.[27] They were also awkward and inconvenient – a 'grievous obstruction in a public thoroughfare'. Only if they gave way could new standards of public comfort and convenience be achieved – and only then, when no longer a liability, might they be susceptible to sentimentalization.[28]

It was, literally, a choice between darkness and light – as we can see, returning again to the City of London, in the public discussion over 'metropolitan improvements' in the particularly torrid period from the 1860s to the 1880s when infrastructural and commercial development pressures converged on the heart of the capital. Intense pressure, and attention, focused on one crucial intersection at the heart of the City, now known by its nearby Underground station 'Bank'.

Photographed in the 1860s, it doesn't look too cramped and congested to modern eyes, but a few years later the construction of Holborn Viaduct (which

---

26   Charles Dellheim, *The Face of the Past: The Preservation of the Medieval Inheritance in Victorian England* (Cambridge: Cambridge University Press, 1982), 58-69; Patrick Joyce, *Visions of the People: Industrial England and the Question of Class, 1840- 1914* (Cambridge: Cambridge University Press, 1991), 172-92. Joyce addresses the issue of the specifically Protestant idea of heritage more directly than does Dellheim, whose focus is on 'provincialism.'

27   Edwin Johnston, letter, *Burnley Gazette* (Jun. 27, 1896), 8.

28   'The Present and the Past,' *Chambers's Edinburgh Journal* 10 (1841), 81-82.

Figure 10.2   **The epicentre of development: the Mansion House and Poultry, photographed in the 1860s. Reproduced by permission of English Heritage**

was channelling more traffic into the City from the west) and other new arteries, as well as new railway stations, and the demand for more office space, were all generating a heightened sensitivity to the age and incommodiousness of the City, even though it had all been rebuilt after the Great Fire of 1666. Narrow Charlotte Row, just beyond the Mansion House, had to be widened substantially – thus creating the Mappin & Webb building on the left (itself the object of a preservation campaign in the late twentieth century) – and even apparently wide and inviting Poultry needed to be widened further and rebuilt, requiring the demolition of the Wren church, St. Mildred Poultry, hiding behind the lamppost on the right. When the reconstruction of Poultry was held up – not by preservationist efforts (there was as yet little organized sentiment on behalf of the Wren churches), but by City bureaucracy and possibly corruption – there was public outcry, and a striking cartoon by Matt Morgan in the radical weekly *Tomahawk*.[29]

Morgan's cartoon counterposes, on the right, dark, crumbling, incommodious Poultry, and on the left, bright visions of the London in the making (including what looks like Holborn Viaduct or Blackfriars Bridge and the new Law Courts).[30] The magazine's figurehead, Tomahawk, is saying to Industria, 'the spirit of improvement', 'You have conjured up a beautiful city; but look at that ugly impeding Poultry; why cumbereth it the ground?' 'Why cumbereth it the ground' is a reference to the parable of the barren fig tree in the Gospel of Luke.[31] It is a finely double-edged parable: the owner of the barren fig tree commands his gardener to cut it down, but the gardener asks for a year's reprieve to have one last attempt at getting the tree to bear fruit. The parable thus embodies, potentially, both ideas of stewardship – the responsibility of the steward to preserve, and the responsibility of the steward to produce. However, Victorian readings of the parable tended to

"*OUT, DAMNED SPOT!*"

Figure 10.3   Matt Morgan on the 'spirit of improvement' in 1860s London, from *Tomahawk*, 13 Nov. 1869 © The British Library Board, All Rights Reserved, pp 5272 b

29   On *Tomahawk*, see Christopher Kent, 'The Angry Young Gentlemen of *Tomahawk*,' in Barbara Garlick and Margaret Harris, eds, *Victorian Journalism: Exotic and Domestic* (St. Lucia, Australia: Queensland University Press, 1998), 75-94.

30   On Morgan's typically melodramatic style, see Thomas Milton Kemnitz, 'Matt Morgan of "Tomahawk" and English Cartooning, 1867-1870,' *Victorian Studies* 19 (1975-6), 5-34.

31   Luke 13:6-9. Thanks to Simon Goldhill and Tim Larsen for their thoughts on this passage.

lean towards the latter, more obvious interpretation[32], and the casual reader of this cartoon is of course also encouraged in this direction by the composition and the Shakespearean injunction to erase the blot. Alongside such imaginative invocations of the spirit of improvement can be found more realistic ones, the photography of improvement, which in contrast to the photography of preservation has only begun to be studied (notably in Lynda Nead's *Victorian Babylon*).

The visual depiction of the modernization of London invoked exciting images of human power and ingenuity, the creative destruction of the old to make way for the bold, the grand, and the new, a distinctive type of the urban sublime.[33]

Something we know even less about is the heroization of the developer. Ernest Terah Hooley is known today, if at all, as a great stock-market speculator (who floated among other companies Raleigh bicycles, Singer sewing machines and Dunlop tyres) and most especially as a great stock-market fraudster, bankrupted four times and eventually jailed.[34] But before that he was also celebrated as a land developer – even hailed as the new type of landowner who could replace the clapped-out aristocracy, 'the coming squire', as the *Estates Gazette* called him in 1896.[35] After his disgrace, he resurfaced in a shadier capacity, as the very speculator who had bought up Tattershall Castle in order to try to milk the National Trust. In fact he had tried the same trick the previous year, with another historic building in Northamptonshire. More familiar is the name of Robert McAlpine, the Scottish bricklayer turned property developer who became popularly known as 'Concrete Bob' and built the biggest construction firm in Britain, McAlpine's. He was involved with Singer sewing machines, too, responsible in 1883 for building their factory in Dunbartonshire, then the largest factory in Europe. The firm remains a mainstay of the building business in Britain today. Throughout the twentieth century, a succession of McAlpines were much in the public eye, and much celebrated too for personal qualities, for feats of daring, and for formidable

---

32   A formidable precedent for this was offered by John Bunyan, in Bunyan, *The Barren Fig Tree* (London: J. Robinson, 1688), where the barren fig tree represents the 'fruitless professor.'

33   Compare Gavin Stamp, *The Changing Metropolis: Earliest Photographs of London 1839-79* (Harmondsworth: Viking, 1984), esp. 16-17, which emphasizes the emergence of preservationist groups such as the Society for Photographing Relics of Old London in the 1870s, to Lynda Nead, *Victorian Babylon: People, Streets and Images in Nineteenth-Century London* (New Haven: Yale University Press, 2000), esp. 29-34, which sees preservationist photography and the spirit of improvement as more closely inter-related. Jerry White, *London in the Nineteenth Century* (London: Jonathan Cape, 2007), ch. 2, gives a particularly vivid picture of the urban sublime in mid- and late Victorian London; note his subtitle, 'A Human Awful Wonder of God.'

34   Hooley's financial unscrupulousness is the focus of P.M. Oppenheimer, 'Hooley, Ernest Terah (1859-1947),' rev., in Matthew and Harrison, *Oxford DNB* (http://www.oxforddnb.com/view/article/37564).

35   'The Coming Squire' (leader), and 'The Future of English Land: An Interview with Mr. Ernest Terah Hooley,' *Estates Gazette* (Nov. 7, 1896), 674; 669-70.

**Figure 10.4   The photography of improvement: building St. Bride Street
in the City of London, c.1871. Reproduced by permission of
English Heritage**

technical achievements – the building of the old Wembley Stadium, for example,
now gone under the wrecking ball but at the time reckoned one of the wonders
of the world.[36] The heroization of this kind of technological achievement, which
we associate with Samuel Smiles's *Lives of the Engineers* and the mid-Victorian
boom, clearly didn't end then any more than Britain's economic growth did.[37]

To the contrary, there is evidence that a new level of public interest in and
approval for modern developments was reached in and after the 1890s (just the
time when the standard narrative of preservation says that public interest in and
approval for preservation was also taking off). As Bernhard Rieger has argued in a
recent book, across Europe the period from the late 1890s to the 1920s witnessed
a new public fascination with 'modern wonders' of unprecedented scale, power
and technical complexity – skyscrapers and ocean liners, motor-cars and airplanes,
telephone, cinema and radio, electrical applications of all sorts. These induced

---

36   Iain F. Russell, 'McAlpine family (*per. c.* 1870-1967),' in Matthew and Harrison,
Oxford DNB (http://www.oxforddnb.com/view/article/49301).

37   David Edgerton, *Science, Technology and the British Industrial "Decline" 1870-
1970* (Cambridge: Cambridge University Press, 1996).

anxiety – about the dwarfing of human scale, the loss of control – but, as Rieger also shows, by the 1930s these anxieties had been managed to the extent that people generally came to embrace the new experiences generated by such unfathomable means.[38] More easily celebrated than 'modern (technological) wonders' were less wonderful but equally modern developments in the urban streetscene – the hotels and restaurants, department stores, chains and multiples that began to construct the high street that we know today.[39] Both sets of developments wrought considerable change on the historic and natural environment, so much so that James Winter in his study of Victorian environmentalism says that the real story begins not with the actually quite mild technologies of coal and steam but with the advent of these newly invasive technologies at the beginning of the twentieth century. As these new development pressures unfolded, they did undoubtedly stimulate the preservationist movement, but they also stimulated countervailing forces. The modern cult of the automobile is reasonably well understood because it persists today; but contemporaneous cults which did not outlast their novelty are more likely forgotten. A case in point is provided by the electricity pylon. The basic design for the British pylon, PL1, was commissioned from the architect Sir Reginald Blomfield in 1928, and Blomfield, naturally enough, was among the first to stake an aesthetic claim for it: 'Anyone who has seen these strange masts and lines striding across the country, ignoring all obstacles in their strenuous march, can realise without a great effort of imagination that [they] have an element of romance of their own.'[40] Here we find the same language not only of comfort and convenience, but again also of beauty, that we encountered in mid-Victorian arguments for development, though now stripped of their scriptural warrant. Similar claims were made for the redevelopment of the urban streetscape necessary to accommodate the new demands of leisure and commerce. John

---

38   Bernhard Rieger, *Technology and the Culture of Modernity in Britain and Germany 1890-1945* (Cambridge: Cambridge University Press, 2005). See also Bernhard Rieger, 'Envisioning the Future: British and German Reactions to the Paris World Fair in 1900,' in Martin Daunton and Bernhard Rieger, eds, *Meanings of Modernity: Britain from the Late-Victorian Era to World War II* (Oxford: Berg, 2001), 145-64.

39   On this commercial celebration of modern life in the interwar period, see Alison Light, *Forever England: Femininity, Literature, and Conservatism Between the Wars* (London: Routledge, 1991), esp. ch. 2, on Agatha Christie's England. For some interesting comments on the recognition by even the more conservative press barons that they could not oppose such developments outright without risking the loss of their mass constituency, see Adrian Bingham, *Gender, Modernity, and the Popular Press in Inter-War Britain* (Oxford: Clarendon Press, 2004), 55-56. But the brevity of such comments points to the difficulty faced by the historian in studying the inner workings of modernization, when the most explicit and articulate historical traces are left by those organizing to resist it.

40   Bill Luckin, *Questions of Power: Electricity and Environment in Inter-War Britain* (Manchester: Manchester University Press, 1990), 98-99. See also ibid., 101-2, where, although his focus is the anti-pylon movement, Luckin acknowledges the existence of explicit pro-pylon feeling as well.

Nash's late Georgian Regent Street quadrant, notoriously, was demolished in the 1910s at the behest of the Crown Estate (seeking enhanced rents, especially from a grand hotel planned for the site) and of the tenant lease-holders (seeking more shop-floor space and, especially, huge plate-glass display windows). Although a lot has been written about preservationist horror at the demolition of Nash's Regent Street – it was one of the founding moments of Georgian preservationism[41] – equally revealing is the torrid debate between architects, landlord and tenants over the extent of the commercial accommodation, which was the crucial debate in determining the Regent Street we have today.[42]

In drawing attention to popular enthusiasm for technology, and to popular enjoyment of consumer culture that added further to development pressure, I am not suggesting that we replace a whiggish history of preservation with a whiggish history of development and destruction. In fact, of course, as contemporaries appreciated, the two were closely linked. 'Two things are characteristic of our age, and more particularly of our island', Trevelyan wrote in 1931. 'The conscious appreciation of natural beauty, and the rapidity with which natural beauty is being destroyed. No doubt it is partly because the destruction is so rapid that the appreciation is so loud.'[43] Instead, what I am suggesting is that we need to consider the social, economic and political forces behind – and also the cultural meanings of – development just as closely as we have plumbed the forces behind and the meanings of preservation, precisely because we will not make full sense of the one without the other. We have studies of preservationists but not developers. We have studies of national parks but hardly any of motorways. We even have studies of the anti-pylon movement, but not of the pro-pylon movement. We can't accept any longer the preservationists' own view that their side was the only side – that against them were ranged not people, or ideologies, or hopes and needs, but only dark, impersonal forces.

Elements of these two movements – for development and for preservation – were always aware of each other, and sometimes alert to opportunities for compromise, or common ground. As early as the 1870s, *The Builder* newspaper – far from the grim utilitarian sheet one might imagine from the title, full of adverts for cheap bricks and new masonry techniques – was adumbrating its own vision of a renewed and recreated city of wide boulevards, the latest transport technologies and excellent public services, but also with carefully preserved historic buildings and open spaces. (They were disappointed that the redevelopment of Poultry didn't create better spaces for people.) Although *The Builder*'s prescriptions

---

41 See 'Epilogue' (on the 'afterlife' of Georgian London), in John Summerson, *Georgian London* (London: Pimlico, 1991), esp. 292-99.

42 For a generally dispassionate and well-documented account of the affair, see 'The Rebuilding of Piccadilly Circus and the Regent Street Quadrant,' in F.H.W. Sheppard, ed., *Survey of London: vols. 31 and 32: St. James Westminster, Part 2. North of Piccadilly* (London: Athlone, 1963), 85-100.

43 Trevelyan, *Call and Claims*, 8.

were rarely heeded in the late Victorian period, its importance lies partly in its awareness that both preservation and development sought to serve real needs and potent dreams. Even *Country Life*, which practically invented the idea of national heritage in the interwar period, was alive to the competing demands of sentiment, tourism, residence, aesthetic enjoyment, historical appreciation, the modernization of agriculture and the rights of private property (to name just a few). Part of the appeal of the planning idea in the 1930s was that it offered a way to adjudicate these competing demands, especially during the Slump, when as one Treasury civil servant insisted, 'it is [essential] for the country not to waste its resources on schemes which, however desirable from the aesthetic point of view, do not enrich the country or add to its commercial equipment', or, from a slightly later, more welfarist standpoint, as a Ministry of Health civil servant queried, 'if Housing, Milk for Mothers and a cancer scheme are to have priority', and, he added, 'it seems difficult to avoid the conclusion that they should'.[44] In response to such demands for prioritization, by 1937, Trevelyan had softened his rhetoric markedly. 'I am not a fanatic in these matters. I fully realize that this is a small island full of folk, and that the prime needs of industry, housing, and defence have to be met, often at the expense of natural beauty. But in disputes and bargains between these rival interests, the interest of amenity is unduly handicapped. It is not officially represented in Government departments ... '[45] Patrick Abercrombie and other founders of the Council for the Preservation of Rural England increasingly recognized the interdependence of town and country and sought, for example, to bend motor tourism into preservationist channels.[46] Even that metaphor, however, views the relationship too much from one point of view. We will not understand properly the origins of the preservationist movement until we stop thinking of it that way, and think instead of preservationism's place in a social, cultural, political and economic field which held, as well, many other players, many other interests, many other meanings.

---

44   E.C. Cleary to M. Fass, Oct. 4, 1924, The National Archives, London (henceforth TNA) T 162/262/E24215/1; E.J. Maude, Jul. 24, 1937, TNA HLG 52/709.

45   George M. Trevelyan, 'Amenities and the State,' in Clough Williams-Ellis, ed., *Britain and the Beast* (London: J.M. Dent, 1937), 184-85.

46   John Sheail, *Rural Conservation in Inter-War Britain* (Oxford: Clarendon Press, 1981) gives a sensitive and well-documented account which seeks to vindicate interwar planning.

Bibliography

Manuscript Collections

BBC Written Archives Centre, Caversham.
Borden Mss, National Archive of Canada/Archive Nationale du Canada, Ottawa.
Admiral Mark Bristol Papers, Library of Congress, Washington.
Papers of James, Viscount Bryce, Bodleian Library, Oxford.
Ronald Burrows Papers, King's College, London.
Papers of the 27th Earl of Crawford and Balcarres, National Library of Scotland, Edinburgh.
Davidson Papers, Lambeth Palace Library, London.
Marquis of Dufferin and Ava Papers, Public Record Office of Northern Ireland, Belfast.
H.A.L. Fisher Papers, Bodleian Library, Oxford.
Geheimes Staatsarchiv Preussischer Kulturbesitz, Berlin.
Gregory Family Papers, Manuscript. Archives and Rare Books Library, Emory University.
Incorporated Society for the Promotion of the Enlargement and Construction of Churches and Chapels Archives, Lambeth Palace Library.
Israel State Archives.
Médiathèque de l'Architecture et du Patrimoine, Archives, Paris.
Montagu Papers, Trinity College, Cambridge.
National Trust Archives, Swindon.
Charles Eliot Norton Papers, Library of Congress, Washington, DC.
Perris Papers, Stanford University Library.
Riksarkivet (National Archives of Sweden), Stockholm:
        Fornminnesvårdskommitténs arkiv.
        Svenska museimannaföreningens arkiv.
St. Sophia Redemption Committee Papers, Anglican & Eastern Churches Association, London.
Society for the Protection of Ancient Buildings Archives, London.
Stanmore Papers, British Library.
Sulgrave Manor Trust Archive.
Templewood Papers, University Library, Cambridge.
The National Archives, London:
    Cabinet Office Papers.
    Colonial Office Papers.
    Foreign Office Papers.

Housing and Local Government Papers.
Treasury Papers.
Office of Works Papers.
Wrench Mss, British Library.

Unpublished Theses

Adamson, Jeremy Elwell. 'Frederic Edwin Church's "Niagara:" The Sublime as Transcendence' PhD diss., University of Michigan, 1981.
Aldrich, Megan. 'Thomas Rickman (1776-1841).' MA thesis, University of Toronto, 1983.
Björkroth, Maria. *Hembygd i samtid och framtid 1890-1930: en museologisk studie av att bevara och förnya*, PhD diss. Umeå universitet, 2000.
Bradley, Simon. 'The Gothic Revival and the Church of England.' PhD diss., University of London, 1996.
Evans, Gail Edith Hallett. 'Storm over Niagara: A study of the interplay of cultural values, resource politics, and environmental policy, 1670s-1950.' PhD diss., University of California Santa Barbara, 1991.
Fernlund, Siegrun. *'Ett Herranom värdigt tempel': kyrkorivningar och kyrkobyggen i Skåne 1812-1912 [Kirchenabrisse und Kirchenbau in Skåne 1812-1912]*, PhD diss., Lunds universitet, 1982.
Geijer, Mia. *Makten* över *monumenten: restaurering av vasaslott 1850-2000*, Nordiska museets förlag, PhD diss., Stockholm Kungliga tekniska högskolan, 2007.
Hillström, Magdalena. *Ansvaret för kulturarvet: studier i det kulturhistoriska museiväsendets formering med särskild inriktning på Nordiska museets etablering 1872-1919*, PhD diss., Linköpings universitet, 2006.
Miele, Chris. 'The Gothic Revival and Gothic Architecture: The Restoration of Medieval Churches in Victorian Britain.' PhD diss., New York University, 1992.
Nissan, Elizabeth. 'The Sacred City of Anuradhapura: Aspects of Sinhalese Buddhism and Nationhood.' PhD diss., The London School of Economics and Political Science, 1985.
Parry, Tim. 'The Incorporated Church Building Society, 1818-1851.' DPhil diss., Trinity College, Oxford, 1984.
Pettersson, Richard. *Fädernesland och framtidsland: Sigurd Curman och kulturminnesvårdens etablering*, PhD diss., Umeå universitet, 2001.
Reid, C. 'American Internationalism: Peace Advocacy and International Relations, 1895-1916.' PhD diss., Boston College, 2005.
Swenson, Astrid. 'Conceptualising 'Heritage' in Nineteenth- and Early Twentieth-Century France, Germany and England.' PhD diss., Cambridge University, 2007.

Wetterberg, Ola. *Monument & miljö. Perspektiv på det tidiga 1900-talets byggnadsvård i Sverige,* Institutionen för arkitekturhistoria, PhD diss., Göteborg Chalmers tekn. Högsk, 1992.

Newspapers, Magazines and Other Serials

American Architect and Building News.
American Scenic and Historic Preservation Society Annual Report.
Atalanta.
Boston Evening Transcripts.
British Architect.
British and Foreign Review.
British Parliamentary Papers.
Burnley Gazette.
Chambers's Edinburgh Journal.
Die Denkmalpflege.
Ecclesiologist.
Estates Gazette.
Huddersfield Daily Chronicle.
Journal Officiel. Paris: Imprimerie du Journal Officiel.
L'Ami des Monuments.
Le Journal des Débats.
Le Temps.
League of Nations Permanent Mandate Commission.
Manchester Guardian.
Morning Post.
National Trust for Places of Historic Interest and Natural Beauty: Annual Reports.
Northamptonshire Notes and Queries.
Pall Mall Gazette.
Penny Illustrated News.
Scribner's Magazine.
Society for the Protection of Ancient Buildings: Annual Reports.
St. Nicholas.
The Builder.
The Christian East.
The Church Times.
The Ecclesiologist.
The Economist.
The Nation.
The New York Times.
The Picayune.
The Times.
The Treasury.

Books and Scholarly Articles

A Description of Westminster Abbey, its Monuments and Curiosities. Brentford, n.d. [1770].

Abel, Vincent and F.-M. *Bethléem: Le Sanctuaire de la Nativité*. Paris: J. Gabalda, 1914.

Adamson, Jeremy Elwell. *Niagara: Two Centuries of Changing Attitudes, 1697-1901*. Washington, DC: Corcoran Gallery of Art, 1984.

Ad patriam illustrandam: hyllningsskrift till Sigurd Curman 30 april 1946. Uppsala, 1946.

Aldrich, Megan. 'Gothic Architecture Illustrated: The Drawings of Thomas Rickman in New York,' *The Antiquaries Journal*, 65 (1985), 427-33.

Almond, Philip. *The British Discovery of Buddhism*. Cambridge: Cambridge University Press, 1988.

Anderson, Benedict. *Imagined Communities: Reflections on the Spread and Origin of Nationalism*. London and New York: Verso, 2006.

Anderson, M.S. *The Rise of Modern Diplomacy, 1450-1919*. London: Longman, 1993.

Anderson, Stuart. *Race and Rapprochement: Anglo-Saxonism and Anglo-American Relations, 1895-1904*. East Brunswick: Associated University Press, 1981.

Andersson, Fredric B., Henrik, O. 'Lilljekvist, Johan Fredrik', *Svenskt biografiskt lexikon*, 23, 1980.

Andersson, Lilian. *Mellan byråkrati och laissez faire: en studie av Camillo Sittes och Patrick Geddes stadsplaneringsstrategier* (Beyond bureaucratic and economic rationality: a study of Camillo Sitte's and Patrick Geddes' town planning strategies). Göteborg: Acta Universitatis Gothoburgensis, 1989.

Arne, Ture J. *Våra fornminnens vård, Studentföreningen Verdandis tidningsartiklar 193*, Uppsala, 1908.

Aronsson, Peter. *Historiebruk: att använda det förflutn*. Lund: Studentlitteratur, 2004.

Ashbee, C.R. *A Palestine Notebook*. New York: Doubleday, 1923.

Ashbee, C.R., ed. *Jerusalem 1918-1920*. London: John Murray, 1921.

Ayres, Philip J. *Classical Culture and the Idea of Rome in Eighteenth-century England*. New York: Cambridge University Press, 1997.

Baker, E.P., ed. *Visitation Returns for the Archdeaconry of Oxfordshire in the Year of 1854*. vol. 35. Oxon Record Society, 1954.

Bar-Yosef, Eitan. *The Holy Land in English Culture 1799-1917*. Oxford: Clarendon Press, 2005.

Barnum, P.T. *Strategies and Triumphs, or Forty Years' Recollections*. Buffalo: Warren, Johnson & Co., 1872.

Barrington, Lewis. *Historic Restorations of the Daughters of the American Revolution*. New York: R.R. Smith, 1941.

Bastiampillai, Bertram. *The Administration of Sir William Gregory, Governor of Ceylon 1872-1877*. Dehiwala: Tisara Prakasakayo, 1968.

Baxandall, Michael. *Painting and Experience in Fifteenth Century Italy*. 2nd Edition. Oxford: Oxford University Press, 1988.

Bell, George K. *Randall Davidson, Archbishop of Canterbury*. 2 vols. Oxford: Oxford University Press, 1935.

Bell, Bethia N. and Heather M. Bell. *H.C.P. Bell: Archaeologist of Ceylon and the Maldives*. Clywd: Archetype Publications, 1993.

Bentley, Michael. *Lord Salisbury's World: Conservative Environments in Late Victorian Britain*. Cambridge: Cambridge University Press, 2001.

Bentwich, Norman. 'The Antiquities Law of Palestine,' *Journal of Comparative Legislation and International Law*, 3rd series, vi:1 (1924), 1-6.

Bercé, Françoise. *Dès Monuments historiques au patrimoine du XVIIIe siècle à nos jours ou 'Les égarements du coeur et de l'esprit.'* Paris: Flammarion, 2000.

Berghoff, Hartmut and Dieter Ziegler, eds. *Pionier und Nachzügler? Vergleichende Studien zur Geschichte Großbritanniens und Deutschlands im Zeitalter der Industrialisierung. Festschrift für Sidney Pollard zum 70. Geburtstag*. Bochum: Brockmeyer, 1995.

Bericht über den VIII. *Internationalen Architekten-Kongress, Wien 1908*. Vienna: Schroll, 1909.

Betänkande med förslag till lag angående kulturminnesvård samt organisation av kulturminnesvården. 1, Historik, memorial angående minnesvårdens nuvarande ståndpunkt, utländsk lagstiftning samt bilagor. Stockholm, 1922.

Bingham, Adrian. *Gender, Modernity, and the Popular Press in Inter-War Britain*. Oxford: Clarendon Press, 2004.

Biörnstad, Arne, ed., *Skansen under hundra* år. Wiken: Höganäs. 1991.

Birnbaum, Charles and Mary V. Hughes, eds. *Design With Culture: Claiming America's Landscape Heritage*. Charlottesville: University of Virginia Press, 2005.

Blackburn, Anne M. *Locations of Buddhism: Colonialism and Modernity in Sri Lanka*. Chicago: University of Chicago Press, 2010.

Boulting, Nikolaus. 'The Law's Delays: Conservationist Legislation in the British Isles,' in Jane Fawcett, ed., *The Future of the Past: Attitudes to Conservation, 1174-1974*. q.v.

Brand, Vanessa, ed. *The Study of the Past in the Victorian Age*. Oxford: Oxford Books, 1998.

Breuilly, John. *Nationalism and the State*. New York: St Martin's Press, 1982.

Breuilly, John. 'Modernisation as Social Evolution: the German Case, c1800-1880,' *Transactions of the Royal Historical Society*, 6th series, xv (2005), 117-48.

Briggs, Martin S. *Goths and Vandals: A Study of the Destruction, Neglect and Preservation of Historical Buildings in England*. London: Constable, 1952.

British-American Peace Centenary Committee. *The British Home of the Washingtons: Sulgrave Manor and Its Associations*. London, 1912.

British-American Peace Centenary Committee. *British-American Peace Centenary: Official Report of the International Conference held at New York, USA, May V to*

IX, MCMXIII, including Proposals for International Programme and Manifesto to the Nations. London, 1913.

Brittain, Sir Harry. *Pilgrims and Pioneers.* London: Hutchinson & Co., 1946.

Brooks, Chris. *Gothic Revival.* London: Phaidon Press, 1999.

Brooks, Chris and Andrew Saint, eds. *The Victorian Church. Architecture and Society.* Manchester: Manchester University Press, 1995.

Brown, Gerald Baldwin. *The Care of Ancient Monuments: An Account of the Legislative and Other Measures Adopted in European Countries for Protecting Ancient Monuments and Objects and Scenes of Natural Beauty and for Preserving the Aspect of Historical Cities.* Cambridge: Cambridge University Press, 1905.

Brown, Robert C. *Robert Laird Borden: A Biography.* 2 vols. Toronto: Macmillan, 1975-77.

Brown, Robert C. and Ramsay Cook. *Canada, 1896-1921: A Nation Transformed.* Toronto: Mclelland and Stewart, 1974.

Bryce, James. *Constantinople.* New York: Harpers & Brothers, 1878.

Bryce, James. 'The Nature of Public Opinion,' in *The American Commonwealth,* vol. 2, 1900, New York: Macmillan, 1960.

Bryce, James. University and Historical Addresses: Delivered during a residence in the United States as Ambassador of Great Britain. London: Macmillan, 1913.

Bunyan, John. *The Barren Fig Tree.* London: J. Robinson, 1688.

Burns, Arthur. The Diocesan Revival in the Church of England, c.1800-1870. Oxford: Clarendon Press, 1999.

Burrows, Stephen M. Report on Archaeological Work in Anuradhapura and Pollonnaruwa. Colombo: Government Printer, 1886.

Butler, Richard and Russell, Roslyn, eds. *Giants of Tourism.* Wallingford: CABI, 2010.

Calhoun, Craig ed., *Habermas and the Public Sphere.* Cambridge, Mass: MIT Press, 1992.

Cannadine, David. *Ornamentalism: How the British Saw Their Empire.* New York: Oxford University Press, 2001.

Ceadel, Martin. *Semi-Detached Idealists: The British Peace Movement and International Relations, 1854-1945.* Oxford: Oxford University Press, 2000.

Celebrations of the 100 Years of Peace among the English-Speaking Peoples: Report of Public Meeting at the Mansion House, December 18th, 1912. London, 1912.

Chadwick, Owen. *The Secularization of the European Mind in the Nineteenth Century.* Cambridge: Cambridge University Press, 1995.

Champion, Timothy. 'Protecting the Monuments: Archaeological Legislation from the 1882 Act to PP16,' in Michael Hunter, ed., *Preserving the Past: The Rise of Heritage in Modern Britain.* q.v.

Chastel, André. 'La Notion de patrimoine,' in Pierre Nora, ed., *Les Lieux de mémoire.* vol. 1, q.v.

Chippendale, Christopher. 'The Making of the First Ancient Monuments Act, 1882, and Its Administration Under General Pitt-River,' *Journal of the British Archaeological Association* 136 (1983), 1-55.

Choay, Françoise. *L'Allégorie du patrimoine*. Paris: Seuil, 1992.

Choay, Françoise. *The Invention of the Historic Monument*. Cambridge: University of Cambridge Press, 1998.

Clark, Christopher and Wolfram Kaiser, eds. *Culture Wars. Secular-Catholic Conflict in Nineteenth-Century Europe*. Cambridge: Cambridge University Press, 2003.

Clason, Isak Gustaf. 'Våra byggnadsminnens vård,, *Arkitektur* 2 (1910), 26-28.

Coates, Colin M., ed. *Imperial Canada, 1867-1917*. Edinburgh: University of Edinburgh Centre of Canadian Studies, 1997.

Coetzee, Frans. *For Party or Country: Nationalism and the Dilemmas of Popular Conservatism in Edwardian England*. Oxford: Oxford University Press, 1990.

Cohn, Bernard S. *Colonialsm and Its Forms of Knowledge: The British in India*. Princeton: Princeton University Press, 1996.

Collier, Peter and Geyer-Ryan, Helga, eds. *Literary Theory Today*. Cambridge: Polity, 1990.

Collini, Stefan, ed. *History, Religion and Culture: British Intellectual History, 1750-1950*. Cambridge: Cambridge University Press, 2000.

Conn, Steven. *Museums and American Intellectual Life, 1876-1926*. Chicago: University of Chicago Press, 1998.

Connell, Philip. 'Death and the Author: Westminster Abbey and the Meanings of the Literary Monument.' *Eighteenth-Century Studies* 38 (Summer 2005), 557-85.

Coxe, Richard C. *A Cursory Survey of the Churches and Church Buildings within the Archdeaconry of Lindisfarne. A Charge Founded on the Reports of the Rural Deans*. London: Francis & John Rivington, 1860.

Craven, Lady Elizabeth. *A Journey through the Crimea to Constantinople*. London: G.G.J. & J. Robinson, 1789.

Crawford, Alan. 'Supper at Gatti's: The SPAB and the Arts and Crafts Movement,' in Chris Miele, ed. *From William Morris: Building Conservation and the Arts and Crafts Cult of Authenticity, 1877-1939*, q.v.

Cullingworth, Barry and Vincent Nadin. *Town and Country Planning in the UK*. 13th Edition. London: Routledge, 2002.

Curman, Sigurd: 'Byggnadsminnen i privatägo,' *Arkitektur* 12 (1908), 133-35.

Cust, L.G.A., ed. *The Status Quo in the Holy Places*. Jerusalem: Ariel, 1980.

Dalton, Kathleen. *Theodore Roosevelt, A Strenuous Life*. New York: A.A. Knopf, 2002.

Dansey, W. *A Letter to the Archdeacon of Sarum on Ruridecanal Charges*. London, 1840.

Daunton, Martin. 'Britain and Globalisation since 1850: I. Creating a Global Order, 1850-1914.' *Transactions of the Royal Historical Society*, 6th Series, xvi (2006), 1-38.

Daunton, Martin and Bernhard Rieger, eds. *Meanings of Modernity: Britain from the Late-Victorian Era to World War II*. Oxford: Berg, 2001.

Davis, Richard W. '"We are all Americans Now!" Anglo-American Marriages in the Later Nineteenth Century,' *Proceedings of the American Philosophical Society*, vol. 135, no. 2 (1991), 140-99.

Dawson Damer, Mrs. G.L. *Diary of a Tour in Greece, Turkey, Eqypt, and the Holy Land*. 2 vols. London: Henry Colburn, 1841.

De Vere, Aubrey. *Picturesque Sketches in Greece and Turkey*. 2 vols. London: Richard Bentley, 1850.

Delafons, John. *Politics and Preservation: A Policy History of the Built Heritage, 1882-1996*. London: Spon, 1997.

Dellheim, Charles. *The Face of the Past: The Preservation of the Medieval Inheritance in Victorian England*. Cambridge: Cambridge University Press, 1982.

Denslagen, Wim. *Architectural Restoration in Western Europe: Controversy and Continuity*. Amsterdam: Architectura & Natura Press, 1994.

Di Giovine, Michael. *The Heritage-Scape: UNESCO, World Heritage, and Tourism*. Lanham, MD: Lexington Books, 2009.

Dobson, Michael. *The Making of the National Poet: Shakespeare, Adaptation and Authorship, 1660-1769*. Oxford: Oxford University Press, 1992.

Dorsheimer, William. *The Life and Public Service of the Honorable Grover Cleveland*. Philadelphia, 1884.

Douglas, J.A. 'The Importance of St. Sophia.' *The Christian East*, vol. 1, no. 2 (June 1920).

Douglas, J.A. *The Redemption of St. Sophia: An Historical and Political Account of the Subject*. London: Faith Press, 1st ed. 1919, 4th ed. 1920.

Dow, Charles Mason, ed. *Anthology and Bibliography of Niagara Falls*. 2 vols. Albany: J.B. Lyon, 1921.

Dow, Charles Mason. *The State Reservation at Niagara, a History*. Albany: J.B. Lyon, 1914.

Dreisziger, N.F. 'The Campaign to Save Niagara Falls and the Settlement of United States-Canadian Differences, 1906-1911.' *New York History* 55 (Oct. 1974), 437-58.

Dryden, Sir H. 'The Washington House at Sulgrave.' *Northamptonshire Notes and Queries* April (1885).

Dubrow, Gail Lee and Jennifer B. Goodman. *Restoring Women's History through Historic Preservation*. Baltimore: Johns Hopkins University Press, 2003.

Dunn, D.H., ed. *Diplomacy at the Highest Level: The Evolution of International Summitry*. London and New York, 1996.

Edenheim, Ralph and Martinus, Eivor, eds. *Skansen*. London and Stockholm: Scala, 1995, trans. Neil Smith.

Edgerton, David. *Science, Technology and the British Industrial "Decline" 1870-1970*. Cambridge: Cambridge University Press, 1996.

Edwards, C. 'The English Home of the Washingtons.' *Atalanta* (January 1, 1898).

Edwards, Penny. *Cambodge: The Cultivation of a Nation, 1860-1945*. Honolulu: University of Hawai'i Press, 2007.

Encylopædia Britannica. 29 vols. New York: Encylopædia Britannica, 11th Edition, 1910.

Ensor, R.C.K. 'The English Countryside,' in Lucian Oldershaw, ed., *England: A Nation*, q.v.

Erder, Cevat. *Our Architectural Heritage: From Consciousness to Conservation*. Paris: UNESCO, 1986.

Ernstrom, Adele. '"Why Should We Always Look Back?" "Christian Art" in Nineteenth-Century Historiography in Britain.' *Art History*, vol. 22, no. 3 (Sept. 1999), 421-35.

Esher, Lionel. *A Broken Wave. The Rebuilding of England, 1940-1980*. London: Allen Lane, 1981.

Everett, Nigel. *The Tory View of Landscape*. New Haven and London: Yale University Press, 1994.

Fataburen: Kulturhistorisk tidskrift, Nordiska museet. Stockholm, 1914.

Fawcett, Jane. *The Future of the Past: Attitudes to Conservation*. London: Thames & Hudson, 1976.

Fawcett, Jane. 'A Restoration Tragedy: Cathedrals in the Eighteenth and Nineteenth Centuries,' in Jane Fawcett, ed. *The Future of the Past: Attitudes to Conservation, 1174-1974*, q.v.

Fellows, Will. *A Passion to Preserve: Gay Men as Keepers of Culture*. Madison: University of Wisconsin Press, 2005.

Fergusson, James. *A History of Architecture in All Countries*. 3 vols. London: John Murray, 3rd ed. 1893.

Findlen, Paula. 'The Museum: Its Classical Etymology and Renaissance Geneology,' *Journal of the History of Collections* 1:1 (1989), 59-78.

Finn, James. *Stirring Times or Records from Jerusalem Consular Chronicles of 1853 to 1856*. London: C. Kegan Paul & Co., 1878.

Firth, Ian. 'The Blue Ridge Parkway: Road to the Modern Preservation Movement,' in Charles Birnbaum and Mary V. Hughes, eds. *Design with Culture: Claiming America's Landscape Heritage*, q.v.

Fisher, John and Antony Best, eds. *On the Fringes of Diplomacy: Influences on British Foreign Policy, 1800-1945*. Aldershot: Ashgate, 2011.

Fisher, Lewis F. *Saving San Antonio: The Precarious Preservation of a Heritage*. Lubbock: Texas Tech University Press, 1996.

Ford, Colin. *The Cameron Collection: An Album of Photographs by Julia Margaret Cameron Presented to Sir John Herschel*. New York: Van Nostrand Reinhold, 1975.

Foreign Relations of the United States: Paris Peace Conference. 13 vols. Washington, DC: Government Printing Office, 1942-47.

Forrest, Craig, *International Law and the Protection of Cultural Heritage*. London and New York: Routledge, 2010.

Francioni, Francesco. The 1972 *World Heritage Convention, A Commentary.* Oxford: Oxford University Press, 2008.

Freeman, Edward A. *Principles of Church Restoration.* London: Joseph Masters, 1846.

Freeman, Michael J. *Railways and the Victorian Imagination.* New Haven and London: Yale University Press, 1999.

Fry, Michael, Goldstein, Erik and Langhorne, R.T.B. *Guide to International Relations and Diplomacy.* London: Continuum, 2002.

Gage, John. *Color in Turner: Poetry and Truth.* New York: Praeger, 1969.

Garlick, Barbara and Margaret Harris, eds. *Victorian Journalism: Exotic and Domestic.* St Lucia, Australia: Queensland University Press, 1998.

Garstang, John. 'Eighteen months' work of the Department of Antiquities for Palestine July, 1920-December, 1921.' *Palestine Exploration Fund Quarterly Statement* (1922).

Gaselee, Stephen. *Stories from the Christian East.* London: Sidgwick & Jackson, 1918.

Gellner, Ernest. *Nations and Nationalism.* Ithaca: Cornell University Press, 1983.

George, David Lloyd. *The Rural Land Problem: What It Is; A Speech.* London: Liberal Publication Department, 1913.

Gibbon, Edward. *The Decline and Fall of the Roman Empire.* 6 vols. New York: Alfred Knopf, 1994.

Gibson, Shimon. 'British Archaeological Institutions in Mandatory Palestine, 1917-1948,' *Palestine Exploration Quarterly* 131 (1999), 115-43.

Gilbert, Martin. *Jerusalem in the Twentieth Century.* London, Chatto & Windus, 1996.

Gilmour, David. *Curzon.* London: John Murray, 1994.

Glendinning, Miles. 'The Conservation Movement: a cult of the modern age,, *Transactions of the Royal Historical Society,* 6th series, vi (2003), 359-76.

Goldstein, Erik. 'Great Britain and Greater Greece, 1917-1920,' *Historical Journal,* vol. 32, no. 2 (1989), 339-56.

Goldstein, Erik. 'Holy Wisdom and British Foreign Policy, 1918-1922: the St. Sophia Redemption Agitation.' *Byzantine and Modern Greek Studies* 15 (1991), 36-64.

Goldstein, Erik. 'The Origins of Summit Diplomacy,, in D.H. Dunn, ed., *Diplomacy at the Highest Level: The Evolution of International Summitry,* q.v.

Goodwin, Godfrey. 'Gaspare Fossati di Morcote and his Brother Giuseppe.' *Environmental Design: Journal of the Islamic Environmental Research Centre* (1990).

Gournay, Isabelle and Elliott Pavlos. 'Americans in Paris,' *Journal of Architectural Education,* vol. 38, no. 4. (Summer, 1985), 22-26.

Governors' Addresses in the Legislative Council of Ceylon and Legislative Council of Ceylon Replies to Governors' Addresses.

Gow, A.S.F. 'Sir Stepen Gaselee, K.C.M.G.,' *Proceedings of the British Academy,* 29 (1943), 441-61.

Grandien, Bo. 'Grogrunden', in Arne Biörnstad, ed., *Skansen under hundra* år. q.v.

Gregory, Augusta, ed. *Mr. Gregory's Letter-box, 1813-1830*. London: Smith, Elder, & Co., 1898.

Green, E.H.H. *The Crisis of Conservatism: The Politics, Economics and Ideology of the British Conservative Party, 1880-1914*. London: Routledge, 1996.

Greenblatt, Steven. 'Resonance and Wonder', in Peter Collier and Helga Geyer-Ryan, eds, *Literary Theory Today*. q.v.

Grusin, Richard. *Culture, Technology, and the Creation of America's National Parks*. Cambridge: Cambridge University Press, 2004.

Guha-Thakurta, Tapati. *Monuments, Objects, Histories: Institutions of Art in Colonial and Postcolonial India*. New York: Columbia University Press, 2004.

Gwynn, Stephen, ed. *The Letters and Friendships of Sir Cecil Spring Rice*. 2 vols. Boston: Houghton Mifflin, 1929.

Habermas, Jurgen. *Strukturwandel der* Öffentlichkeit*: Untersuching zu einer Kategorie der bürgerlichen Gesellschaft. Frankfurt, 1962*. (The Structural Transformation of the Public Sphere: An Inquiry into a Category of Bourgeois Society, trans. Thomas Burger Cambridge, Mass: MIT Press, 1988).

Hadfield, Miles. *A History of British Gardening*. Harmondsworth: Penguin Books, 1985.

Haig, Alan. *The Victorian Clergy*. London: Croom Helm, 1984.

Hall, Melanie. 'The Politics of Collecting: The Early Aspirations of the National Trust, 1883-1913,' *Transactions of the Royal Historical Society*, 6th series, vi (2003), 345-58.

Hall, Melanie. 'Affirming Community Life: Preservation, National Identity and the State, 1900,' in Chris Miele, ed. *From William Morris: Building Conservation and the Arts and Crafts Cult of Authenticity, 1877-1939*, q.v.

Hall, Melanie and Goldstein, Erik. 'Writers, the Clergy, and the 'Diplomatization' of Culture: Sub-structures of Anglo-American Diplomacy, 1820-1914,' John Fisher and Antony Best, eds, *On the Fringes of Diplomacy: Influences on British Foreign Policy*, q.v.

Hamilton, R.W. *The Church of the Nativity Bethlehem*. Jerusalem: Government of Palestine, Department of Antiquities, 1947.

Hammer, Felix. *Die geschichtliche Entwicklung des Denkmalrechts in Deutschland*. Tübingen: Mohr, 1995.

Hammarlund-Larsson, Cecilia. 'I denna tid af slapp nationalkänsla: om Artur Hazelius, vetenskapen och nationen,' in: *Samhällsideal och framtidsbilder: perspektiv på Nordiska museets dokumentation och forskning*. q.v.

Hans, Günter. *Denkmalschutz in Baden im 19. und 20. Jahrhundert*. Freiburg: Univ. Hochschulschrift, 1985.

Hardinge, Lord of Penshurst, *Charles. Old Diplomacy*. London: J. Murray, 1947.

Harringron, Paul. 'Holst and Vaughan Williams: Radical Pastoral,' in Christopher Norris, ed., *Music and the Politics of Culture*. q.v.

Harris, John. *Moving Rooms: The Trade in Architectural Salvages*. New Haven and London: Yale University Press, 2007.

Harvey, William. *The Preservation of St. Paul's Cathedral and Other Famous Buildings*. London: The Architectural Press, 1925.

Harvey, William. *Structural Survey of the Church of the Nativity, Bethlehem*. London: Oxford University Press, 1935.

Harvie, Christopher. 'Ideology and Home Rule: James Bryce, A.V. Dicey and Ireland, 1880–1887,' *English Historical Review*, xci (1976), 298-314.

Haskell, Francis and Penny, Nicholas. *Taste and the Antique, the Lure of Classical Sculpture, 1500-1900*. New Haven and London: Yale University Press, 1981.

Hastings, Adrian. *The Construction of Nationhood, Ethnicity, Religion and Nationalism*. Cambridge: Cambridge University Press, 1997.

Hays, Samuel P. *Conservation and the Gospel of Efficiency, The Progressive Conservation Movement 1890-1920*. Cambridge, Mass: Harvard University Press, 1969.

Helfert, Josef Alexander Freiherr von. *Denkmalpflege:* Öffentliche *Obsorge für Gegenstände der Kunst und des Altertums nach dem neuesten Stande der Gesetzgebung in den verschiedenen Culturstaaten*. Vienna, Leipzig: Braumüller, 1897.

Hellal, Salima. 'Les Etudes de législations comparées du patrimoine en France, 1887-1913,' *Mémoire de Maîtrise*, Université Paris-IV, 2000.

Hendrick, Burton J. *The Life and Letters of Walter Hines Page*. 3 vols. London: W. Heinemann, 1926.

Hendrick, Burton J. *The Training of an American: The Earlier Life and Letters of Walter H. Page*. Boston: Houghton, 1928.

Henry, David. *An Historical Description of Westminster Abbey, its Monuments and Curiosities ... Designed Chiefly as a Guide to Strangers*. London, 1753 [1778].

Herrick, Francis H. 'Gladstone and the Concept of the "English-Speaking Peoples."' *Journal of British Studies*, vol. 12, no. 1 (1972), 150-56.

Hewison, Robert. *The Heritage Industry: Britain in a Climate of Decline*. London: Methuen London, 1987.

Hewison, Robert. *Ruskin on Venice: 'The Paradise of Cities'*. New Haven and London: Yale University Press, 2009.

Hill, Rosemary. *God's Architect: Pugin and the Building of Romantic Britain*. London: Allen Lane, 2007.

Hilton, Boyd. *The Age of Atonement: The Influence of Evangelicalism on Social and Economic Thought, 1785-1865*. Oxford: Clarendon Press, 1988.

Hinsley, F.H., ed. *British Foreign Policy under Sir Edward Grey*. Cambridge: Cambridge University Press, 1977.

Hirtzel, Sir Arthur. *The Church, The Empire, and The World*. London: S.P.C.K., 1919.

Hobsbawm, Eric and Terence Ranger, eds. *The Invention of Tradition*. Cambridge: Cambridge University Press, 1983.

Hobsbawm, Eric. 'Mass-Producing Traditions: Europe, 1870-1914,' in Eric Hobsbawm and Terence Ranger, *Invention of Tradition*, q.v.

Hobsbawm, Eric. *Nations and Nationalism Since 1780 Programme, Myth, Reality*. Cambridge: Cambridge University Press, 1992.

Holleran, Michael. *Boston's 'Changeful Times': Origins of Preservation and Planning in America*. Baltimore: Johns Hopkins University Press, 1998.

Hood, Jean. *Trafalgar Square: A Visual History of London's Landmark Through Time*. London: Batsford, 2005.

Hooper-Greenhill, Eilean. *Museums and the Shaping of Knowledge*, London: Routledge, 1992.

Horsman, Reginald. 'Origins of Racial Anglo-Saxonism in Great Britain Before 1850.' *Journal of the History of Ideas*, vol. 37, no. 3 (1976), 387-410.

Hoskins, William G. *The Making of the English Landscape*. Harmondsworth: Penguin Books, 1973.

Hosmer, Charles B., Jr. *Presence of the Past: A History of the Preservation Movement in the United States Before Williamsburg*. New York: Putnam, 1965.

Hosmer, Charles B., Jr. *Preservation Comes of Age: From Williamsburg to the National Trust, 1926-1949*. Charlottesville: University Press of Virginia, 1981.

Howe, Barbara J. 'Women in the Nineteenth-Century Preservation Movement,' in Gail Lee Dubrow and Jennifer B. Goodman, *Restoring Women's History through Historic Preservation*, q.v.

Hughes, Matthew, ed. *Allenby in Palestine: The Middle East Correspondence of Field Marshal Viscount Allenby, June 1917-October 1919*. Stroud: Sutton, for the Army Records Society, 2004.

Hunt, Tristram. *Building Jerusalem: The Rise and Fall of the Victorian City*. London: Weidenfeld & Nicolson, 2004.

Hunter, Michael, ed. *Preserving the Past: The Rise of Heritage in Modern Britain*. Phoenix Mill: Alan Sutton Publishing, 1996.

Huse, Norbert, ed. *Denkmalpflege. Deutsche Texte aus drei Jahrhunderten*. Munich: Beck, 1996.

Hutton, William Holden, ed. *Letters of William Stubbs*. London: Archibald Constable, 1904.

Ievers, R.W. *Manual of the North Central Province, Ceylon*. Colombo: Government Printer, 1899.

Ions, Edmund. *James Bryce and American Democracy, 1870-1922*. London: Macmillan, 1968.

Impey, Oliver and MacGregor, Arthur. *The Origin of Museums: The Cabinet of Curiosities in Sixteenth and Seventeenth Century Europe*. Oxford: Clarendon Press, 1985.

Irwin, William. *The New Niagara, Tourism, Technology, and the Landscape of Niagara Falls 1776-1917*. University Park: University of Pennsylvania Press, 1996.

Jackson, Basil, ed. *Recollections of Thomas Graham Jackson, 1835-1924*. London: Oxford University Press, 1950.

Jackson, Sir Thomas Graham. *Byzantine and Romanesque Architecture*. Cambridge: Cambridge University Press, 1913.

Jackson, W. Turrentine. 'The Creation of Yellowstone National Park.' *Mississippi Valley Historical Review*, vol. 29, no. 2 (1942), 187-206.

James, Henry. Portraits of Places. Boston: Houghton, 1884.

Jasanoff, Maya. *Edge of Empire: Lives, Culture, and Conquest in the East, 1750-1850*. New York: Alfred A. Knopf, 2005.

Jenkins, Brian. *Sir William Gregory of Coole: The Biography of an Anglo-Irishman*. Gerrards Cross: Colin Smythe, 1986.

Jenkins, Jennifer and Patrick James. *From Acorn to Oak Tree: The Growth of the National Trust, 1895-1994*. London: Macmillan, 1994.

Jenkyns, Richard. *The Victorians and Ancient Greece*. Cambridge, Mass: Harvard University Press, 1980.

Jokilehto, Jukka. *A History of Architectural Conservation*. Oxford: Butterworth-Heinemann, 1999.

Jones, Eugene. *Native Americans as Shown on the Stage, 1753-1916*. Metuchen: Scarcrow Press, 1988.

Joyce, Patrick. *Visions of the People: Industrial England the Question of Class, 1840-1914*. Cambridge: Cambridge University Press, 1991.

Kauffmann, Eric. '"Naturalizing the Nation:" The Rise of Naturalistic Nationalism in the United States and Canada,' *Comparative Studies in Society and History*, vol. 40, no. 4 (1998), 666-95.

Kay, Jane Holtz. *Lost Boston*. Boston: Houghton Mifflin, 1980.

Keenleyside, Hugh L., ed. *The Growth of Canadian Policies in External Affairs*. Westport: Greenwood Press for Duke University, Commonwealth-Studies Center, 1960.

Kelvin, Norman, ed. *The Collected Letters of William Morris*. 4 vols. Princeton: Princeton University Press, 1984.

Kemnitz, Thomas Milton. 'Matt Morgan of "Tomahawk" and English Cartooning, 1867-1870,' *Victorian Studies* 19 (1975-6), 5-34.

Kemper, Steven. *The Presence of the Past: Chronicles, Politics, and Culture in Sinhala Life*. Ithaca: Cornell University Press, 1991.

Kennedy Paul M. and Anthony Nicholls, eds. *Nationalist and Racialist Movements in Britain and Germany before 1914*. London: Macmillan, 1981.

Kennet, Wayland. *Preservation*. London: Temple Smith, 1972.

Kenyon, John P. *The History Men: The Historical Profession in England since the Renaissance*. Pittsburgh: University of Pittsburgh Press, 1984.

Kertzer, D.I. *Ritual, Politics, and Power*. New Haven and London: Yale University Press, 1988.

King, Richard. *Orientalism and Religion: Post-Colonial Theory, India and the Mystic East*. London: Routledge, 1999.

Kinross, Lord. *Hagia Sophia*. New York: Newsweek, 1972.

Kirk, Clara Marburg. *W.D. Howells and Art in His Time*. New Brunswick: Rutgers University Press, 1965.

Kohte, Julius. *Die Pflege der Kunstdenkmäler in Italien*. Berlin: Ernst, 1898.

Kongl: Mayst:tz Placat och Påbudh, Om Gamble Monumenter och Antiquiteter. Printed by Georg Hantsch, 1666.

Koshar, Rudy. *From Monuments to Traces: Aspects of German Memory, 1870-1990*. Berkeley: University of California Press, 2000.

Kowsky, Francis. 'Municipal Parks and City Planning: Frederick Law Olmsted's Buffalo Park and Parkway System,' *Journal of the Society of Architectural Historians* 46 (1987), 49-64.

Krass, Peter. *Carnegie*. Hoboken: J. Wiley & Sons, 2002.

Kungl. Vitterhets-, historie- och antikvitetsakademiens månadsblad., Kongl. Vitterhets, historie och antiqvitetsakademien, Stockholm, 1903.

Kulturfastighetsutredningen. Bilagor. Stockholm: Statens fastighetsverk, 2009.

Ladd, Brian. *Urban Planning and Civic Order in Germany, 1860-1914*. Cambridge, Mass: Harvard University Press, 1990.

LaFeber, Walter. *The Panama Canal: The Crisis in Historical Perspective*. New York: Oxford University Press, 1978.

Laity, Paul. *The British Peace Movement, 1870-1914*. Oxford: Clarendon Press, 2001.

Lambert, Andrew. *Nelson: Britannia's God of War*. London: Faber, 2004.

Lambert, Robert A. *Contested Mountains: Nature, Development and Environment in the Cairngorms Region of Scotland, 1880-1980*. Cambridge: White Horse Press, 2001.

Langhorne, R.T.B. 'The Development of International Conferences, 1648-1830', *Studies in History and Politics* iii (1981-2), 65-75.

Langhorne, R.T.B. *The Coming of Globalization, its Evolution and Contemporary Consequences*. Basingstoke: Palgrave, 2001.

Lagercrantz, Bo. 'Nordiska museet, Skansen och hembygdsrörelsen,' in Arne Biörnstad, ed. *Skansen under hundra år*. q.v.

Leatherbarrow, J. 'The Victorians Look at Their Churches.' *Transactions of the Worcester Archaeological Society*, 3rd series, vol. 9 (1984), 87-94.

Lectures on Art Delivered in Support of the Society for the Protection of Ancient Buildings. London, 1882.

Lectures delivered before the Young Men's Christian Association, in Exeter Hall, from November, 1862, to February, 1863. London: James Nisbet & Co., 1863.

Leerssen, Joseph. *National Thought in Europe: A Cultural History*. Amsterdam: Amsterdam University Press, 2006.

Lees-Milne, James, ed. *The National Trust: A Record of the First Fifty Years*. London: B.T. Batsford, 1945.

Leggo, William. *The History of the Administration of the Right Honorable Frederick Temple, Earl of Dufferin, K.P., G.C.M.G., K.C.B., F.R.S., Late Governor General of Canada*. Montreal: Lovell, 1878.

Lemire, Eugene, ed. *The Unpublished Lectures of William Morris*. Detroit, 1969.

Leniaud, Jean-Michel. *Les Archipels du passé. Le patrimoine et son histoire*. Paris: Fayard, 2002.

Leniaud, Jean-Michel. *L'Utopie française. Essay sur le patrimoine*. Paris: Mengès, 1992.

Léon, Paul. *La Vie des monuments français. Destruction, restauration*. Paris: Picard, 1951.

Lethaby, William R. and Harold Swainson. *The Church of Sancta Sophia: A Study in Byzantine Building*. London and New York: Macmillan & Co., 1894.

Levine, Lawrence. *Highbrow, Lowbrow: The Emergence of Cultural Hierarchy in America*. Cambridge, Mass: Harvard University Press, 1988.

Levine, Philippa. *Amateur and the Professional: Antiquarians, Historians and Archaeologists in Victorian Britain*. Cambridge: Cambridge University Press, 1986.

Lewis, George R. *Illustrations of Kilpeck Church, Herefordshire*. London: G.R. Lewis and William Pickering, 1843.

l'Hôpital, W. de, *Westminster Cathedral and its Architect: vol. I, The Building of the Cathedral*. New York: Dodd, Mead and Company, n.d.

Ljungström, Lars. *Aendnu gamblare: Fredrik Lilljekvists restaurering av Gripsholms slott och 1890-talets restaureringsdebatt*. Nyköping: Södermanlands museum, 1987.

Light, Alison. *Forever England: Femininity, Literature, and Conservatism Between the Wars*. London: Routledge, 1991.

Linden-Ward, Blanche. *Silent City on a Hill: Landscapes of Memory and Boston's Mount Auburn Cemetery*. Columbus: Ohio State University Press, 1989.

Lindgren, James M. *Preserving Historic New England: Preservation, Progressivism, and the Remaking of Memory*. New York: Oxford University Press, 1995.

Lindgren, James M. *Preserving the Old Dominion: Historic Preservation and Virginia Traditionalism*. Charlottesville: University Press of Virginia, 1993.

Lindsay, Lord. *Sketches in the History of Christian Art*. London: J. Murray, 1847.

Logan, Thomas. 'The Americanization of German Zoning,' *Journal of the American Institute of Planners* 42 (1976), 377-85.

Lopez, Donald S., ed. *Curators of the Buddha: The Study of Buddhism Under Colonialism*. Chicago: University of Chicago Press, 1995.

Lord Curzon in India, 1898-1905. London: Macmillan, 1906.

Loufenberg, H. von. *That Night in May ... Lament on the Capture of Saint Sophia by the Turk. Translated from the modern Greek of G. Drosini by G.R. Woodward, air by H. von Loufenberg, harmonies by Dr. C. Wood, etc*. London: Faith Press, 1919.

Lowenthal, David. *The Heritage Crusade and the Spoils of History*. Cambridge: Cambridge University Press, 1998.

Lowenthal, David. *The Past is a Foreign Country*. Cambridge: Cambridge University Press, 1985.

Lucas, Bernard. *The Empire of Christ*. London: Macmillan, 1907.

Luckin, Bill. *Questions of Power: Electricity and Environment in Inter-War Britain*. Manchester: Manchester University Press, 1990.

Lynch, Kevin. *The Image of the City.* Cambridge, Mass: Joint Center for Urban Studies, 1960.

MacCarthy, Fiona. *William Morris: A Life for Our Time.* London: Faber and Faber, 1994.

MacGill, Thomas. *Travels in Turkey, Italy, and Russia.* 2 vols. London: John Murray and Archibald Constable, 1808.

Mackail, J.W. *The Life of William Morris.* 2 vols. New York: Dover Publications, 1995.

Mackail, J.W. and Guy Wyndham. *Life and Letters of George Wyndham.* 2 vols. London: Hutchinson & Co., [1925].

Mandler, Peter. 'Against "Englishness:" English Culture and the Limits to Rural Nostalgia, 1850-1940,' *Transactions of the Royal Historical Society*, 6th Series, vii (1997), 155-75.

Mandler, Peter. *Fall and Rise of the Stately Home.* New Haven and London: Yale University Press, 1997.

Mandler, Peter. '"Race" and "Nation" in Mid-Victorian Thought,' in Stefan Collini, ed. *History, Religion and Culture: British Intellectual History, 1750-1950.* q.v.

Mandler, Peter and Swenson, Astrid, eds. 'From Plunder to Preservation. Britain and the Heritage of Empire, 1800-1950,' *Proceedings of the British Academy*, Oxford: Oxford University Press, (forthcoming, 2011).

Markland, James. *Remarks on the Sepulchral Memorials of the Past ... with Some Suggestions for Improving the Condition of Our Churches, a Letter Addressed to the Oxford Architectural Society.* Oxford: John Henry Parker, 1840.

Marsden, Joshua. 'Thoughts on the Present Crisis of the Turkish Empire.' *Imperial Magazine*, vol. 10, no. 114 (June 1828).

Marsh, Jan. *Back to the Land: The Pastoral Impulse in Victorian England from 1880 to 1914.* London: Faber and Faber, 1982.

Martin, Christopher. *A Glimpse of Heaven: Catholic Churches of England and Wales.* Swindon: English Heritage, 2006.

Martin, Edward S. *The Life of Joseph Hodges Choate.* 2 vols. London: Constable, 1920.

Mason, Randall. 'Historic Preservation, Public Memory, and the Making of New York City,' in Max Page and Randall Mason, eds. *Giving Preservation a History: Histories of Historic Preservation in the United States*, q.v.

Matthew, H.C.G. and Brian Harrison, eds. *Oxford Dictionary of National Biography.* Oxford: Oxford University Press, 2004.

McCue, George. *The Octagon: Being an Account of a Famous Washington Residence: Its Great Years, Decline & Restoration.* Washington: American Institute of Architects Foundation, 1976.

McGann, Jerome, ed. *Lord Byron: The Complete Poetical Works.* vol. 2: *Childe Harold's Pilgrimage.* Oxford: Clarendon Press, 1979.

McKinsey, Elizabeth. *Niagara Falls, Icon of the American Sublime.* Cambridge: Cambridge University Press, 1985.

Merle, Gabriel. *Emile Combes.* Paris: Fayard, 1995.

Messamore, Barbara J., 'British Imperialists and Canadian "Nationalists,"' in Colin M. Coates, ed., *Imperial Canada, 1867-1917*. q.v.

Metcalf, Thomas. *An Imperial Vision: Indian Architecture and Britain's Raj*. New Delhi: Oxford University Press, 2002.

Miele, Chris. '"A Small Knot of Cultivated People:" William Morris and the Ideologies of Protection.' *Art Journal* 54 (1995), 73-79.

Miele, Chris. '"Their Interest and Habit:" Professionalism and the Restoration of Medieval Churches, 1837-1877,' in Christopher Brooks and Andrew Saint, *The Victorian Church. Architecture and Society*, q.v.

Miele, Chris. 'The First Conservation Militants,' in Michael Hunter, ed., *Preserving the Past: The Rise of Heritage in Modern Britain*, q.v.

Miele, Chris. 'Real Antiquity and the Ancient Object: The Science of Gothic Architecture and the Restoration of Medieval Buildings,' in Vanessa Brand, ed., *The Study of the Past in the Victorian Age*, q.v.

Miele, Chris, ed. *From William Morris: Building Conservation and the Arts and Crafts Cult of Authenticity, 1877-1939*. New Haven and London: Yale University Press, 2005.

Miele, Chris. 'Conservation and the Enemies of Progress?,' in Chris Miele ed. *From William Morris* q.v.

Miele, Chris. 'Morris and Conservation,' in Chris Miele ed. *From William Morris,* q.v.

Miles, Margaret. *Art as Plunder: The Ancient Origins of Debate about Cultural Property*. Cambridge: Cambridge University Press, 2008.

Millman, Richard. *Britain and the Eastern Question, 1875-1878*. Oxford: Clarendon Press, 1979.

Mires, Charlene. *Independence Hall in American Memory*. Philadelphia: University of Pennsylvania Press, 2002.

Mock, W. 'The Function of "Race" in Imperialist Ideologies: The Example of Joseph Chamberlain,' in Paul M. Kennedy and Anthony Nicholls, eds, *Nationalist and Racialist Movements in Britain and Germany before 1914*, q.v.

Morley, Judy Mattivi. *Historic Preservation and the Imagined West: Albuquerque, Denver and Seattle*. Lawrence, Kansas: University of Kansas Press, 2006.

Morison, Elting, ed. *The Letters of Theodore Roosevelt*. 8 vols. Cambridge, Mass: MIT Press, 1951-54.

Morris, William. *Letters Sent Home*. London: Frederick Warne, 1875.

Mosse, George L. *The Nationalization of the Masses: Political Symbolism and Mass Movements in Germany from the Napoleonic Wars Through the Third Reich*. New York: H. Fertig, 1975.

Mullin, John R. 'American Perceptions of German City Planning at the Turn of the Century.' *Urbanism Past and Present* 3 (1976).

Mulvey, Christopher. *Anglo-American Landscapes, A Study of Nineteenth-century Anglo-American Travel Literature*. Cambridge: Cambridge University Press, 1983.

Mulryne, J.R. and Goldring, E. *Court Festivals of the European Renaissance: Art, Politics and Performance*. Aldershot: Ashgate, 2002.

Mulryne, J.R., O'Kelly, Helen Watanabe and Shewring, Margaret, eds. *Europa Triumphans: Court and Civic Festivals in Early Modern Europe*, 2 vols, Aldershot: Ashgate, 2004.

Murphy, Graham. *Founders of the National Trust*. London: National Trust, 2002.

Murray, Tim. 'The History, Philosophy and Sociology of Archaeology: The Case of the Ancient Monuments Protection Act 1882,' in Valerie Pinsky and Alison Wylie, eds, *Critical Traditions in Contemporary Archaeology*. q.v.

Murtagh, William J. *Keeping Time: The History and Theory of Preservation in America*. Hoboken: John Wiley, 2006.

Nead, Lynda. *Victorian Babylon: People, Streets and Images in Nineteenth-Century London*. New Haven and London: Yale University Press, 2000.

Neary, Peter. 'Grey, Bryce, and the Settlement of Canadian-American Differences, 1905-1911,' *Canadian Historical Review*, vol. 49, no. 4 (1968), 357-80.

Nelson, Robert and Olin, Margaret eds. *Monuments and Memory Made and Unmade*. Chicago: University of Chicago Press, 2003.

Nelson, Robert. *Hagia Sophia, 1850-1950: Holy Wisdom Modern Monument*. Chicago: University of Chicago Press, 2004.

Nicolson, Harold. *Diplomacy*. London, New York: Oxford University Press, 1950.

Nora, Pierre. *Les Lieux de mémoire*. Paris: Gallimard, 1994.

Normand, Charles. *Ministère du Commerce, de l'Industrie et des Colonies. Exposition universelle internationale de 1889. Direction générale de l'exploitation. Congrès international pour la protection des œuvres d'art et des monuments, tenu à Paris du 24 au 29 juin 1889. Procès-verbaux sommaires. Rédigés par le secrétaire général Charles Normand, architecte diplômé par le gouvernement, directeur de l'Ami des monuments, secrétaire générale de la Société des amis des monuments parisien*. Paris: Imprimerie Nationale, 1889.

Norris, Christopher, ed. *Music and the Politics of Culture*. London: Lawrence & Wishart, 1989.

Norton, Sara and M.A. DeWolfe Howe, eds. *Letters of Charles Eliot Norton*. Boston: Houghton Mifflin, 1913.

Oechelhaeuser, Adolf von. *Wege, Ziele und Gefahren der Denkmalpflege: Festrede bei dem Feierlichen Akte des Rektoratswechsels an der Großherzoglich-Technischen Hochschule Fridericiana zu Karlsruhe am 20. November 1909*. Karslruhe: Braun, 1909.

Oldershaw, Lucian, ed. *England: A Nation*. London: R. Brimley Johnson, 1904.

Ousby, Ian. *The Englishman's England: Taste, Travel and the Rise of Tourism*. Cambridge: Cambridge University Press, 1990.

Page, Max. *The Creative Destruction on Manhattan, 1900-1940*. Chicago: University of Chicago Press, 1999.

Page, Max and Mason, Randall eds. *Giving Preservation a History: Histories of Historic Preservation in the United States*. New York: Routledge, 2004.

Pardoe, Miss. *The City of the Sultan*. 2 vols. London: Henry Colburn, 1837.

Pelizzari, Antonella, ed. *Traces of India: Photography, Architecture, and the Politics of Representation, 1850-1900*. Montréal: Canadian Centre for Architecture and New Haven: Yale University Press, 2003.

Perkin, Harold. *The Rise of Professional Society: England since 1880*. London: Routledge, 1989.

Perkins, Bradford. *Castlereagh and Adams: England and the United States, 1812-1823*. Berkeley: University of California Press, 1964.

Perkins, Bradford. *The Great Rapprochement: England and the United States, 1895-1914*. London: Gollancz, 1969.

Pevsner, Nikolaus and Bridget Cherry. *The Buildings of England: Northamptonshire*. London: Yale University Press, 1973.

Pevsner, Nikolaus and Jane Fawcett, eds. *The Future of the Past: Attitudes to Conservation, 1174-1974*. London: Thames & Hudson, 1975.

Phillips, Robert B. *Letter to the Landowners of the Diocese of Hereford*. Hereford: J. Head, 1847.

Pinsky, Valerie and Alison Wylie, eds. *Critical Traditions in Contemporary Archaeology*. Cambridge: Cambridge University Press, 1989.

Poole, Randall. 'Religion, War, and Revolution: E.N. Trubetskoi's National Liberal Construction of Russian National Identity, 1912-20,' *Kritika: Explorations in Russian and Eurasian History*, vol. 7, no. 2 (Spring 2006), 195-240.

Port, Michael H. *Six Hundred New Churches: The Church Building Commission 1818-1856*. Reading: Spire Books, Ltd, 2006.

Porter, Roy ed. *Myths of the English*. Cambridge: Polity Press, 1992.

Poulot, Dominique. *Patrimoine et musée: L'institution de la culture*. Paris: Hachette, 2001.

Prestige, G.L. *The Life of Charles Gore: A Great Englishman*. London: Heinemann, 1935.

Price, Nicholas, Talley, M. Kirby Jr., and Vaccaro, Alessandra (eds). *Historical and Philosophical Issues in the Conservation of Cultural Heritage*. Los Angeles: The Getty Conservation Institute, 1996.

Priestland, Jane, ed. *Records of Jerusalem, 1917-1971*. Cambridge: Archive Editions, 2002.

Prince Trubetskoi. *Saint Sophia Russia's Hope and Calling*. London: Faith Press, 1916.

Pringle, Allan. 'Albert Bierstadt in Canada,' *American Art Journal*, vol. 17, no. 1 (1985), 2-27.

Pugh, R.K., ed. *The Letter-books of Samuel Wilberforce, 1843-68*. vol. 47. Buckinghamshire and Oxfordshire County Record Societies, 1970.

Quinault, Roland. 'The Cult of the Centenary, c. 1784-1914,' *Historical Research*, vol. 76, no. 176 (1998), 303-23.

Quinault, Roland. 'Westminster and the Victorian Constitution,' *Transactions of the Royal Historical Society*, 6th series, ii (1992), 79-104.

Ramsay, Sir William Mitchell. *The Imperial Peace: An Ideal in European History*. Oxford: Clarendon Press, 1913.

Reinsch, Paul S. 'International Unions and their Administration,' *American Journal of International Law*, vol. I, no.3 (1907), 597-623.

Rentzhog, Sten. *Friluftsmuseerna. En skandinavisk idé erövrar världen.* Stockholm, 2007.

Rhodes, Albert. *Jerusalem As It Is.* London: J. Maxwell, 1865.

Ricard, Serge. 'Théodore Roosevelt: Principes et pratique d'une politique étrangère.' Aix-Marseilles: Aix-en-Provence Publ. de l'Univ. de Provence, 1991.

Richards, Henry E. *The Progress of International Law and Arbitration: An Inaugural Lecture delivered at the University of Oxford.* Oxford: Clarendon Press, 1911.

Rieger, Bernhard. 'Envisioning the Future: British and German Reactions to the Paris World Fair in 1900,' in Martin Daunton and Bernhard Rieger, eds, *Meanings of Modernity: Britain from the Late-Victorian Era to World War II*, q.v.

Rieger, Bernhard. *Technology and the Culture of Modernity in Britain and Germany 1890-1945.* Cambridge: Cambridge University Press, 2005.

Riegl, Alois. *Der modern Denkmalkultus sein Wesen, seine Entstehung.* Vienna: W. Braumuller 1903.

Riley, Athelstan. *Give Back Saint Sophia: a report of the speeches delivered at the opening meeting of the crusade for the redemption of Saint Sophia, held at the Cannon Street hotel, on Jan. 23rd, 1919.* London: Faith Press, 1919.

Ritvo, Harriet. *The Dawn of Green: Manchester, Thirlmere, and Modern Environmentalism.* Chicago: University of Chicago Press, 2009.

Ritvo, Harriet. 'Fighting for Thirlmere – The Roots of Environmentalism.' *Science*, vol. 300, no. 5625 (Jun. 6, 2003), 1510-511.

Robbins, Keith. 'History and Politics: The Career of James Bryce.' *Journal of Contemporary History*, vol. 7, no. 3-4 (1972), 37-52.

Roessel, David. *In Byron's Shadow: Modern Greece in the English and American Imagination.* Oxford: Oxford University Press, 2002.

Roosevelt, Theodore. *Biological Analogies in History: The Romanes Lecture 1910.* Oxford: Clarendon Press, 1910.

Rubens, Godfrey. *William Richard Lethaby: His Life and Work, 1857-1931.* London: The Architectural Press, 1986.

Runte, Alfred. *National Parks, The American Experience.* Lincoln: University of Nebraska Press, 1997.

Ruskin, John. *The Seven Lamps of Architecture.* Smith, Elder, and Co., 1849.

Ruskin, John. *The Seven Lamps of Architecture.* New York: Farrar Strauss and Giroux, 1971.

Ruskin, John. *The Stones of Venice.* London: Smith, Elder, and Co., 1851-53.

Rydell, Robert. *All the World's a Fair: Visions of Empire at American International Expositions 1876-1916.*(Chicago: University of Chicago Press, 1987.

Rydin, Yvonne. *Urban and Environmental Planning in the UK.* London: Macmillan, 2003.

Rydjold, John. *Indian Place Names, Their Origin, Evolution and Meanings.* Norman: University of Oklahoma Press, 1968.

Said, Edward. *Orientalism.* New York: Vintage Books, 1979.

St. Clair, William. *That Greece Might Still be Free; the Philhellenes in the War of Independence.* London: Oxford University Press, 1972.

'St. Sophia, Constantinople.' *The Builder*, vol. 68, no. 2720 (Mar. 23, 1895) and no. 2721 (Mar. 30, 1895).

Salmon, Frank. *Building on Ruins: The Rediscovery of Rome and English Architecture.* Aldershot: Ashgate, 2000.

Samhällsideal och framtidsbilder : perspektiv på Nordiska museets dokumentation och forskning. 2004.

Samuel, Raphael. *Theatres of Memory.* London: Verso, 1994.

Saunders, Andrew Downing. 'A Century of Ancient Monuments Legislation 1882-1992,' *Antiquaries Journal* 63 (1983), 11-33.

Schama, Simon. *Landscape and Memory.* New York: Vintage Books, 1996.

Schroeder, P.W. 'The 19th Century System: Balance of Power or Political Equilibrium?' *World Politics* xxxix (1986-7), 135-53.

Schroeder, P.W. *The Transformation of European Politics, 1763-1848.* Oxford, New York: Oxford University Press, 1994.

Schück, Henrik. *Kgl. Vitterhets historie och antikvitets akademien: dess förhistoria och historia.* Stockholm: Wahlström & Widstrand, 1932-1944.

Schultz, R. Weir, ed. *The Church of the Nativity at Bethlehem.* London: B.T. Batsford, 1910.

Scott, George G. *A Plea for the Faithful Restoration of Our Ancient Churches.* London: John Henry Parker, 1850.

Scott, James B., ed. *President Wilson's Foreign Policy: Messages, Addresses, Papers.* New York: Oxford University Press, 1918.

Scott-Harper, Queenie. 'The Betrayal of Saint Sophia.' *The Christian East*, vol. 2, no. 1 (Apr. 1921).

Searle, G.R. 'The "Revolt from the Right" in Edwardian Britain,' in Paul M. Kennedy and Anthony Nicholls, eds, *Nationalist and Racialist Movements in Britain and Germany before 1914.* q.v.

Seibel, George A. *Ontario's Niagara Parks, One Hundred Years, A History.* Ontario: Niagara Parks Commission, 1985.

Semmel, Bernard. *Imperialism and Social Reform.* London: G. Allen and Unwin, 1960.

Shand-Tucci, Douglass. *Ralph Adams Cram: Life and Architecture. vol. 1: Boston Bohemia, 1881-1900.* Amherst: University of Massachusetts Press, 1995.

Sharp, Frank C. 'Exporting the Revolution: The Work of the SPAB outside Britain, 1878-1914,' in Chris Miele, ed., *From William Morris: Building Conservation and the Arts and Crafts Cult of Authenticity, 1877-1939.* q.v.

Sharp, Paul. 'For Diplomacy: Representation and the Study of International Relations.' *International Studies Review*, vol. 1, no. 1 (1999), 33-57.

Sheail, John. *Rural Conservation in Inter-War Britain*. Oxford: Clarendon Press, 1981.

Sheppard, F.H.W., ed. *Survey of London: vols. 31 and 32: St James Westminster, Part 2. North of Piccadilly.* London: Athlone, 1963.

Siegel, Michael. *Denkmalpflege als öffentliche Aufgabe. Eine ökonomische, institutionelle und historische Untersuchung.* Göttingen: Vandenhoeck & Ruprecht, 1985.

Simsarian, James. 'The Diversion of Waters Affecting the United States and Canada,' *American Journal of International Law*, vol. 32, no. 3 (1938), 907-21.

Singh, Upinder. *The Discovery of Ancient India: Early Archaeologists and the Beginnings of Archaeology*. Delhi: Permanent Black, 2004.

Sivasundaram, Sujit. *Nature and the Godly Empire: Science and Evangelical Mission in the Pacific, 1795-1850*. Cambridge: Cambridge University Press, 2005.

Slade, Adolphus. *Records of Travels in Turkey, Greece, &c.* 2 vols. London: Saunders and Otley, 1833.

Slater, Samuel ed. *The Correspondence of Emerson and Carlyle*. New York: Columbia University Press, 1964.

Smith, Goldwin. *Three English Statesmen: A Course of Lectures on the Political History of England*. London: Macmillan, 1868.

Smith, Goldwin. *The United States: An Outline of Political History*. New York: Macmillan, 1983.

Smith, H. Clifford. *Sulgrave Manor and the Washingtons: A History and Guide to the Home of George Washington's Ancestors*. New York: Macmillan, 1933.

Smither, James. G. *Architectural Remains: Anuradhapura, Ceylon; Comprising the Dagabas and Certain Other Ancient Ruined Structures*. Colombo: The Ceylon Government, 1894.

Society for the Liberation of Religion from State-Patronage and Control. *The Ecclesiastical Commission, or the Bishops as Church Reformers*. London, 1860.

Society for the Protection of Ancient Buildings. *Annual Report*. London: SPAB, 1919.

Southgate, Donald. *The Passing of the Whigs, 1832-1886*. London: Macmillan, 1965.

Speitkamp, Winfried. *Die Verwaltung der Geschichte. Denkmalpflege und Staat in Deutschland 1871-1933*. Göttingen: Vandenhoeck & Ruprecht, 1996.

Spender, J.A. *Weetman Pearson, First Viscount Cowdray, 1856-1917*. London: Cassell, 1930.

Stafford, Barbara M. *Voyage into Substance, Art, Science, Nature, and the Illustrated Travel Account, 1760-1840*. Cambridge, Mass: MIT Press, 1984.

Stamp, Gavin. *The Changing Metropolis: Earliest Photographs of London 1839-79*. Harmondsworth: Viking, 1984.

Stead, William T. *The Splendid Paupers: A Tale of the Coming Plutocracy.* London: Review of Reviews, 1894.

Stenographische Berichte über die Verhandlungen des Preußischen Hauses der Abgeordneten. Berlin: Moeser, 1901.

Stenographische Berichte über die Verhandlungen des Preußischen Herrenhauses. Berlin: Sittenfeld, 1914.

Stoklund, Bjarne. 'Between Scenography and Science. Early Folk Museums and Their Pioneers,' *Ethnologia Euroaea: Journal of Ethnography*, vol. 3 (2003), 21-36.

Stonehouse, W.B. *The Office of the Rural Dean ... A Sermon ... Being the Annual Visitation of the Venerable Archdeacon of Stow.* London, 1841.

Storrs, Ronald. *Orientations.* London: Ivor Nicholson & Watson, 1937.

Stradling, Robert and Meirion Hughes. *The English Musical Renaissance, 1860-1940: Construction and Deconstruction.* London: Routledge, 1993.

Strong, Roy. *Art and Power, Renaissance Festivals 1450-1650.* Berkeley: University of California Press, 1984.

Summerson, John. *Georgian London.* London: Pimlico, 1991.

Svenska fornminnesföreningens tidskrift, Stockholm, 1871-72, 1902.

Swift, Emerson. *Hagia Sophia.* New York: Columbia University Press, 1950.

Swift, Emerson. *Roman Sources of Christian Art.* New York, 1951.

Tambiah, Stanley J. *Buddhism Betrayed?* Chicago: University of Chicago Press, 1994.

Tanner, Tony. *The Reign of Wonder, Naivety and Reality in American Literature.* Cambridge: Cambridge University Press, 1965.

Teague, Francis. *Shakespeare and the American Popular Stage.* Cambridge: Cambridge University Press, 2006.

The Dardanelles: Their Story and Significance in the Great War. London: Andrew Melrose, 1915.

'The Editor's Causerie.' *The Balkan Review*, vol. 1, no. 2 (Mar. 1919).

Thomas, John. 'Sir Thomas Graham Jackson and the Church of the Hagia Sophia,' *Architectural History* 25 (1982), 98-101.

Thompson, Andrew S. *Imperial Britain: The Empire in British Politics, c. 1880-1932.* London: Longman, 2000.

Thompson, Edward P. *William Morris: Romantic to Revolutionary.* New York: Pantheon Books, 1975.

Trachtenberg, Alan. *Shades of Hiawatha. Staging Indians, Making Americans.* New York: Hill and Wang, 2004.

Trevelyan, George M. *The Call and Claims of Natural Beauty.* London: University College London, 1931.

Trollope, Anthony. *North America.* 2 vols. London, 186.

Trubetskoi, E.N. *Saint Sophia: Russia's Hope and Calling.* London: Faith Press, 1916.

Tschudi-Madsen, Stephen. *Restoration and Anti-Restoration.* Oslo: Universitretsforlaget, 1976.

Tuckerman, Henry T. *Book of the Artists*. New York, 1867.

Tulloch, Hugh. *James Bryce's American Commonwealth: The Anglo-American Background*. Woodbridge: Boydell Press, 1988.

Underdånig skrifvelse i restaureringsfrågan, Stockholm, 1902.

Upmark, Gustaf. *Svensk byggnadskonst 1530-1760*. Stockholm: Norstedt, 1904.

Vance, Aylmer. *William Morris: His Art, His Writings, and His Public Life – A Record*. London: G. Bell, 1897.

Verhandlungen der ersten und zweiten Kammer der Landstände des Großherzogtums Hessen. Darmstadt: Leste, 1902.

Viollet-le-Duc, Eugène-Emmanuel. *Discourses on Architecture*. Boston: James R. Osgood, 1875.

Viollet-le-Duc, Eugène-Emmanuel. *The Foundations of Architecture: Selections from the Dictionnaire Raisonné*. New York: George Braziller, 1990.

von Bogdandy, Armin, Wolfrum, Rüdiger, von Bernstorff, Jochen, Dann, Philipp and Goldmann, Matthias, eds. *The Exercise of Public Authority by International Institutions Advancing International Institutional Law*. Heidelberg: Springer, 2010.

von Heidenstam, Verner. *Modern Barbarism: några ord mot restaurerandet af historiska byggnader*. Stockholm: Bonnier, 1894.

Wallace, Elisabeth. 'Goldwin Smith on England and America,' *American Historical Review*, vol. 59, no. 4 (1959), 884-94.

Walton, John K. 'Thomas Cook', in Richard Butler and Roslyn Russell, eds. *Giants of Tourism* q.v.

Waterson, Merlin. *The National Trust: The First Hundred Years*. London: BBC Books, 1994.

Watkin, David. *The English Vision, The Picturesque in Architecture, Landscape and Garden Design*. London: John Murray, 1982.

Weardale, Lord P. Stanhope. *British-American Peace Centenary: Visit of the British Delegation to the United States, May MCMXIII*. London, 1913.

Webster, Christopher. *R.D. Chantrell (1793-1872) and the Architecture of a Lost Generation*. Reading: Spire Books Ltd, 2010.

Webster, Christopher and John Elliott, eds. *'A Church As It Should Be:' The Cambridge Camden Society and Its Influence*. Stamford: Shaun Tyas, 2000.

Wenzel, F., S. Szaktilla, and H. Pliett. *The Roof of the Church Of Nativity in Bethlehem*. Paris: UNESCO, 2000.

West, Patricia. *Domesticating History: The Political Origins of America's House Museums*. Washington: Smithsonian Institution Press, 1999.

Westlund, Per-Olof. 'Byggnadsminnen',Iin: *Ad patriam illustrandam: hyllningsskrift till Sigurd Curman 30 April 1946*. q.v.

Wetterberg, Ola, *Monument & miljö*; *Kulturfastighetsutredningen. Bilagor*, Stockholm: Statens fastighetsverk, 2009.

Wickremeratne, Ananda. *The Genesis of An Orientalist: Thomas William Rhys Davids and Buddhism in Sri Lanka*. Delhi: Motilal Banarsidas, 1984.

Wickremeratne, L.A. 'The Rulers and the Ruled in British Ceylon: A Study of the Function of Petitions in Colonial Government,' *Modern Ceylon Studies*, vol. 1, no. 2 (1970).

Wiener, Martin J. *English Culture and the Decline of the Industrial Spirit, 1850-1980.* Cambridge: Cambridge University Press, 1981.

Williams-Ellis, Clough, ed. *Britain and the Beast.* London: J.M. Dent, 1937.

Williams-Ellis, Clough. *England and the Octopus.* London: Geoffrey Bles, 1928.

Wilson, Richard Guy. *The Colonial Revival House.* New York: Harry N. Abrams, 2004.

Wilton-Ely, J. '"Classic Ground:" Britain, Italy, and the Grand Tour,' *Eighteenth-Century Life*, vol. 28, no. 1 (2004), 136-65.

Winter, James. *Secure from Rash Assault: Sustaining the Victorian Environment.* Berkeley: University of California Press, 1999.

White, Jerry. *London in the Nineteenth Century.* London: Jonathan Cape, 2007.

Whitemore, Thomas. *The Mosaics of St. Sophia at Istanbul: Preliminary Report on the Year's Work, 1931-1932.* Paris: Byzantine Institute of America, 1933-1952.

Womack, John. *Zapata and the Mexican Revolution.* Harmondsworth: Penguin Books, 1972.

Wrangel, Ewert. 'Våra fornminnen och deras vård,' *Föreningen Heimdals populärvetenskapliga tidningsartiklar* 9, Uppsala, 1908.

Wussow, A. von. *Die Erhaltung der Denkmäler in den Kulturstaaten der Gegenwart. Im Auftrage des Herrn Ministers der geistlichen, Unterrichts- und Medizinalangelegenheiten nach amtlichen Quellen dargestellt.* Berlin: Heymann, 1885.

Young, George. *Corps de droit ottoman.* Oxford: Clarendon Press, 1905.

Young, G.M. 'The Manor House,' in James Lees-Milne, ed., *The National Trust: A Record of the First Fifty Years,* q.v.

Zaslow, Morris ed. *The Defended Border: Upper Canada and the War of 1812.* Toronto: Macmillan, 1964.

Websites

'Conservation Principles,' *English Heritage,* May, 2008. Online at http://www. english-heritage.org.uk.

English Heritage: Heritage Matters. Online at http://www. english-heritage.org. uk.

UNESCO World Heritage Centre, *World Heritage: Challenges for the Millennium.* 2007. Online at http://whc.unesco.org/documents/publi_millennium_en.pdf.

Index